Test Procedures for the Blood Compatibility of Biomaterials

Test Procedures for the Blood Compatibility of Biomaterials

edited by

S. Dawids

Konstruktionsteknik/DTH,
Instituttet for Konstruktionsteknik,
Danmarks Tekniske Højskole,
Lyngby, Denmark

Kluwer Academic Publishers

Dordrecht / Boston / London

Library of Congress Cataloging-in-Publication Data

```
Test procedures for the blood compatibility of biomaterials / edited
  by S. Dawids.
       p.    cm.
   Includes index.
   ISBN 0-7923-2107-3 (HB : acid-free paper)
   1. Biocompatibility--Laboratory manuals.  2. Blood--Effect of
implants on.   I. Dawids, S. G.
R857.M3T47  1993
610'.28--dc20                                            92-41726
```

ISBN 0-7923-2107-3

Published by Kluwer Academic Publishers,
P.O. Box 17, 3300 AA Dordrecht, The Netherlands.

Kluwer Academic Publishers incorporates
the publishing programmes of
D. Reidel, Martinus Nijhoff, Dr W. Junk and MTP Press.

Sold and distributed in the U.S.A. and Canada
by Kluwer Academic Publishers,
101 Philip Drive, Norwell, MA 02061, U.S.A.

In all other countries, sold and distributed
by Kluwer Academic Publishers Group,
P.O. Box 322, 3300 AH Dordrecht, The Netherlands.

Printed on acid-free paper

Printed in the Netherlands

T A B L E OF C O N T E N T S

VII   TEST METHODS FOR THE DETECTION OF THROMBOSIS, FIBRINOLYSIS AND EMBOLI FORMATION

VIII   TEST METHODS FOR THE DETECTION OF TOXICITY

IX GLP-GMP RULES FOR TESTING OF POLYMERS

FOREWORD

This book represents the first European effort to provide a collection of test descriptions used in evaluation of the compatibility of biomaterials in contact with tissues and blood. The urge to compile this book arose from the fact that it is the properties of the material which ultimatively seem to determine the functional outcome of a medical device, almost regardless of how ingenious the construction of the very device is. The longer the exposure is, the more important these basic properties become. Unfortunately only a small part of the interactive phenomena is fully elucidated and understood. This challenge reflects itself in an effort to cover numerous aspects of testing, beginning with fundamental analysis of the material, continuing with the mechanical properties, the resistance to degradation and the analysis of surface and chemical properties including adsorption patterns of proteins ending with test on cell cultures, ex vivo and in vivo.

A number of the tests which are generally accepted as being important are already described as official requirements (primarily Pharmacopeas). These official requirements are not included in order to limit the size of the book.

It is the aim of this book to present the tests like a recipe in a uniform way to ease the reader in finding his/her way through the material and to present it as a kind of "cook-book" in an order to provide an easy access to copy the procedures. This has unfortunately not been possible in all circumstances. The editors regret this but do, on the other hand, accept the contributions as it is an unusual way for scientists to publish their material. The difficulties in bringing a uniform format to the presentations have unfortunately delayed the book, but it is with respect and gratitude that the editors take the opportunity to thank contributors and fellow scientists for their efforts.

This book is the offspring of a decade of coordinated research which has been supported by the European Community through the European Concerted Action "EUROBIOMAT" of the 4th Medical Research Programme. It is in some ways a continuation of a long series of publications within the topics of biomaterial. The financial aid from the Community has enabled the European scientifist to form a network, and we are very grateful for the Commission's foresight in sponsoring this important field.

We thank the present project leader dr. Willy Lemm whose never failing help has been of considerable importance. Finally I wish to thank Mrs. Vibeke Munck, head of the secretariat. Without her meticulous retyping of edited versions and her ability to keep the logic of the book in sight in spite of sometimes insurmountable difficulties, this book had not been born. The Institute of Engineering Design, Technical University of Denmark, has devoted many ressources into redrawing and improving the figures to provide a result of uniform high quality.

Steen Dawids
editor

Chapter I

GENERAL INTRODUCTION

Chapter ed.: S. Dawids

Institute of Engineering Design, Biomedical Section, Technical University of Denmark,
DK - 2800 Lyngby

INTRODUCTION

When biomaterials are exposed to blood a rapid sequence of events occur. Many of these have not been fully elucidated and their roles for e.g. the patency of a vascular graft is also unknown. The chapter will provide a brief introduction on questions of haemocompatibility.

Testing for haemocompatibility is always extensively carried out on animals prior to clinical use. This chapter therefore also provides some guidelines for selecting animal species for such tests.

S. Dawids (ed.), Test Procedures for the Blood Compatibility of Biomaterials, 1.
© 1993 *Kluwer Academic Publishers. Printed in the Netherlands.*

Haemocompatibility, What Does It Mean?

Author: S. Dawids

Institute of Engineering Design, Biomedical Section, Technical University of Denmark,
DK - 2800 Lyngby

The greatest problem encountered when foreign materials are inserted into the blood is usually the rapid formation of visual thrombus material on the foreign surface. Much has been done to prevent this formation including extensive research on surface properties, interface phenomena and efforts to mimic the natural thrombolytic mechanisms of the body. Nevertheless, the fact that one third of the population in the Western world eventually succumb to vascular diseases indicates that even natures own haemocompatibility and repair processes fall short with time and show us that haemocompatibility is always a time limited feature.

It is important to realise that the extent of properties for adequate haemocompatibility differ depending on the application. A catheter or guide wire for short term inwelling should "only" provide prevention of clot formation for minutes up to an hour. Long term exposure, on the other hand, requires continuous replentishment of the relevant abilities to prevent clotformation. For decades research has focused on preventing the clotting in the last step(s) of fibrin formation. It is now becoming clear that the early processes, far before fibrin formation occurs, are more important as well as the ability to dissolve clots. Even if all toxic and tissue incompatibilities are absent, no biomaterial for vascular purpose has in any way matched natures own solutions.

NORMAL VASCULATURE

In the ageing organism it is generally the arterial blood supply which determines the vitality of the tissue. Although the structure of arteries varies according to size and localization, the therapeutical problems are localised in the large and medium size vessels. Without going into details, it should be mentioned that the arterial wall is in itself a vascularised organ where the inner surface is coated with

3

S. Dawids (ed.), Test Procedures for the Blood Compatibility of Biomaterials, 3–11.
© 1993 *Kluwer Academic Publishers. Printed in the Netherlands.*

endothelial cells which continuously play an active role in the defense of the vascular surface integrity. Under the basal membrane of the vessel wall, the tissue receives its own vascular supply (vasa vasorum) and has a surrounding lymph drainage system. Fluid and solutes move accross the endothelial lining into the vessel wall. The diffusive properties vary considerably in different tissues. In the arteries of the brain the endothelial lining is very tight with a low escape of solutes compared to e.g. arteries in the liver.

The varying properties of the endothelial lining are believed to be a result of an interplay with the environment (i.e. substances from the blood and from the subendothelia cells). All the tissue components in the wall are continuously replaced. The turnover of tissue is generally unknown but must be very high some places which explains why e.g. heart valves function all life.

It should be realized that infarctions occurring in organs such as brain, heart and kidney most often are the integrated results of the occlusion of an artery together with a pre-existant low tissue perfusion pressure. In young persons possessing a good tissue perfusion pressure there is adequate collateral vascular supply to replace the tissue perfusion lost by the compromised artery. Thus occlusion of middle size arteries in young, healthy persons is rarely accompanied by clinical symptoms.

The metabolism of endothelial cells has been extensively investigated, and it is evident that endothelial cells can provide very different expressions, depending on the environmental conditions. The turnover of endothelial cells varies depending on the site, but is considered very rapid. Efforts to provide proper conditions for endothelial cells on biomaterials is still an unfulfilled challenge (*Lelkes & Samet, 1991*).

BODILY REACTIONS TOWARDS BLOODCOMPATIBLE MATERIALS

Several interacting defense mechanisms or systems are activated when a polymer is exposed to blood. These systems have traditionally been divided into the coagulation system, the mutually complement system, the kinin-kallikrein system and cellular systems. They are all coordinated towards eliminating the invading foreign body by isolation (fibrin) and by attack (fagocytosis combined with liberation of enzymes and reactive radicals). The fibrinolytic system plays an important role in the normal physiological condition by limiting the clot formation to the relevant site and resolve the activity with time. This activation is, inter alia, obtained with enzymes from the underlying tissue (*Fig.1*). In synthetic materials this tissue factor is largely absent or reduced. Resolution of thrombus material will proceed very slowly or not occur.

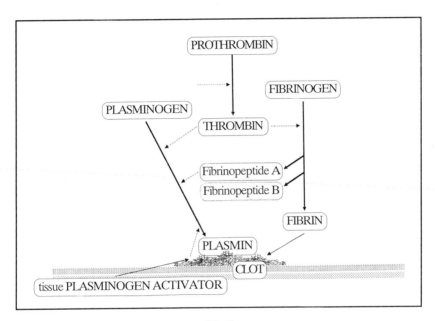

Fig. 1

Fibrin formation shown in a simplified diagram with the final steps involved in fibrin-clot formation and its dissolution. The pivoting protein thrombin is generated by factors in the coagulation cascade (not shown). The generated thrombin splits fibrinogen which then polymerizes through several steps (not shown) into a fibrin clot. Circulating plasminogen will preferentially bind to the fibrin. It is activated by several activators leading to the gradual dissolution of fibrin. The major activator under physiological conditions is tissue plasminogen activator although others play a role (Fibrinopeptide A and Thrombin).

Research emphasis has been put on the different aspects of the defense systems disregarding what is believed to be important. This is understandable, as the obvious formation of clot and emboli should be avoided because only a limited number of known factors are easily measurable for research purposes. The relative importance of the different aspects are still not quite known.

POLYMER EXPOSURE TO BLOOD

When a blood polymer-interface is established a rapid sequence of processes occur. These processes have not been fully elucidated, mainly because the steps replace each other quickly and cannot at present be studied in detail in vivo. The use of in vitro models have provided a number of clues, although the relevance of the in vitro observations is doubtful in most cases (for survey, see *Missirlis & Lemm, 1991*). However, it is generally accepted that the processes can be divided arbitrarily into the following groups pf events (which partly occur simultaneously):

1. Adsorption of plasma proteins onto the polymer surface.
2. Activation of the systems of complement, kinin/kallikrein, blood cells and coagulation initiated by adsorbed proteins from the systems.
3. Adhesion of cell components (thrombocytes, granulocytes and monocytes) to the protein coating.
4. Formation of fibrin onto the surface.

ad 1. Adsorption of specific proteins from the plasma rapidly takes place. The proteins adsorbed initially are those present in relatively high concentrations in the plasma including species such as albumin, fibrinogen and IgG. However, these are subsequently replaced and covered by other protein molecules. This effect of absorption and desorption (the Vroman effect) was initially believed to provide a source of informaiton on the compatibility of the material. There seems, however, to be a very low correlation to specific adsorption pattern and in vivo evaluation. The affinity of protein molecules to the polymer surface is quite dependent on the property of the surface. The adsorption of proteins is at first reversibel, but gradually becomes irreversibel due to increasing multivalent binding to sites on the polymer surface. This may induce deformation of the protein molecules leading to irreversibel changes (denaturation). It is believed that adsorption on hydrophilic polymers takes place at "hydrophilic" sites on the protein, and is mainly ionogenic. Hydrophobic polymers bind proteins at their "lipophilic" sites with hydrogen bonds. The relative contents in the polymer surface of hydrophobic and hydrophilic sites play an unpredictable role. Many antibodies of IgG type adsorbed to the surface enhance activation in vitro.

ad 2. Several proteins from the systems of complement, kinin/kallikrein, coagulation etc. are activated by the binding to the adsorbed proteins on the polymer surface. They may directly or by subsequently binding to other proteins display strong activating properties leading locally to activation of the systems (thrombus formation) or generally to inflamation through liberation of fragments of split molecules into the blood (*Fig. 2*).

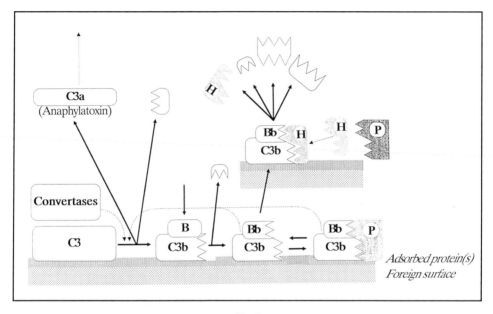

Fig. 2

Complement activation shown in a simplified diagram. C3 binds to the adsorbed proteins, and is split by convertases. The fragment C3b binds from plasma factor B which is split. The remaining complex C3b-Bb activates C3 leading to accelerated activation of C3. The complex is broken down by contact with factor H or stabilised by binding to protein P.

Activation of the systems of complement, kallikrein-kinin, coagulation and other defense systems then follow. This phenomenon is called contact activation and serves to start the major systems involved in the protection against physiological injuries. Of the many possible proteins, four are believed to be major initiators: Factor XII (Hageman factor), Factor XI, Prekallikrein and High Molecular Weight Kininogen (HMWK)(*Fig. 3*).

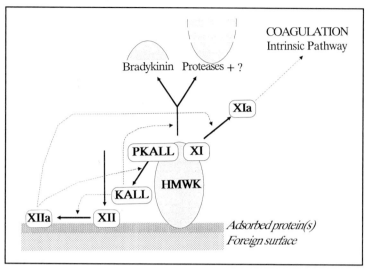

Fig. 3

Surface activation shown in a simplified diagram. High Molecular Weight Kininogen binds to the adsorbed proteins. It will here bind prekallikrein where it is converted to kallikrein by activated factor XII (Hageman factor). This, in turn, activates prekallikrein leading to accelerating activation. Factor XI is likewise activated by factor XII leading to coagulation. HMWK is broken down and the molecular fragments activate, inter alia, cellular systems.

The first three mentioned are zymogen forms of serine proteases, while the fourth acts as a co-factor in the activation. Although the steps in the activation processes have been analysed in detail, the actual steps that trigger activation have not been fully elucidated (although a number of hypotheses exists). The role of other systems (not mentioned here) with yet immeasurable roles in the bodily defenses such as the co-activation and influence of the renin/angiotensin system on the coagulation and complement systems is not known. One established activation of the complement system occurs through adsorption of C3 together with "convertases". The speed of complement activation/deactivation of a polymer surface has been found to depend on its ability to bind protein B (which stabilizes the surfaces bound complexes) or to deactivate the complement system by its relative ability to bind inhibiting factors such as factor H (which promotes the breakdown of complexes). This seems to be of clinical relevance.

ad 3. Adhesion of cell components takes place when the surface has been coated with proteins. The adhesion of thrombocytes occurs within seconds and subsequently they are activated. While it is believed that the platelets gradually are repelled and shed from normal endothelial lining, they will adhere to the polymer surface and subsequently release factors with stimulatory activation (serotonin, Factor V, von Willebrand factor, fibronectin, thrombospondin etc.) as well as others with inhibitory and mitogenic effects. Adhesion of platelets on polymers has been shown to generate progressing activation of other thrombocytes, of the coagulation system and of release of factors generating general reactions. The counteracting systems are less activated (e.g. fibrinolysis). Adhesion of neutrophils to the polymer surface is mediated, inter alia, by the complement system, even though it also seems to occur independently. Activation of the neutrophils is mediated by all the mentioned systems. In turn, they release proteases (local action), lactoferrin, interleukins etc. (generalised action) and initiate phagocytoses (with liberation of reactive radicals). The monocytes also adhere to the surface and are likewise activated through the influence of factors from all the mentioned systems as well as by interleukins which provide a generalised activation of the immune systems. The release of proteases and oxygen radicals are believed to play an important role in the degradation of polymers.

ad 4. The generation of fibrin clots on the polymer surface is the visible results of the above complex and rapid sequences of interlaced processes. The clot formation can be compared to that which occurs at sites where the endothelium has been damaged. However, under normal physiological conditions the stabilised fibrin clot will gradually break down by the action of plasmin. The fibrin preferentially adsorbs circulating plasminogen which is then locally activated by tissue activators promoting lysis of the clot. On a polymer surface the tissue factor is not present, and thus making the removal of fibrin very slow compared to a "normal" situation. This difference may, to some extent, explain why cumulative fibrin formation occurs in artificial vessels and prevents the use of small size calibers.

While prevention of clot formation has not been achieved in spite of countless approaches, it is, as indicated above, ambiguous which factors are of most importance. It is doubtful whether the microscopic surface geometry (as is commonly believed) is the most important factor. The distribution of binding sites on the surface is probably more important as this interferes with the surface binding of proteins. It must be realised that prevention of clot formation requires a continuous active counteraction in the dynamic environment of blood. This requires continuous supply of chemical energy/replacement of enzymes which obviously perform more successful from a biological point of view compared to the isolated effort to hamper clot formation on a polymer surface. Once a fibrin coating has occurred, it will, if resolution of the fibrin has not taken place, become organized and converted into fibrous-like tissue.

The approaches to render a polymer biocompatible and non-thrombogenic can be divided arbitrarily into 4 groups:

1. Coating the surfaces with known antithrombogenic substances such as heparin or heparin-like radicals has been tried in numerous ways. The results have rarely been promising in vivo, not even if in vitro demonstrations have shown the properties of the treated material to be much superior to those of the untreated surface. The antithrombogenic property may in reality last just a few minutes, either because of rapid release of the surface bound heparin or because of gradual coating with protein(s).

2. Coating the polymer surface with substances which selectively bind proteins which in vitro have the ability to inhibit coagulation. Elastomeric materials can be coated to provide selective adsorption of e.g. albumin. No successes have been reported. One of the main difficulties is perhaps that the surface produced is not rugged enough to handle or inadvertently is modified during sterilization.

3. Slow release of active substances (heparin, prostacyclin, plasminogen activators etc.) has been tried on elastomeric polymers which can absorp large quantities of heparin e.g. and release it slowly over periods of days. A number of published attemps are very difficult to reproduce, perhaps because of biological variations or other overlooked factors.

4. Coating of polymers with endothelial cells has long been believed to be a key solution to acceptable blood compatibility. In this respects most polymers do not inhibit the overgrowth, but the morphology and function of the seeded endothelium (pseudoendothelium) is

morphologically and behaviorally different from normal endothelium and has not yet provided convincing results.

Thus new polymer surfaces still continue to generate mural clots which can generate emboli or which gradually add successive layers of pseudointima with interposed debris leading to narrowing of the lumen. Only large caliber vessels can remain patent and avoid thrombosis under these circumstances.

As a consequence most vascular surgeons have given up replacing medium size vessels with artificial vessels and are to day replacing medium size arteries (in the effort of tissue revascularisation of e.g. of limbs) using in situ venous by-pass. Such veins have a very high patency rate, presumably attributed to the normal endothelial lining and existance of subendothelial organised tissue. The veins are quickly arterialised with thickening of the walls maintaining the structure.

REFERENCES

Lelkes, P.I. & Samet, M.M. (1991) "Endothelialization of the luminal sac in artificial cardiac prostheses: A challenge for both biologists and engineers", J. Biomed. Eng. 113, 132-142.

Missirlis, Y.F. & Lemm, W. (1991) Modern Aspects of Protein Adsorption on Biomaterials, Kluwer Academic Publishers, Dordrecht.

Guidelines For Selecting The Proper Animal Species For A Particular Test

Authors: J. Belleville & R. Eloy
Unité 37, INSERM, F - 69500 Bron

INTRODUCTION

The haemostatic mechanism of most animals differs significantly from that of humans. Crucially important protein systems also vary between species. Nevertheless animal models are of value for the evaluation of biomaterial haemocompatibility because until now there is no in vitro experimental procedures which can mimic the human response so closely. Comparative blood constituents have been studied and focus has been made on those involved in blood-material interaction with the aim to justify the selection of animal models and to identify the limitations of such experiments in the extrapolation to clinical human situation.

COMPARATIVE BIOCHEMICAL, HAEMATOLOGICAL AND HAEMOSTATIC VALUES

Comparative biochemical and haematological values.

Proteins are present in human blood at the concentration of 6.8 ± 4 g/l. This does not significantly differ in most species apart from the pig. Values are given in *Table I*. The albumin/globulin ratio is generally higher in humans than in most animals which may be of importance knowing the role of albumin in passivating the surface against platelets. Serum calcium levels in rabbit, dog and cattle are similar to human levels, and their blood is commonly used for investigation of compatibility. Values are given in *Table II*.

Values on erythrocyte, haemoglobin, haematocrit and leucocyte distributions from different animals are shown in *Table III* and *Table IV*.

S. Dawids (ed.), Test Procedures for the Blood Compatibility of Biomaterials, 13–34.
© 1993 *Kluwer Academic Publishers. Printed in the Netherlands.*

Table I

Serum proteins of normal animals (Mitruka & Rawnsley, 1977).

ANIMAL	TOTAL PROTEIN g/dl	ALBUMIN g/dl	ALBUMIN/ GLOBULIN
Mouse	6.25 ± 0.75	2.52 / 4.84	0.56 / 1.30
Rat	7.61 ± 0.50	2.70 / 5.10	0.72 / 1.21
Hamster	6.94 ± 0.32	2.63 / 4.10	0.58 / 1.24
Guinea Pig	5.60 ± 0.28	2.10 / 3.90	0.72 / 1.34
Rabbit	6.90 ± 0.36	2.42 / 4.05	0.68 / 1.15
Chicken	5.92 ± 0.82	2.10 / 2.45	0.58 / 1.30
Cat	6.10 ± 0.32	2.20 / 3.20	0.60 / 1.20
Dog	7.10 ± 0.20	2.12 / 4.00	0.50 / 1.68
Monkey-Rhesus	7.20 ± 0.44	1.80 / 4.60	0.61 / 1.65
Pig	8.90 ± 0.48	1.80 / 5.60	0.47 / 1.19
Sheep	6.80 ± 0.30	2.70 / 4.55	0.70 / 1.60
Cattle	6.97 ± 0.45	2.45 / 4.20	0.70 / 1.15
MAN	6.80 ± 0.40	3.50 / 4.70	1.15 / 2.17

Table II

Seric calcium level. Atomic absorption spectrophotometric assay (mean ± SD). (Mitruka & Rawnsley, 1977).

ANIMAL	Mean (mg/dl)	SD
Mouse	5.60	0.40
Rat	12.00	0.80
Hamster	9.55	0.95
Guinea Pig	9.80	0.60
Rabbit	10.00	1.10
Chicken	14.50	5.00
Cat	10.00	0.86
Dog	10.20	0.50
Monkey-Rhesus	11.20	0.69
Pig	9.60	1.00
Sheep	11.40	0.30
Cattle	10.70	1.15
MAN	9.80	1.10

Table III

Erythrocytes parameters of normal animals. (Mean and standard deviations from 145 adult males). (Mitruka & Rawnsley, 1077).

ANIMAL	Number x 10^6/mm^3		Haemoglobin g/dl		MCV µ^3		Haematocrit ml/%	
Mouse	9.30	1.20	11.3	0.10	49.0	0.75	41.5	4.20
Rat	8.95	0.40	14.6	0.58	53.8	2.00	47.4	1.50
Hamster	7.50	1.40	16.8	1.20	70.0	3.19	52.5	2.30
Guinea Pig	5.60	0.62	14.4	1.38	77.0	3.00	42.0	2.50
Rabbit	6.70	0.62	13.9	1.75	62.5	2.00	41.5	4.25
Chicken	3.50	0.30	10.3	1.40	102.0	1.75	35.6	4.82
Cat	7.40	0.20	10.7	0.40	56.6	1.10	41.4	1.20
Dog	6.98	0.60	17.1	0.90	77.2	2.25	53.9	2.70
Monkey-Rhesus	5.15	0.91	13.2	1.35	80.6	9.98	41.5	4.20
Pig	7.09	0.80	13.0	0.25	60.0	1.00	43.2	0.50
Sheep	10.30	0.42	10.9	0.45	31.0	0.50	31.7	0.92
Cattle	8.10	1.15	11.5	1.45	50.0	1.75	40.0	3.50
MAN	5.20	1.00	14.5	1.50	89.0	5.00	45.0	7.00

Table IV

Leucocytes values of normal animals. (Mean and standard deviations from 145 adult males). (Mitruka & Rawnsley, 1977).

ANIMAL	Leucocyte count 10³/mm³		Neutro. %		Eosino. %		Baso. %		Lympho. %		Mono. %	
Mouse	14.20	0.87	17.4	2.1	2.09	0.36	0.00	0.15	72.6	5.0	2.35	0.06
Rat	9.92	0.96	24.4	9.6	0.35	0.14	0.20	0.20	75.4	9.0	0.23	0.21
Hamster	7.62	1.30	22.1		0.90		1.00		73.5	9.4	2.50	0.80
Guinea Pig	11.50	3.00	42.0		4.00		0.70		49.0		4.30	
Rabbit	9.00	1.70	46.0		2.00		5.00		39.0		8.00	
Chicken	20.40	5.30	31.5		8.00		4.25		55.8		0.50	
Cat	17.40	1.20	58.3		6.90		0.02		32.0		2.80	
Dog	12.90	2.40	61.2		4.60		0.30		29.7		2.30	
Monkey-Rhesus	10.60	1.20	37.6		1.80		0.20		58.8		1.60	
Pig	15.50	2.90	39.0		4.50		1.20		52.1		3.30	
Sheep	7.80	1.20	35.7		2.50		0.40		56.9		4.50	
Cattle	9.25	1.40	30.5		9.50		0.50		51.4		8.10	
MAN	7.70	3.00	66.0		3.00		0.50		35.0		6.40	

Haemostasis.

Platelets.

Platelets vary in size, survival time and in number among different species as shown in *Table V*. Very high platelet counts are found in the rat. Platelet membrane glucoproteins involved in platelet substrate or platelet-proteins interactions shown remarkable interspecies similarities as shown in *Table VI* (*Nurden, 1976*). The release of dense granules from platelets is an important step of normal platelet activation. In *Table VII* notable differences are seen among the various species. Human dense granule platelet release is characterized by a high level of ADP and Ca^{2+} and a low level of serotonin. Alphagranules contain a large number of proteins which induce inflammatory and coagulative or repair functions. Fibrinogen and platelet factor IV (which exhibits an antiheparin activity) have been studied in animals such as cat, rat, rabbit, sheep, cow, pig, horse, monkey and dog where they are present in amounts similar to those in human. Apart from that, little has been investigated in the animals.

Platelet adhesion in vitro on glass is specific to human platelets (*Belleville, 1966; Sinakos, 1967*), but in vivo platelet adhesion to exposed subendothelium is similar in humans and animals and is more relevant. In all mammals platelet aggregation is stimulated by thrombin and collagen, provided that adequate concentrations of extracellular Ca^{2+} and fibrinogen exist. Following exposure to these initiating activators, the platelets may themselves generate agonists by at least three other mechanisms (*Fig. 1*). These are:

1. The arachidonic acid pathway (AA) where the liberation and metabolism of AA produce thromboxane A2 and prostaglandin endoperoxides.
2. The dense granule pathway which leads to the release of serotonin and ADP from the dense granules.
3. The platelet activating factor (PAF) pathway generated from phospholipids in some species (guinea pig, rabbit, horse, cattle, rat).

It should be noted that rabbit and porcine platelets aggregate without the presence of fibrinogen, although fibrinogen potentiates their aggregation responses in contrast to human platelets which require external fibrinogen for aggregation. Pig, rabbit and rat platelets are more dependent upon extracellular Ca^{2+} compared to human platelets. They do not exhibit significant release reactions, until Ca^{2+} is added to the medium.

Table V

Characteristics of platelet from different species.

SPECIES	Platelet count x 10^9/l	Size (fl)	Survival days (51 Cr)
Mouse	232 ± 80	3.3	
Rat (Wistar)	1350 ± 250	4.7	
Hamster	410 ± 75		
Guinea Pig	500 ± 120	4.5	
Rabbit	480 ± 88	3.3	2.8
Chicken thrombocytes	31 ± 9		
Cat	235 ± 65	9.9	
Dog	257 ± 52	7.3	5.4
Monkey-Rhesus	144 ± 15		5.6
Pig	300 ± 34	6.9	
Sheep	350 ± 52	4.4	
Cattle	350 ± 87	4.5	10-12
Horse	250 ± 50	5.1	
MAN	150 - 400	8.3	8-9

Table VI

Platelet membrane glycoproteins. Nature and molecular weights of the platelet membrane glycoproteins from different species. (Nurden, 1976).

SPECIES	GLYCOPROTEINS		
	I	II	III
Baboon	153,000	130,000	103,000
Dog	148,000	125,000	98,000
Rabbit	170,000	132,000	108,000
Cat	absence	130,000	100,000
MAN	155,000	135,000	103,000

Table VII

Dense granules. Thrombin induced release of selected constituents (expressed in $\mu Mol/10^{11}$ platelets). (Meyers, 1986).

SPECIES	Serotonin	ADP	Ca^{2+}
Cat	0.9	0.9	2.7
Cow	2.3	0.8	3.5
Dog	0.6	0.7	2.0
Horse	0.6	0.5	3.0
Pig	1.2	2.0	4.8
Rabbit	4.0	0.7	2.4
MAN	0.3	3.2	12.5

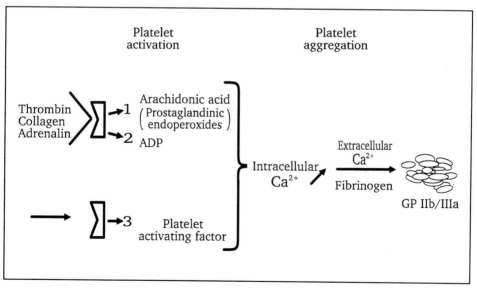

Fig. 1

Model of platelet aggregation in animals.

The arachidonic acid pathway shows considerable variations in the amount of TXB^2 generated in platelets. Humans, dogs, mice and rats produce high amounts while horse, pig and cow have a lower release (*Meyers, 1980*). The arachidonic acid pathway is well developed in platelets from humans and from the common laboratory animals. Between dogs there are considerable differences in the response of platelets to AA ranging from no aggregation to reversible aggregation to irreversible aggregation. Irreversible aggregation induced by AA occurs in approximately 33% of beagles and 50% in mongrels.

The dense granule pathway uses ADP as a universal agonist of mammalian platelets although there are species differences in the ADP concentration required from maximal aggregation and in the occurence of biphasic aggregation as shown in *Table VIII*. Serotonin is not a potent agonist of most mammalian platelets, but is the case for the dog.

The platelet activation factor pathway has been identified in all species as a potent aggregating agent, although there are conflicting results whether platelets can, or can not, synthesize significant amounts of PAF-acether when they are stimulated by a physiological agonists. The rank order in sensitivity to PAF for aggregation is: guinea pig > rabbit = horse = cattle > human > rat (*Meyers, 1986*).

It should be noted that platelets from most mammalians species readily aggregate when exposed to ADP, thrombin and collagen. On the other hand they are relatively unresponsive to adrenalin, AMP, dipyridamole, bovine factor VIII and ristocetin (*Dodds, 1978*). Human platelets, however, are aggregated by ristocetin. The sensitivity of platelets appear to be modulated by factors outside the platelet, notably prostacyclin (PGI_2) and adrenalin. Platelets from all species appears sensitive to PGI_2 at concentrations of 0.5 - 3.5 ng/ml, and binding of PGI_2 to platelet membranes has been reported in rats, dogs, cows and humans. Adrenaline exerts a more species related effect: it aggregates and potentiates platelet aggregation in humans, cats, dogs and mice, but is without effect on platelets from horses, cows, sheep and pigs. In rats adrenaline causes either an inhibition or a potentiation of platelet aggregation depending on other factors.

Coagulation.

Blood coagulation is the result of a cascade of reactions involving trace coagulation factors shown in *Table IX*. They are present in plasma as inert precursors apart from one of them. Two activation paths exist, the extrinsic mediated by cell thromboplastin production and the intrinsic initiated by factor XII activated by

Table VIII

Platelet from different species responses to ADP (Meyers, 1986).

SPECIES	ADP concentration for maximal aggregation (μM)	Biphasic Aggregation
Monkey-Rhesus	14.0	−
Guinea Pig	2.0	+
Mouse	5.0	−
Rabbit	4.9	−
Rat	3.4	−
Cat	5.0	+
Dog	5-10	+/−
Horse	1.5	+/−
Cow	5.0	−
Pig	50.0	−
Sheep	1.5/14	−
MAN	22 - 3.5	+/−

Table IX

Blood Coagulation factors in human.

FACTOR	Common name	Molecular Weight (K Dalton)	Plasma concentration (mg/dl)
Factor I	Fibrinogen	340	200 - 400
Factor II	Prothrombin	72	10
Factor III	Tissue Thromboplastin		0
Factor IV	Calcium		4 - 5
Factor V	Proaccelerin	330	1
Factor VII	Proconvertin	48	0.05
Factor VIII:C	Antihaemophilic factor	242	1 - 2
Factor IX	Christmas factor	57	0.3
Factor X	Stuart-Prower factor	59	1
Factor XI	Plasma Thromboplastin antecedent	185	0.5
Factor XII	Contact factor Hageman factor	80	3
Factor XIII	Fibrin stabilizing factor	326	1 - 2
High Mol. Weight kininogen	Fitzgerald, Flaujeac Williams factor	160	6
Prekallikrein	Fletcher factor	83	5

contact with foreign surfaces. This leads to activation of a thrombin which, in turn, activates the soluble fibrinogen into insoluble fibrin as shown in *Fig. 2*. The physiological inhibition of activated clotting factors is ensured by inhibitors such as antithrombin III, alpha-2-macroglobulin, C_1 inactivator and protein C with its cofactor protein S. In all vertebrates, fibrinogen molecules are composed of three pairs of non identical chains.

Conversion to fibrinogen consists of proteolysis at the N-terminal region releasing fibrinopeptides A and B which differ among species (*Doolittle, 1988*). The speed of fibrinogen clotting in different mammals varies with the origin of thrombin as shown in *Table X*. With the same thrombin all the fibrinogens react differently. This is also the case with other factors as demonstrated by the efficacy of autologous versus heterologous thromboplastin, which earlier was used as quantitative coagulation initiater as shown in *Table XI*. Only considering mammal coagulation factors, the cofactors V and VIII are generally present in lower amounts compared to humans. It has, on the other hand, been demonstrated that kininogen seem to exhibit different molecular weights and functions according to the species (*Muller-Esterl, 1987*).

In this respect, a third kind of kininogen has been found in rat plasma, termed T-kininogen. This is of intermediate molecular weight, lying between high and low molecular weight kininogens. It plays a role during acute inflammation. The inhibitors, protein C, alpha-2-macroglobulin and antithrombin III are present in all vertebrates, and they all exhibit heparin-dependent anti-Xa and anti-IIa activities. Molecular species variations in man, monkey, dog and rabbit lead to variable immunoreactivity towards human antithrombin III antibodies (*Fareed et al., 1985*). Heparin is physiologically present in various tissues, especially in the liver of many species, but not in the tissue of rabbit.

Comparative blood coagulation parameters in laboratory animals are known from humans, primates, rabbits and dogs (*Fareed et al., 1985*). Tests in monkeys and humans show similar values in prothrombin time (PT), activated partial thromboplastin time (APTT), thrombin time (TT) and reptilase time. Rabbit and dog exhibit considerably shorter PT and APTT, indicating a faster and/or higher thrombin generation. Reptilase time and thrombin time are similar in the four species.

Many hereditary disorders have been described in animals (*Dodds, 1974*) encompassing:

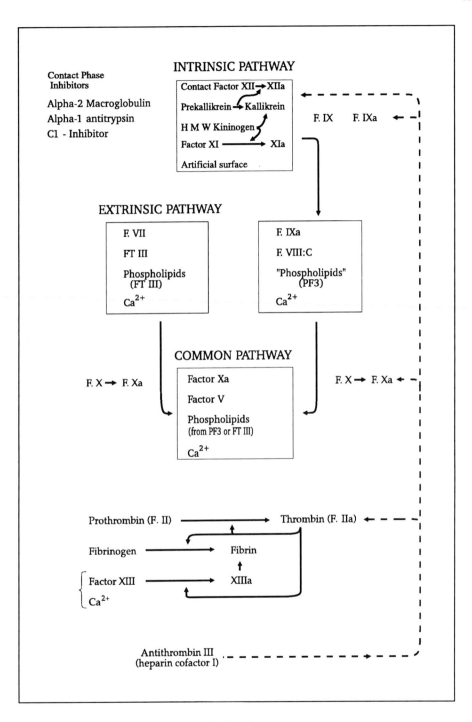

Fig. 2

Coagulation activating pathways and their inhibitors.

Table X

Species differences in the clottability of fibrinogen (thrombin time). (Chandraseklar & Laki, 1968).

THROMBIN	Man	Monkey	Dog	Sheep	Rabbit
FIBRINOGEN	Rabbit	Rabbit	Sheep	Rabbit	Dog
	Sheep	Dog	Rabbit	Man	Man
	Dog	Sheep	Man	Sheep	Rabbit
	Man	Man	Dog	Dog	Sheep
	Monkey	Monkey	Monkey	Monkey	Monkey

This table shows fibrinogen arranged in order to the speed with which they were clotted by thrombin. The fastest reactions are given at the top of the columns.

Table XI

Prothrombin time (sec). Comparison between results obtained with heterologous brain extracts thromboplastin (from human) and homologous thromboplastins. (Leroy, 1987).

SPECIES	Man	Dog	Sheep	Rabbit	Rat
Man	12	14	16	14	17
Dog	28	11	17	12	
Sheep	44	22	15	23	25
Rabbit	12	10	14	9	13
Rat	20	15	17	13	10

- congenital deficiency of fibrinogen in goats and dogs,
- hereditary factor VIII deficiency in beagles,
- haemophilia A (factor VIII:C deficiency) and haemophilia B (factor IX deficiency) in most breeds of dogs,
- severe haemophilia in horses,
- von Willebrand's disease in pigs and dogs as in humans.

In estimating animal clotting factors II, V, VII, VIII, IX, X, XI and XII human factor-deficient plasma is used. In primates all clotting factors seem to range from 75% - 150% of that in human plasma. In rabbits and dogs factors V, VIII:C, IX and XI are considerably higher. Application of chromogenic and coagulation assays for antithrombin III in humans, monkeys, rabbits and dogs (*Fareed et al., 1985*) shows an identical heparin co-factor.

Fibrinolysis.
Fibrinolysis (*Fig. 3*) is achieved by the proteolysis of fibrin by activated plasmin. The inert precursor, plasminogen in plasma is converted by specific plasminogen activators (PAs). Intrinsic PAs include factor XII, factor XI and prekallekrein. They are activated when blood comes into contact with the foreign surface. Extrinsic PAs are liberated in almost all tissues including vascular endothelium (tissue plasminogen activator or t-pA) and urine (urokinase or U-PA). Like other factors it appears to be present in normal circulated blood as a proenzyme (pro-urokinase). Plasmin is inhibited mainly by alpha-2-antiplasmin, but substances such as alpha-2-macroglobulin, antithrombin III and C_1 INH appear to be active in vitro.

Fear stimulates in general the fibronolytic activity in animals and particularly in monkey, dog, pig and sheep. Animals should therefore receive tranquilizing doses (e.g. phencyclidine hydrochloride, 1 - 1.5 mg/kg) before venepucture. For anaesthesia sodium pentobarbitane is recommended as it does not influence coagulation or fibrinolytic tests. Gazeous anaesthetics should be avoided. Many animals including bovidae have no measurable fibrinolytic activity in blood.

The range of plasminogen levels is similar in human and animals. Plasminogen activators similar to urokinase are found in urine from baboons, but not in gorillas and orangutans. Low levels could be detected in concentrated samples of urine from mice. Most human and mammalian endothelial cells synthesize and secrete tissue plasminogen activators. Human, bovine and rabbit endothelial cell synthesize in vitro both tissue plasminogen activator and urokinase. Fibrinolytic inhibitors

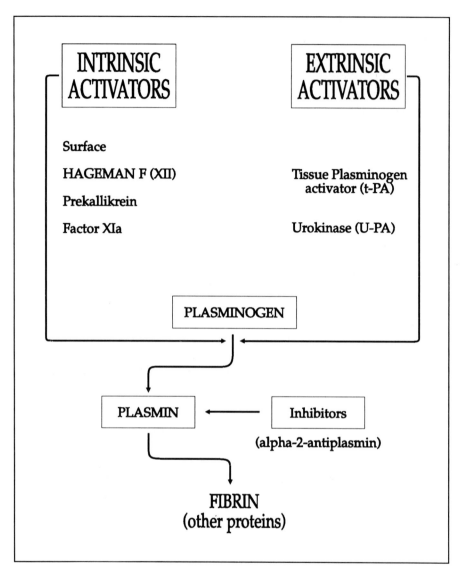

Fig. 3

The fibrinolytic system.

detected by global fibrinolytic inhibitor tests show that the level in monkeys and carnivores is similar to that in many other experimental animal models.

COMPARATIVE HAEMOSTASIS BIOLOGICAL TESTING

In the evaluation of haemocompatibility, commonly available standard reagents may be used in mammals, because their coagulation steps and pathways are similar to those of humans. As an example, the results of cardiovascular implant evaluations are given for different species summarized in *Tables XII, XIII* and *XIV*. Various coagulation tests and factor assays have been tested in humans, bamboons, dogs and rats. Bamboons seem to be most similar to humans despite e.g. increased factor XII activity and decreased levels of fibrinogen, kallekrein activity and alpha-2-macroglobulin together with an increased fibrinolytic activity. Haemocytological comparisons of bamboons and humans are shown in *Table XV*.

SELECTION OF ANIMAL SPECIES FOR IN VIVO HAEMOCOMPATIBILITY TESTING

In vivo testing of cardiovascular devices requires relatively large species (dogs, calves, sheep, goats or pigs). Species with favourable anatomy for certain cardiopulmonary surgical interventions are calf and sheep. Smaller animals such as rabbit, cat, guinea pig and rat are only useful under special circumstances.

Variations in drug sensitivity imply some general rules:

- The evaluation of low molecular weight heparin effect should only be performed on primates since rabbits, dogs and rats are generally not considered as "good responders",
- The testing of antiaggregants should only be made on primates due to individual variations even between inbred and animals.
- The evaluation of fibrinolytic drugs should be avoided in dogs (spontaneous fibrinolysis) and in rabbits (absence of streptokinase induced activation).

Table XII

Comparison of various coagulation tests (mean and range values) in Wistar rats (n = 20), in human and in mongrel dogs (n = 30).

TEST or FACTOR	Rat		Man		Dog	
	M	Range values	M	Range values	M	Range values
Platelet count x 10^9/l	1.590	1130 - 2050	254	120 - 400	257	200 - 300
APTT (sec)	21	14 - 25	25	22 - 28	12	10 - 14
Prothrombin time (sec)	11.1	10 - 12	13	11.5 - 14	6.5	4 - 7.5
Thrombin time (sec)	24	21 - 26	16	14 - 20	17	14 - 19
Reptilase time (sec)	20	18 - 24	18	16 - 20	18	16 - 20
Fibrinogen (g/l)	1.9	1 - 2.6	3	2 - 4	3	1.6 - 5
125^IFibrinogen half-time survival (day)			2.5	2.4 - 2.6	4	3 - 5
Factor II (%)	104	71 - 120	-	60 - 100	110	75 - 150
Factor V (%)	284	160 - 380	-	60 - 100	600	500 - 800
Factor VII + X (%)	186	110 - 260	-	60 - 100	260	150 - 300
Factor VIII:C (%)	60	40 - 150	108	50 - 200	250	150 - 400
Factor IX (%)	75	50 - 150	105	60 - 140	480	350 - 600
Factor XI (%)	61	45 - 106	-	55 - 180	290	150 - 350
Factor XII (%)	241	90 - 440	-	40 - 150	78	40 - 180
Euglobulin lysis time (hour)	9.3	6 - 12	2.1	2 - 3	1.4	0.5 - 3
AT III (%)	110	90 - 140	105	80 - 150	100	80 - 120
PDF (μg/ml)	< 10		< 10		< 10	

Table XIII

Comparison of the bleeding time and some platelet characteristics (mean ± SD) in humans and mongrel dogs.

TEST	Mongrel Dog			Man
	n	M	SD	M
Bleeding time (min) (Ivy)	30	5	± 2	6.30
Platelet count x 10^9/l	100	257	± 52	254 ± 53
In vitro platelet retention (%) (Hellem)	30	52	± 6	36
In vivo platelet retention (%) (Borchgrevink)	30	51	± 6	44
Half-time survival ^{51}Cr (days)	9	5.4		9
^{75}Seleno-methionine platelet survival	9	8.8		11.5

The dog model is widely used for haemocompatibility evaluation in spite of an increased platelet activity, thrombin generation and fibrinolytic activation. This model has also been validated in catheters (*Eloy, 1987*) and vascular prosthesis as predictive of clinical performance of biomaterials in humans (*Didisheim, 1984*). According to NIH guidelines (*NIH Publication, 1985*), the genetics, origin, health and nutrition status must be strictly controlled together with environmental factors in accordance with existing regulations for animal research facilities.

Selection of the test system.

If possible, human blood should be preferred for in vitro testing. Animals are necessary for in vivo studies and useful for some specific in vitro tests. Specific homologous reagents may be recommended for e.g. thromboplastin. All chronometric coagulation tests or chromogenic substrates can be employed in all mammals. Tests using antigen-antibody techniques may be human specific and can not be used on animal species.

Table XIV

Statistical comparison of various coagulation tests in human and baboon (n = 7) whole blood or plasma.

COAGULATION TESTS	Baboon		Man		Signifi-cance
	M	SD	M	SD	
Platelet count (x 10^9/l)	290.0	57.00	254.0	53.00	NS
APTT (sec)	29.4	4.60	29.4	0.90	NS
Prothrombin time (sec)	12.9	0.70	13.0	0.14	NS
Thrombin time (sec)	20.6	5.00	22.0	1.80	NS
Reptilase time (sec)	20.1	1.29	18.2	2.10	NS
Fibrinogen (g/l)	<u>1.9</u>	0.38	<u>3.1</u>	1.10	$p < 0.05$
Factor II (sec)	17.8	1.40	18.4	1.20	NS
Factor V (sec)	18.0	5.30	18.6	1.20	NS
Factor VII (sec)	18.1	0.90	20.2	0.90	NS
Factor X (sec)	39.1	2.70	41.9	0.40	NS
Factor VIII:C (sec)	40.4	3.60	36.7	5.20	NS
Factor IX (sec)	50.0	3.70	46.4	1.90	NS
Factor XI (sec)	32.4	1.80	35.3	6.60	NS
Factor XII (sec)	<u>30.6</u>	3.10	<u>39.4</u>	5.80	$p < 0.05$
Total Kallikrein activity (%)	<u>70.6</u>	17.40	<u>101.7</u>	3.70	$p < 0.05$
Antithrombin III (%)	106.7	12.90	103.7	13.00	NS
Whole blood lysis (hour)	<u>10.8</u>	8.20	$\geq$24.0		$p < 0.01$
PDF (mg/l)	< 10		< 10		
Total antiplasmin activity (%)	75.0	17.00	101.0	15.00	NS
α2-Macroglobulin (mg/dl)	118.0	7.90	149.0	10.00	$p < 0.05$

Table XV

Haemocytology of the baboon as compared to human.

| | Baboon normal values | | Human normal range | |
	M	SD	male	female
Leucocytes x 10^9	10.0	1.79	4-8	
Erythrocytes x 10^{12}	5.3	0.24	4.5-5.5	4.5
Haemoglobin (g/l)	142.6	5.80	145-165	125-145
Haematocrit %	41.5	2.20	40-54	37-47
MCV (fl)	78.0	1.60	82-98	
MCH (pg)	25.9	0.40	23-32	
MCHC (g/l)	335.8	6.90	320-350	

MCV   =   Mean Corpuscular Volume (fl)
MCH   =   Mean Corpuscular Haemoglobin (pg)
MCHC =   Mean Corpuscular Haemoglobin Concentration (gl)

Table XVI

Common anticoagulants for blood sampling.

Anti-coagulant	Composition		Amount required ml/ml of blood	Useful for
Sodium citrate	Trisodium Citrate	3.8 g	0.15	Coagulation tests, (e.g. Prothrombin time, APTT ...)
	Distilled water	100 ml		
EDTA	Potassium salt of ethyldiamide tetra-acetate	1.0 g	0.10	Platelet, white and blood cells counts, morphology
	0.07% saline solution (in liquid or dry form)	100 ml		

Handling and sampling techniques.

The animals should not experience fear in connection with anaesthesia, sampling and surgery since it may mobilize systems of coagulation and fibrinolysis as well as influence platelet in reducing number and function and lead to a refractory state. The choice of anaesthetic agent is of importance and should be selected to each specific study. As an example, ether induces fibrinolysis in the dog, but does not modify platelet function in the rat. The choice of anticoagulant is important since:

- heparin may interfere with platelets in rabbit and dog
- exposure of sodium citrate reduces the plasmatic level of Ca^{2+} thereby impairing the platelet responses to agonists in dog, rat, pig and rabbit.

The usual anticoagulants are given in *Table XVI*. However, some specific anticoagulant solutions should be selected such as:

a. For FpA determination in blood sampling:

Sodium citrate	0.11 M
Heparin	1000 IU/ml
Aprotinin	1 TIU/ml
Sodium Azoture	1 g/l

b. For determination of ß-thromboglobulin and platelet factor IV:

PGE 1	0.33 μg/ml
Theophyllin	10 mM
Na_2 EDTA	78 mM (pH 7.4)

Blood sampling depends on the animal species. In larger animals blood samples may be carefully drawn from superficial veins, using 21 gauge butterfly needles (dogs, primates, rabbits) and a double syringe technique (the blood sample collected in the first syringe is discarded and only the sample of the second syringe is used for analysis). Blood should be drawn slowly to avoid collapse of the vein. Anticoagulants may be introduced either in the syringe or in the test tube. If it is technically feasible, a direct sampling in the test tube is preferable.

The major clinical problems with prosthetic biomaterials in contact with blood are: thrombosis, embolism, excessive pseudointimal proliferation, infection and calcification all leading to failure of the prosthesis.

CONCLUSIONS

Animal models are not justified for screening of materials, but are "irreplacable" for predicting long term behaviour in man, providing simulated conditions of use: rheological conditions, geometry of the device, non anticoagulated whole blood environment etc. Extrapolation of experimental in vivo results to humans must, however, be accompanied by a careful analysis of the specific interaction of each animal under the given experimental conditions. Nevertheless, there is sufficient evidence for the need, usefulness and relevance of animal models before entering into clincal studies.

REFERENCES

Belleville, J., Thouvez, J.P., Mikaeloff, P. & Descotes, J. (1966) Path. Biol., 14, 41-55.

Chandrasekhar, N. & Laki, K. (1968) in "Fibrinogen" (Laki, K. ed.) pp. 117-130, Marcel Dekker, New York.

Didisheim, P., Devanjee, M.K., Frisk, C.S., Kaye, M.P. & Fass, D.N. (1984) in "Host response-clinical applications - New technology and legal aspects" (Boretos, J.W. & Eden, M., eds.) pp. 132-179, Noyes Publication, Park Ridge, USA.

Dodds, W.J. (1974) in "Progress in Haemostasis and thrombosis" (Spaet, T., ed.), vol. 2, pp. 215-247, Grune et Stratton, New York and London.

Dodds, W.J. (1978) in "Platelet: A multidisciplinary approach" (de Gaetano, G. & Garatini, S., eds.) pp. 45-59, Raven Press, New York.

Doolittle, R.F. (1983) Ann. N.Y. Acad. Sci., 408, 13-27.

Eloy, R., Belleville, J., Paul, J., Pusineri, C., Baguet, J., Rissoan, M.C., Cathignol, D., French, P., Ville, D. & Tartullier, M. (1987) Thromb.Res., 45, 223-233.

Fareed, J., Kumar, A., Rock, A., Walenga, J. & Davis, P. (1985) Sem. Thromb. Haemost., 11, 2, 138-154.

Leroy, J. (1987) in "Le sang et les vaisseaux" (Caen, J. ed.) pp. 215-233, Hermann, Paris.

Meyers, K.M. (1986) in "Platelet response and metabolism" (Holmsen, H. ed.) vol. 1, pp. 209-234, CRC Press, Boca Raton, Fl., USA.

Meyers, K.M., Katz, J.B., Clemmons, R.M., Smith, J.B. & Holmsen, H. (1980) Thromb.Res., 20, 13-20.

Mitruka, B.M. & Rawnsley, H.M. (1977) Clinical, biochemical and haematological reference values in normal experimental animals, 272 p., Masson Pub., New York, Paris.

Muller-Esterl, W. (1987) Sem. Thromb. Haemost., 13, 1, 115-126.

NIH Publication (1985) "Guidelines for blood material interactions", 85, 2185.

Nurden, A.T., Butcher, P.D. & Hawkey, C.M. (1977) Comp. Biochem. Physiol., 5613, 407-413.

Sinakos, Z. & Caen, J. (1967) Thromb. Diath. Haemorrh., 17, 99-111.

Chapter II

TEST METHODS FOR THE DETECTION
OF MECHANICAL PROPERTIES

Chapter eds.: W. Lemm[1] & S. Dawids[2]

[1]   Department of Experimental Surgery, Rudolf-Virchow-Clinic, Location Charlottenburg,
D - 1000 Berlin 19

[2]   Institute of Engineering Design, Biomedical Section, Technical University of Denmark,
DK - 2800 Lyngby

INTRODUCTION

Mechanical failures of implanted prostheses are serious events and may lead to unpleasant and deleterious consequences for the patient (ruptures in vascular grafts, broken heart valves or leakages in the diaphragm of an implanted blood pump). A material which is used for the fabrication of medical devices must thus basically comply with the maximum mechanical demands.

In other engineering technologies the components which are highly stressed are regularly replaced by new parts to prevent untimely or severe breakdowns. In general such a replacement is not feasible on implantable materials because of additional risks for the patients. In medicine the components must therefore be designed to comply fully with the maximal range of physiological stresses of the tissue which they replace. Requirements for the socalled biofunctionality of an artificial implant often exceeds the mechanical demands of comparable technical components. As examples heart valves, vascular grafts and similar prostheses should work safely for at least ten years without risks of failures. In practice it is expected to function for the rest of the expected life time of the patient.

This makes it obvious that basic information and data of the mechanical behaviour of prosthetic materials are of tremendous importance with regard to the choice of the material and to the design of the device. The evaluation of the mechanical stability should include, inter alia, investigations of the abrasion, the rate and mode

35

S. Dawids (ed.), Test Procedures for the Blood Compatibility of Biomaterials, 35–43.
© 1993 *Kluwer Academic Publishers. Printed in the Netherlands.*

of degradation of synthetic materials which are exposed continuously to the aggressive biological environment.

Increased and prolonged stress on a material will make it yield. In this respect elastomeric materials are especially vulnerable. Mechanical tests of medical elastomers have to include at least the following methods which are described in this chapter:

I.   Test for short time stress and strain.

II.   Test for tensile stress.

III.   Test for tear strength.

IV   Test for dynamic stress and strain.

V.   Test for material fatigue.

VI.   Test for creep, relaxation and set.

VII.   Tests for material compliance.

VIII.   Test for porosity of vascular grafts.

On national and European levels an increasing number of test procedures have been standardized for the mechanical characterization and for the quality control of flexible materials. All the standards may be obtained from the national standard bureaus. A list of these are provided at the end of the chapter. They are also described in the relevant literature (see *References* at the end of this introduction).

The International Standard Organization (ISO) coordinates the national activities of standardization to equalize national diversities of testing conditions and formulates internationally accepted test descriptions. These international standard methods can be applied on the mechanical characterization of medical polymers as well.

Because of these already existing and precisely described test procedures a recapitulation of these methods in all details can be omitted without conflicting with the basic intention of this book.

All tests mentioned in this chapter are referred to the relevant ISO methods and provided with the exact ISO specification.

Furthermore, the international standardization of mechanical tests should be regarded as a positive model for a desirable standardization of medical test procedures.

Standards of tests.

Within the frame the international standard methods developed for technical purposes can be adapted to the medical fields of application to qualify a new polymer or to control the batches of already established materials to obtain the permission for clinical utilization. But far more internationally coordinated and accepted examination methods are required. The medical demands exceed in many cases those required for the technical specifications.

Within the frame of EC numerous standards are made uniform by the member countries on medical utensils and their testing. Those who want to introduce particular mechanical test procedures will obtain detailed descriptions on any method from the below mentioned national ISO-members.

LIST OF INTERNATIONAL STANDARDS ORGANIZATION MEMBERS

AUSTRALIA
Standards Association of Australia (SSA)
Standards House
80-86 Arthur Street
North Sydney
NSW 2060

AUSTRIA
Österreichisches Normungsinstitut (ON)
Heinestrasse 38
Postfach 130
A - 1021 Wien

BELGIUM
Institut Belge de Normalisation (IBN)
29, Ave. de la Brabanconne
B - 1040 Bruxelles

CANADA
Standards Council of Canada (SCC)
International Standardization Branch
2000, Argentina Road, Suite 2-401
Mississauga, Ontario L5N IV8

CHINA
China State Bureau of Standards (CSBS)
P.O.Box 820
Beijing

CYPRUS
Cyprus Organization for Standards and
Control of Quality (CYS)
Ministry of Commerce and Industry
Nicosia

CZECHOSLOVAKIA
Urad pro Normalizaci a Mereni (CSN)
Václavské Námesti 19
113 47 Praha 1

DENMARK
Dansk Standardiseringsraad (DS)
Aurehøjvej 12
Postbox 77
DK - 2900 Hellerup

FINLAND
Suomen Standardisoimisliitto SFS (SFS)
P.O.Box 205
SF - 00121 Helsinki

FRANCE
Association Francaise de Normalisation (AFNOR)
Tour Europe, Cedex 7
F - 92080 Paris La Défense

GERMANY, F.R.
DIN Deutsches Institut für Normung (DIN)
Burggrafenstrasse 4-10
Postfach 1107
D - 1000 Berlin 30

GREECE
Hellenic Organization for Standardization (ELOT)
Didotou 15
GR - 10680 Athens

HUNGARY
Magyar Szabványügyi Hivatal (MSZH)
1450 Budapest 9
Pf. 24.

INDIA
Indian Standards Institution (ISI)
Manak Bhavan
9, Bahadur Shah Zafar Marg,
New Delhi 110002

IRELAND
National Standards Authority of Ireland (NSAI)
Ballymun Road
IRL - Dublin 9

ISRAEL
Standards Institution of Israel (SII)
42, University Street
Tel Aviv 69977

ITALY
Ente Nazionale Italiano di Unficazione (UNI)
Piazza Armando Diaz 2
I - 20123 Milano

JAPAN
Japanese Industrial Standards Committee (JISC)
c/o Standards Department
Agency of Industrial Science and Technology
Ministry of International Trade and Industry
1-3-1 Kasumigaseki, Chiyoda-ku
Tokyo 100

KOREA, REP. OF
Bureau of Standards
Industrial Advancement Administration (KBS)
2 Chungang-dong Kwach'on-myon
Kyonggi-do 171-11

NETHERLANDS
Nederlands Normalisatie-Instituut (NNI)
Kalfjeslaan 2
P.O.Box 5059
NL - 2600 GB Delft

NEW ZEALAND
Standards Association of New Zealand (SANZ)
Private Bag
Wellington

NORWAY
Norges Standardiseringsforfund (NSF)
Postboks 7020
Hormansbyen
N - 0306 Oslo 3

PORTUGAL
Direcco-Geral da Qualidade (DGQ)
Rua José Estêvao, 83-A
P - 1199 Lisboa Codex

SAUDIA ARABIA
Saudi Arabian Standards Organization (SASO)
P.O.Box 3437
Riyadh-11471

SINGAPORE
Singapore Institut of Standards and
Industrial Research (SISIR)
Maxwell Road
P.O.Box 2611
Singapore 9046

SOUTH AFRICA, Rep, of
South African Bureau of Standards (SABS)
Private Bag X 191
Pretoria 0001

SPAIN
Instituto Espanol de Normalizacion (IRANOR)
Calle Fernandez de la Hoz, 52
E - 28010 Madrid

SWEDEN
Standardiseringskommissionen i Sverige (SIS)
Tegnergatan 11
Box 3295
S - 103 66 Stockholm

SWITZERLAND
Swiss Association for Standardization (SNV)
Kirchenweg 4
Postfach
CH - 8032 Zürich

TURKEY
Türk Standardlari Enstitüsü (TSE)
Necatibey Cad. 112
Bankanliklar
Ankara

UNITED KINGDOM
British Standards Institution (BSI)
2, Park Street
London W1A 2BS

USA
American National Standards Institute (ANSI)
1430 Broadway
New York, N.Y. 10018

USSR
USSR State Committee for Standards (GOST)
Leninsky Prospekt 9
Moskva 117049

REFERENCES

Annis, D. (1987) Protocol to determine the average porosity of vascular grafts. Royal Liverpool Hospital, Univ. of Liverpool.

Brown, R.P. (1979) Physical Testing of Rubbers. Applied Science Publishers Ltd. London.

Danusso, F. (1989) "Mechanical Properties of Polymers" in "Polymers: Their Properties and Blood Compatibility" (ed. S. Dawids), pp.195-212, Kluwer Academic Publishers, Dordrecht.

DIN-Taschenbuch 18 (1971) Materialprüfnormen für Kunststoffe, Kautschuk und Gummi. Beuth-Vertrieb GmbH, Berlin.

Guidon, R., King, M., Marceau, D. & Cardou, A. (1987) J. Biomed. Mat. Res. 21, 65-87.

Murabayashi, S. (1987) "Compliance Effect on Patancy of Small Diameter Vascular Grafts" Report on the National Heart, Lung and Blood Institute, Bethesda, Md USA.

Zartnack, F., Henning, E., Ott, F. & Bücherl, E.S. (1983) "Development and fatigue testing of a new blood pump". Proc. X Ann. Meeting of ESAO, Life Support System I, 13-16.

Test for Short Time Stress and Strain
(Shore Hardness Measurement)

Author: W. Lemm

Department of Experimental Surgery, Rudolf-Virchow-Clinic, Location Charlottenburg,
D - 1000 Berlin 19

INTRODUCTION

Short time stress and strain tests release information about the resistance of a material exposed to a single unrepeated deformation. In some particular cases this deformation will culminate in the destruction of the sample yielding ultimate characteristics.

Name of method:

Shore hardness measurement.

Aim of method:

By applying a sudden mechanical load on the material a reproducible basic information on its initial elasticity is obtained.It is generally known as Shore hardness measurement.

Physical background of the method:

Shore hardness measure the initial elesticity of the material. Many polymers are elastomers, but yield when exposed to strain. This pseudoelasticity is not pronounced at low stress, but increases with force and time of deformation. Methods which provide an objective and fast hardness measurement of an elastomer are used for the adjustment of production parameters, for the quality control of the final product and the effect of e.g. sterilization procedures, ageing and determination of shelf life. The test should be done under standardized conditions especially concerning temperature, because the elastomeric properties are temperature dependent.

S. Dawids (ed.), Test Procedures for the Blood Compatibility of Biomaterials, 45–47.
© 1993 *Kluwer Academic Publishers. Printed in the Netherlands.*

The Shore hardness is defined as the resistance of a material to the indentation of a cone or cylinder.

A rigid ball, cylinder or cone to which a force (spring load) is applied indents into the test sample. Pocket types of Shore hardness meters are provided with a scale on which the hardness in Shore units is indicated. The hardness is given in Shore A (ball), C (cylinder) or D (cone) units in a range between 0 and 100 (ISO 48, ISO 1400, ISO 1818, DIN 53 505).

Scope of method:

The method is commonly used as an industrial routine procedure. Some polyurethane types are indicated commercially with their Shore hardness (Pellethane 2363-80A or Pellethane 2363-55D).

DETAILED DESCRIPTION OF METHOD

Equipment.

Shore hardness meter. Several types of equipment are commercially available.

Outline of method.

The performance of this test requires a Shore hardness meter either available as a pocket type or in a desk version. The Shore A and C meter is provided with a body in the form of a cylinder made of stainless, polished steel. The diameter of the cylinder is 0.79 mm with a spring load force applied. In the Shore D version, a cone with spherical end (diameter 0.1 mm) is used.

Description of procedure.

The device is positionated on the test sample. The surface area must be smooth and flat and at least 30 mm in diameter. The thickness of soft (hard) materials has to be at least 6 mm (3 mm). The reading should be taken 3 sec. after the instrument has been applied to the test piece, in case of creeping of the material 15 sec. after the application. This deviation from the standards has to be declared in the protocol. The body (cylinder) is released and recoils. The recoil is a measure for the Shore hardness. In the desk versions, the fall is 140 mm and the recoil may be up to 100 mm.

The standard method requires at least three tests of each sample performed with an impact distance of more than 5 mm between each and more than 13 mm from the border of the test sample.

Safety aspects.

No special precautions are necessary.

RESULTS

No data are available.

DISCUSSION

This test provides useful but not very detailed results. The method can only indicate the physical or chemical dimensions like e.g. the degree of cross-linking in the material, provided a number of standard measures are made.

Limitations of method.

The test temperature must be 20° ± 2°C, especially for elastomers. In the case of larger deviations, they have to be specified in the test protocol

Test for Tensile Stress

Author: W. Lemm

Department of Experimental Surgery, Rudolf-Virchow-Clinic, Location Charlottenburg,
D - 1000 Berlin 19

INTRODUCTION

Name of method:

Test for tensile stress.

Aim of method:

A specially shaped test piece is slowly stretched in one dimension to measure the force/elongation relationship of the test sample.

Physical background of the method:

This well-defined standard method provides a general clue to the quality of an elastomer with an extended scale of information. It gives an indication of the content of plasticizers or fillers as well as the degree of cross linkages and of the effect of stretching. An applied constant tension stressing and deforming an elastomer causes the polymeric macromolecules to form a high degree of orientation. The distances between the polymer chains are reduced creating a large amount of reversible intra-molecular secondary bonds which improves the tensile strength of the material for further deformation.

A typical stress strain curve (*Fig. 1*) illustrates all the steps of response to increased stress produced in the continuously elongated test sample. At different elongations (50, 100 and 300%) the stress is recorded as well. The ultimate tensile strength and the elongation at break end the graph. The increasing slope of the curve points to the internal reinforcements of the material by generation of additional cross linkages. If the slope flattens out, it indicates creeping or weakening through degradation of the chemical bonds.

S. Dawids (ed.), Test Procedures for the Blood Compatibility of Biomaterials, 49–53.

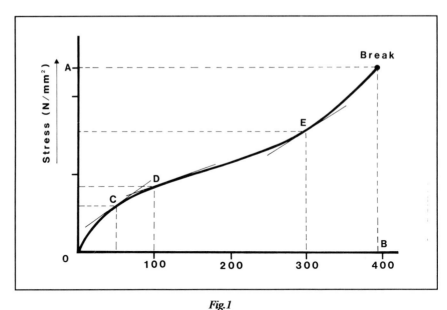

Fig.1

Stress-Strain plot of an elastomer:

A: *ultimate tensile strength*   D: *stress at an elongation of 100%*

B: *elongation at break*   E: *stress at an elongation of 300%*

C: *stress at an elongation of 50%*

Scope of method:

This method is widely used as an industrial procedure to obtain reliable information on the mechanical properties of an elastomer to which all the major international and national standards are basically in agreement.

The tensile strength test is also a valuable tool to quantify the effect of bio-degradation in such cases where synthetic materials are exposed to biological environment for a considerable period of time.

DETAILED DESCRIPTION OF METHOD

Equipment.

Tensile strength machine, INSTRON Company, High Wycombe (London), UK; Canton, Mass., USA.

Outline of method.

Preparation of test samples.
Dumb-bell shaped test pieces whose dimensions are given in *Fig. 2* are strained at a constant rate. The force and the corresponding extensions are continuously recorded. The forces are expressed as stresses related to the original cross-sectional area of the test piece (ISO 37). The thickness of the samples should be between 0.7 and 2.0 mm. Usually the dumb-bell shaped test pieces are stamped out of films using sharp tools if the thickness of the film does not exceed 2.5 mm.

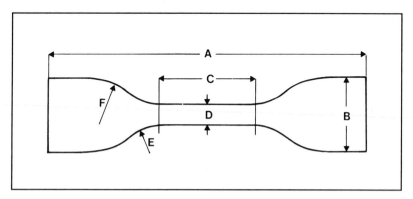

Fig. 2

Dumb-bell shaped test piece (redrawn from "Physical Testing of Rubbers", R.P. Brown, 1986)

Dimensions	Type I (mm)	Type II (mm)
A	115	75
B	25 ± 1.0	12.5 ± 1.0
C	33 ± 2.0	25.0 ± 1.0
D	6 ± 0.4	4.0 ± 0.1
E	14 ± 1.0	8.0 ± 0.5
F	25 ± 2.0	12.5 ± 1.0

Test machine.
The test machines are commercially available and cover the basic demands described in the standard methods (INSTRON Company).

Description of procedure.

The samples are fixed in the clamps of the tensile machine with the extensometer. The tensile properties of a material depend also on the speed of extension and

52

therefore the test pieces have to be stretched under unified conditions: 1% per minute.

Safety aspects.

No special precautions are necessary.

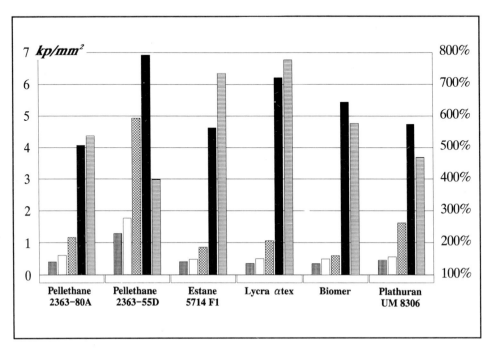

Fig. 3

Comparison of the mechanical properties of some polyurethane elastomers.

1. Column: stress at an elongation of 50%   3. Column: stress at an elongation of 300%
2. Column: stress at an elongation of 100%   4. Column: ultimate tensile strength
5. Column: elongation at break (%)

RESULTS

An important and commonly used characteristic of a material is the modulus of elasticity or Young's modulus. It is the resistance of a material to its deformation or the slope of the stress-strain curve. But in most cases the relation between stress and elongation is not linear. Therefore the stress at an elongation of 50, 100 and 300% is often documented beside the ultimate tensile strength and the elongation at break.

As an example the mechanical properties of some polyurethane elastomers are compared in *Fig. 3*. The results are the average of 8 - 10 individual tests.

DISCUSSION

The clinical field of application requires primarily a mechanical stability of the synthetic material of which a therapeutic device is made. Data obtained by tensile stress tests may attest an initial durability. In contact with body fluids these basic mechanical properties might change. The chapter on *Detection of Biodegradation* describes the effects of altered mechanical properties of biomaterials exposed to a biological environment.

Limitations of method.

There are no limitations of this method if the material is elastomeric and can be mounted in the equipment.

Test for Tear Strength

Author: W. Lemm

Department of Experimental Surgery, Rudolf-Virchow-Clinic, Location Charlottenburg,
D - 1000 Berlin 19

INTRODUCTION

Name of method:

Test for tear strength.

Aim of method:

To determine the tendency to tear and to provide information about an elastomer in the case of damages and latent imperfections of the material.

Physical background of the method:

Deviating from the tensile strength tests, the tear strength tests are performed with test samples carrying a slight discontinuity at the test area which may vary in its shape, geometry and dimension (hole, cavity, cut, incision, puncture, any kind of inhomogeneity etc.). Hence additional information about an elastomer in the case of damages and slight injuries is recognized.

This test procedure is similar to the previously mentioned tensile strength test. The contribution of a geometrical discontinuity to the mechanical failure of the material is precisely detected.

These measurements are performed with a tensile test machine (see: *Test for Tensile Stress*) but with specially shaped tear test samples. An example for the proposed dimensions of such tear test samples is given in *Fig. 1* (ISO R34, ISO 816).

S. Dawids (ed.), Test Procedures for the Blood Compatibility of Biomaterials, 55–57.
© 1993 *Kluwer Academic Publishers. Printed in the Netherlands.*

Scope of method:

The method is used as an industrial routine procedure. Apart from a few generally used standard methods, other test procedures for specific purposes have been developed for example the stich tear test or the needle tear test which has its relevance for biomaterials testing as well (e.g. vascular grafts or surgical patches).

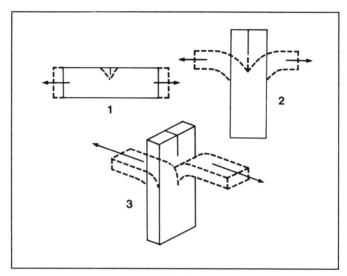

Fig. 1

Three different types of tear test samples (redrawn from "Physical Testing of Rubbers", R.P. Brown, 1986)

DETAILED DESCRIPTION OF METHOD

Equipment.

Tensile strength machine, INSTRON Company, High Wycombe (London), UK; Canton, Mass., USA.

Outline of method.

The test itself is performed in context with other mechanical tests described in this chapter but with specially shaped tear test samples. They are inserted into the clamps of the machine and stretched until break. The thickness of the samples should be between 0.7 and 2.0 mm. All the current standards specify a stretching rate of 500 ± 50 mm/min.

Description of procedure.

The samples are fixed in the clamps of the tensile machine with the extensometer. The tensile properties of a material depend also on the speed of extension and therefore the test pieces have to be stretched under unified conditions: 1% per minute.

Safety aspects.

No special precautions are necessary.

RESULTS

Certainly the results will heavily depend on and vary with the large number of different geometries of the discontinuity in the test piece. Hence the significance of the test results must be carefully evaluated.

DISCUSSION

The clinical application requires primarily mechanical stability of the elastomer from which a therapeutic device is made. Data from test for tear strength should be applied on the designed material to provide a realistic picture of the performance. It should be noted that degradation might create cracks from which tear can spread. Therefore the test can be useful in the design of elastomeric devices.

Limitation of method.

No known limitations exist. However, it is important to test a number of identical specimens to determine the scatter of measurements.

Test for Dynamic Stress and Strain

Author: W. Lemm

Department of Experimental Surgery, Rudolf-Virchow-Clinic, Location Charlottenburg,
D - 1000 Berlin 19

INTRODUCTION

Dynamic properties of elastomers are important in a large number of engineering and medical applications, for example the movements of heart valves or diaphragms of blood pumps, the bending of pace maker leads, the inflation and deflation of intra-aortic balloon pumps, the squeezing of tubes in roller pumps of extracorporeal circuits and similar devices.

In a dynamic test the material is exposed to cyclic and repeated deformations while the stress and strain are monitored. Basically, there are two types of dynamic motions: a free vibration in which the test sample is set in oscillation and the amplitude is allowed to decay according to the damping behaviour of the test material and a forced vibration in which the oscillation is maintained (ISO 2856).

Name of method:

Test for dynamic stress and strain.

Aim of method:

If a test piece is stressed by a forced oscillation with regular periods of stretching and relaxing, and both the force and the elongation are recorded continuously, a hysteresis loop is obtained (*Fig. 1*) which reflects the time related strain and relaxation of the material.

S. Dawids (ed.), Test Procedures for the Blood Compatibility of Biomaterials, 59–62.
© 1993 *Kluwer Academic Publishers. Printed in the Netherlands.*

Physical background of the method:

An ideal and perfect spring would not exhibit hysteresis, but a viscoelastic material needs a certain time to regain its basic dimensions thereby giving rise to a hysteresis loop.

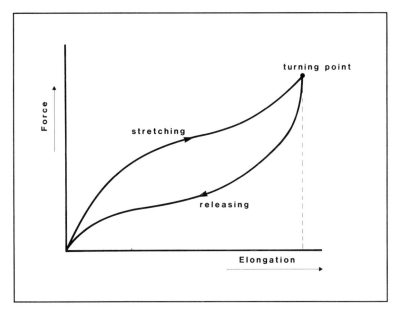

Fig. 1

Hysteresis loop of an elastomer stressed by a forced oscillation.

The long term repetition of such strain cycles may induce progressive changes in the dynamic properties, because they depend strongly on the oscillation frequency and the temperature.

In some cases the material needs an infinite long period of time to recover. It is thus deformed permanently (residual permanent deformation). On molecular level the internal organization of the polymer chains has reached its maximum.

Scope of method:

The method is used as an industrial routine procedure. The relevance for the biomedical field is for example the determination of the residual permanent deformation of the diaphragm in blood pumps. The moving diaphragm is not allowed to touch the housing to avoid haemolysis. On the other hand the moving distance of the diaphragm should be as large as possible because it limits the stroke volume and the output of the pump.

The restoring time to open the full inner cross-section of tubes in roller pumps should be as short as possible to avoid insufficient blood flow.

DETAILED DESCRIPTION OF METHOD

Equipment.

Tensile strength machine, (INSTRON Company, High Wycombe (London) UK; Canton, Mass., USA) adapted to oscillating deformations.

Outline of method.

The performance of the test requires the same dumb-bell shaped test pieces as described in *Test for Tensile Strength*. The tensile test machine must allow a cyclic and variable operation over a considerable period of time. The adapted plotter or computer unit displays the hysteresis loop (*Fig. 1*) and its time depending variations.

Description of procedure.

The samples are fixed in the clamps of the tensile machine with the extensometer. The tensile properties of a material depend also on the speed of extension, and therefore the test pieces have to be stretched under unified conditions: 1% per minute.

Safety aspects.

No special precautions are necessary apart from those arising from the use of the particular machine.

RESULTS

No general results are available.

DISCUSSION

Oscillating deformation at various temperatures (preferably body temperature) can provide information on the dynamic response in relation to body tissue. Long term oscillating deformations can be used in the field of ageing and fatique of the material. Individual tests of specimens are necessary to provide information on a particular production.

Limitation of method.

No particular limitations are given of the method. However, it is important to realise that material may undergo changes due to sterilization, storage and implantation.

Test for Material Fatigue

Author: W. Lemm

Department of Experimental Surgery, Rudolf-Virchow-Clinic, Location Charlottenburg,
D - 1000 Berlin 19

INTRODUCTION

Name of method:

Test for material fatigue.

Aim of method:

The test material is exposed to repeated cyclic deformation until break to determine the fatigue properties.

Physical background of the method:

Fatigue of a material occurs as a change of the properties after a long time of repeated cyclic deformations, preferably the decrease of mechanical strength. It is important to realize that the breakdown of the test sample does not only depend on the material itself. The geometry of the sample, the type of stress and the environmental conditions also influence the results of such tests. Thus the testing conditions are not generally standardized. They should as far as possible simulate the future demands and simulate the exposition in use. Only two international standards exist (ISO 132 and ISO 133) which describe the initiation of cracks and cut growth.

The relation of flex life and the applied stress or strain is given by the "Wöhler curve" (*Fig. 1*). If stress or strain is reduced the flex life will finally increase to infinity.

S. Dawids (ed.), Test Procedures for the Blood Compatibility of Biomaterials, 63–66.
© 1993 *Kluwer Academic Publishers. Printed in the Netherlands.*

Scope of method:

Different methods to test fatigue are used in engineering technologies. Data on flex life and fatique are of immense importance for permanently implanted medical devices or prostheses to predict their liability in vivo. In cases of mechanical failures in vivo, the immediate death of the patient might be the consequence.

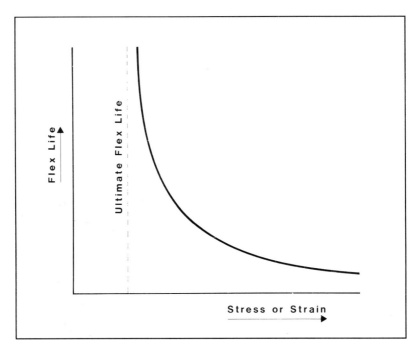

Fig. 1

Wöhler curve describing the dependence of flex life on the stress or strain.

DETAILED DESCRIPTION OF METHOD

Equipment.

There is no standardized test equipment available. The test machine has to simulate the impact on the material in vivo.

Outline of method.

The specimen (e.g. a vascular prosthesis) must be set up to mimic the conditions concerning flow, pressure, biological fluid and temperature. The exposures should be the highest physiological possible (e.g. hypertension, fever). Flexlife and stress/strain can be related in an empirical fashion to the expected life time. Acceptable life times of vascular prostheses should exceed 20 years.

Description of procedure.

A fluid containing enzymes suitable for the test should be made aseptic. The fluid should be circulated at elevated temperatures and replaced at suitable intervals to maintain enzymatic activity. Individual test stands must be made, but it is advisable to provide a set-up which is easily disassembled for individual cleaning and sterilization.

Safety aspects.

No particular safety aspects can be mentioned.

RESULTS

No general results are available. However, an example is given of degradation of vascular prostheses based on glutaraldehyde stabilized collagen bovine vessels. It can be seen that the decay of collagen appears to be progressive.

DISCUSSION

Because of additional influences, as mentioned above, which limit the fatigue life of an elastomeric material, it is recommended to test original devices under the very circumstances these devices are exposed to later in reality.

Materials for the manufacturing of heart valves or blood pumps (artificial hearts or extracorporeal circulatory assist devices) are tested as identical devices in water at 37°C (*Zartnack et al., 1983*).

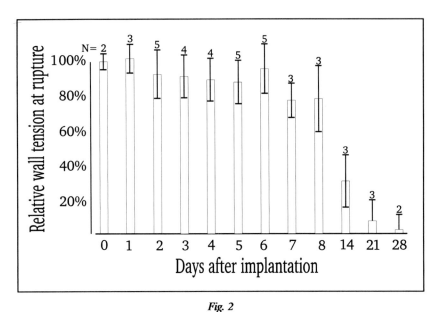

Fig. 2

Tensile strength at rupture for modified collagen vascular material. A progressive weakening of the material occurs simultaneously with a replacement by surrounding tissue build up. (S.Dawids, unpublished data).

Flex life tests are highly time consuming and consequently accelerated test frequencies are generally used. In such way, obtained numbers of flex cycles which were tolerated by the test device until break cannot be converted to an expected life time under realistic conditions.

It was generally found that the flex life in vivo is lower than in vitro. Furthermore, the attack of the blood components must be taken into consideration. An in vivo test would be fatigue tests of original prosthesis material in animal experiments as shown in *Fig. 2*. Additional impacts of the biological environment like calcification, enzymatic attack or ageing by oxidation and similar ill known events on the fatigue life of the moving parts of a prosthesis are included by these in vivo experiments.

Limitation of method.

Because of the large number of effects which influence the results of fatigue and flex life tests, standardized procedures do not exist. The test conditions should adapt as far as possible the environment and circumstances to which the device is finally exposed.

Test for Creep, Relaxation and Set

Author: W. Lemm

Department of Experimental Surgery, Rudolf-Virchow-Clinic, Location Charlottenburg,
D - 1000 Berlin 19

INTRODUCTION

Name of method:

Test for creep, relaxation and set.

Aim of method:

Under a constant force of deformation (stretching), a longtime response of the test material is studied for the properties of creep, relaxation and set.

Physical background of the method:

Creep is the consequence of a longtime effect. Under a constant force the deformation either compression or elongation of a test sample increases. Although the performance of a creep test is very simple no precise international standard method exists (ISO DR 748, British Standard 903, DIN 53 444).

Relaxation is the change of stress under a constant longtime strain. Stress relaxation tests can be performed in compression, shear or tension (ISO DIS 3384),

Set is defined as the recovery after removal of an applied stress or strain. Set is expressed as the percentage of the original thickness of the test piece (ISO 815, ISO 1633, ISO 2285)

Scope of method:

The method is used as an industrial routine procedure. Within the frame of test methods for the blood compatibility of biomaterials creep, relaxation and set have

S. Dawids (ed.), Test Procedures for the Blood Compatibility of Biomaterials, 67–70.
© 1993 *Kluwer Academic Publishers. Printed in the Netherlands.*

no particular importance. Artificial bone material should certainly maintain under compression its original dimensions. As already mentioned the moving diaphragm of a pulsatile blood pump should sustain as well its basic size to avoid haemolysis in the case of touching the housing.

DETAILED DESCRIPTION OF METHOD

Equipment.

Tensile strength machine (INSTRON Company, High Wycombe (London), UK; Canton, Mass., USA) adapted to test creep, relaxation and set.

Outline of method.

Creep, relaxation and set are obtained by applying one test method. The test sample is exposed to a constant force either of tension or compression.

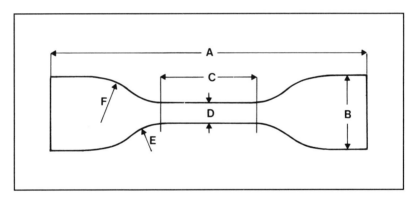

Fig. 1

Dumb-bell shaped test piece (redrawn from "Physical Testing of Rubbers", R.P. Brown, 1986)

Dimensions	Type I (mm)	Type II (mm)
A	115	75
B	25 ± 1.0	12.5 ± 1.0
C	33 ± 2.0	25.0 ± 1.0
D	6 ± 0.4	4.0 ± 0.1
E	14 ± 1.0	8.0 ± 0.5
F	25 ± 2.0	12.5 ± 1.0

A test sample (*Fig. 1*) is fixed in a regular tensile strength test machine and loaded with a permanent force of tension maintaining constant environmental conditions.

Creep is defined as the change in length with time (%).
Relaxation is defined as the change of stress with time (%).

After the permanent force is removed, the test sample is allowed to recover.

Set is defined as the change in length with time after the removal of an applied stress (%).

Description of procedure.

The samples are fixed in the clamps of the tensile machine with the extensometer. The tensile properties of a material depend also on the speed of extension and therefore the test pieces have to be stretched under unified conditions: 1% per minute.

Safety aspects.

No special precautions are necessary.

RESULTS

No standard results are available.

DISCUSSION

The described method is a very conventional test procedure for the evaluation of specimens. Routine control can provide good information of both the uniformity of the material and of the production parameters. It is necessary to build up a base of data for each individual design and use of material. This method can not stand alone, but should be used together with a number of other test procedures described in this book.

Limitation of method.

The test time may vary and should be adapted to later demands. DIN 53 444 suggests to expose the test sample exactly a time of 1 min. The creep, relaxation and set of the parameters are recorded after this period.

A longer test period requires a constant temperature, in general room temperature (23°C) and a constant relative air humidity. Both parameters may seriously influence the reproducibility of results and should not vary more than ± 2°C and ± 3%.

Tests performed under compression need disc test pieces (ISO DIS 3384). They have to be 29 mm in diameter and 12.5 mm thick. This standard method suggests to apply a compression of either 15% or 25% at a temperture of 23°C.

Test for Material Compliance

Author: W. Lemm

Department of Experimental Surgary, Rudolf-Virchow-Clinic, Location Charlottenburg,
D - 1000 Berlin 19

INTRODUCTION

Name of method:

Test for material compliance.

Aim of method:

To detect the compliance (elasticity) of manufactured items or test pieces by applying a sudden and repeated pressure wave on the item (e.g. a vascular graft), its elasticity and in particular its increasing volume or diameter.

Physical background of method:

The compliance is of particular interest for artificial vascular grafts. It is suspected that a change in compliance occurs at the junction between the natural blood vessel and an inserted vascular graft and might cause thromboembolic events, because the normal pulse wave of the blood flow is disturbed.

The mathematical expression for the compliance is:

$$C = \Delta V / \Delta P \ (ml/kPa)$$

This indicates the expansion of a vascular tubing when an increased pressure is applied.

A modified definition for the compliance is given by the ratio of an increasing diameter of the vascular graft as the consequence of a growing pressure:

71

S. Dawids (ed.), Test Procedures for the Blood Compatibility of Biomaterials, 71–73.
© 1993 *Kluwer Academic Publishers. Printed in the Netherlands.*

$$C = \Delta d / \Delta p \quad (\%/kPa)$$

An average compliance value for vascular grafts is suggested as (measured at 100 mm Hg or 13.33 kPa):

$$12.7 \pm 2.0 \times 10^{-2} \quad (\%/mmHg)$$

A standardized and generally accepted procedure to test the compliance does not exist.

Scope of method:

Apart from other properties like thromboresistance, porosity and stability to kinking, the patency of artificial vascular grafts depends on an accurate compliance.

DETAILED DESCRIPTION OF METHOD

Equipment.

No commercial equipment exists. A test bench must be developed individually and designed for specific requirements.

Outline of method.

As the material in use often has visco-elastic properties, it is important to apply repeated sudden pressure waves to the specimen. A slow increase of pressure should be applied to measure the static compliance of the specimen. Although the test bench should be able to generate pressure and flow parameters which exceed physiological levels (to provoke a rupture of the specimen), the long term test should be set at levels at the high end of physiological values.

Description of procedure.

The test procedure is performed according to the mathematical definition for the compliance.

The test object, a piece of a 10 cm long vascular graft is occluded at one end and tightly connected at the other end to a small calibrated glass pipette. The whole equipment is now filled with distilled water and loaded with a pressure of 100 mm Hg (or 13.33 kPa). Air bubbles must be strictly avoided. The expansion of the tubing can be directly taken from the calibration of the glass pipette.

Safety aspects.

No specific precautions are required. However, bacterial or fungal growth should be prevented due to their possible high enzymatic action upon the material.

RESULTS

Until now no results exist.

DISCUSSION

The test protocol should include an extention of this procedure. The device should be tested at various pressure loads to obtain comprehensive information on the compliance behaviour.

Limitation of method.

Only non-porous tubings and vascular grafts are intended to be tested by this method.

Test for Porosity of Vascular Grafts

Author: W. Lemm

Department of Experimental Surgery, Rudolf-Virchow-Clinic, Location Charlottenburg,
D - 1000 Berlin 19

INTRODUCTION

All synthetic vascular prostheses or surgical patches available at present are porous: the woven types made of poly-(ethylene-terephthalate) well known as Dacron® or the fenestrated types made of expanded poly-(tetrafluorethylene) with its trade name Gore-Tex®.

Name of method:

Test for porosity of vascular grafts.

Aim of method:

To measure the leak rate of water through the wall of the graft at a constant hydrostatic pressure (26.66 kPa = 200 mm Hg).

Physical background of the method:

The water porocity test registers the flow rate of water through the wall of the graft at a constant hydrostatic pressure of 200 mm Hg (26.66 kPa).

Scope of method:

The characterization of the porosity is of importance for a safe utilization of vascular grafts. But no standardized test method exists to determine the porosity towards blood.

S. Dawids (ed.), Test Procedures for the Blood Compatibility of Biomaterials, 75–77.
© 1993 *Kluwer Academic Publishers. Printed in the Netherlands.*

DETAILED DESCRIPTION OF METHOD

Equipment.

No commercially available equipment exists. An individual test bench must be developed and designed for specific demands.

Outline of method.

The vascular graft is mounted in the test bench and a specific pressure using pure water, plasma (or a similar viscous substance like dextran solution) can be applied at a constant pressure. The oozing of fluid should be measured repeatedly as a gradual clotting of porosities occurs or the material swells by absorption of fluid.

Description of procedure.

The test object, a piece of a 100 - 150 mm long vascular graft is occluded at one end and tightly connected at the other end to a small calibrated glass pipette. The whole equipment is now filled with distilled water and exposed to a pressure of 200 mm Hg (26.66 kPa). The water penetration through the wall can directly be taken from the calibration of the glass pipette by reading the fallen water level. - For micro-porous grafts the water must be highly purified.

Safety aspects.

No specific precautions are necessary.

RESULTS

There are no general results available.

DISCUSSION

The hydrophilic or hydrophobic properties of the polymer from which the graft is made has a large influence on the water porosity of the micro-porous graft. Micro-porous grafts of poly-(tetrafluorethylene) PTFE are almost totally impervious to water, but once wetted with alcohol they become freely permeable. The water porosity of grafts made of poly-(ether-urethanes) increases progressively with an increasing duration of exposition to water (*Annis, 1987, Guidoin et al., 1987*).

Limitation of method.

It is very obvious that the test issue depends on many influences which limit the suitability of this method. A generally agreed and standardized test procedure is urgently required.

It remains doubtful whether the water porosity measurement is a suitable indicator of the possibility of excessive loss of blood through the wall of the vascular graft.

For highly porous grafts this method may not be suitable because of the pressure drop in the system.

Chapter III

TEST METHODS FOR THE DETECTION OF BIODEGRADATION

Chapter ed.: W. Lemm

Dept. of Experimental Surgery, Rudolf-Virchow-Clinic, Location Charlottenburg,
D - 1000 Berlin 19

INTRODUCTION

Permanent or longtime implants of polymeric material are exposed to a biological
environment which gradually degrades them. Widely used prostheses intended to
remain functioning in the body up to the lifetime of the patient are heart valves,
insulators for pace maker leads, vascular grafts, permanent catheters, implanted
blood pumps either as support systems or as total artificial hearts (as "bridging" to
a transplant).

The gradual alteration of material properties, leading occasionally to the disruption
of an implanted device by the environmental impact of the living organism is called
biodegradation. These changes of the material may have a detrimental influence
on biocompatibility and biofunctionality.

In general all polymers are susceptible to degradation, but the conditions under
which polymers degrade vary within wide ranges. Degraded materials have changed
their chemical, mechanical and biochemical properties and in some cases the
surface quality which implies serious consequences for the blood compatibility.

Cellular blood components adhere preferably on such partly degraded surfaces,
release clotting factors and induce thrombus formation. Subsequent mineralisation
starts from degenerated cells, forming small strongly attached calcium-carbonate
plaques which gradually transform into crystaline calcium-hydroxy-apatite (*Hennig,
1981*).

S. Dawids (ed.), Test Procedures for the Blood Compatibility of Biomaterials, 79–87.
© 1993 *Kluwer Academic Publishers. Printed in the Netherlands.*

Polymeric materials degrade in two ways: either by hydrolytic or by oxidative scissions of the polymer backbone releasing low molecular fragments, oligomers or monomers which might be toxic.

Hydrolysis.

Despite great efforts to study the kinetics of biodegradation, its mechanism in vivo is not yet clearified in all details. While the influence of temperature, moisture, adhered proteins etc. is generally known, the attack of e.g. phagocytes on local spots of a polymer surface is not definitely understood. Some hypotheses try to describe the different phenomena. It is, however, generally accepted that before any significant degradation occurs, the polymer surface has to be hydrophilic. A hydrolysis has to take place at a physiological pH-value and at body temperature.

It is known that some unspecific proteases and esterases are able to catalyse a hydrolysis of condensation polymers such as polyesters and polyamides. Papain, trypsin and chymotrypsin cleaves the urethane bonds in polyether urethanes (*Lemm et al., 1986*). Normally enzymes are known as proteins with highly substrate specific, bio-catalytic properties. Their catalytic effect on synthetic polymers which are insoluble in water cannot be interpreted in a common way, although some polymers are composed of ester-, amid-, urea- and other chemical bonds being potentially susceptible to enzymatic attack. Consequently relatively unstable polymers such as poly-(glycolic acid) and poly-(lactic acid) exhibit a hydrolytic degradation in the presence of enzymes. Enzymatic attack on normally stable polymers requires that they are to some extent hydrophilic and allow the interaction of enzyme molecules with the polymer matrix before this type of degradation process is initiated (*Williams et al., 1977*).

Oxidation.

Beside these hydrolytical aspects of degradation, an oxidative mechanism is observed with poly-ethylene, poly-propylene and poly-styrene (*Smith et al., 1987*).

Other poorly defined factors may also be involved, leading to unexpected degradation mechanisms within the body. Insulators of pace maker leads made of poly-ether-urethane are preferably destroyed around the area where the electrode penetrates the wall of the blood vessel. In case of an infected surrounding of an artificial implant, the process of degradation is highly accelerated.

Leaching.

A third mechanism is a simple leaching of low molecular additives or oligomers by the body fluids, creating first defective areas on which further degradation proceeds.

Swelling.

Apart from these pure chemical attacks on primary covalent bonds within the backbone of the polymer, the simple incorporation of low molecular biological compounds like lipids, organic acids and metabolites or water weaken the secondary bonds within the polymer structure like plasticizers. This enlarges the distance between the polymer chains and causes the polymer to swell. The consequences are reduced mechanical strength and increased flexibility and softness. Stress-strain tests of polymer samples in a standardized shape illustrate these effects (*Fig. 1*) (See also chapter: *Detection of Mechanical Properties*).

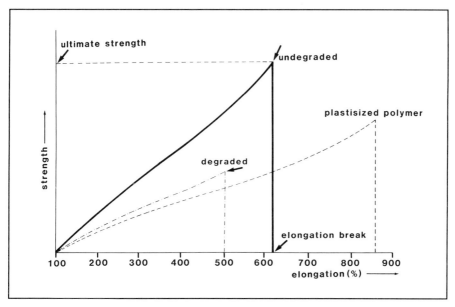

Fig. 1

Stress-Strain-Plot of elastomers. Degraded and undegraded materials.

The reasons for a reduced tensile strength are thus either scissions of the primary bonds within the polymer chain or a lower degree of crosslinkages or both. An increased elongation break point indicates an incorporation of plasticizing compounds.

Scissions of the polymer chain are reflected in a reduction of the average molecular weight or a changed molecular weight distribution of the polymer. In general, synthetic polymers never have a uniform molecular weight.

The best method to study both aspects simultaneously is the gel permeation chromatography (GPC) or exclusion chromatography. *Fig. 2* shows the molecular weight distribution of a polymer. M_n is called the number average molecular weight of the sample and represents the most frequent molecular weight (maximum of the distribution curve). M_w is called the weight average molecular weight. Experimental details on the GPC method will be described in the methodical part.

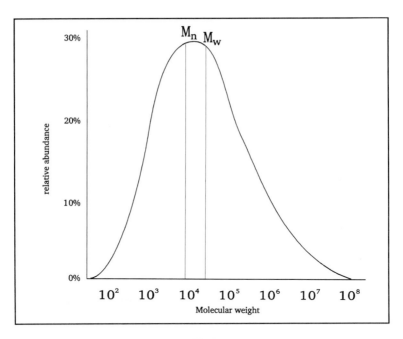

Fig. 2

Molecular weight distribution and average molecular weight.

As an example a polyurethane (Pellethane 2363-80A) shows the relationship between the average molecular weight and the ultimate mechanical properties, elongation at break and ultimate tensile strength. The polyurethane can be degraded slowly in solution (400 days at 55°C) and the process can be monitored by taking samples at distinct time intervals for preparation of thin films for testing (see chapter: *Detection of Mechanical Properties*). In *Fig. 3* the changes are shown in the three parameters comparing their interaction. The decreasing viscosity of the polymer solution provides a sensitive indication of degradation as well.

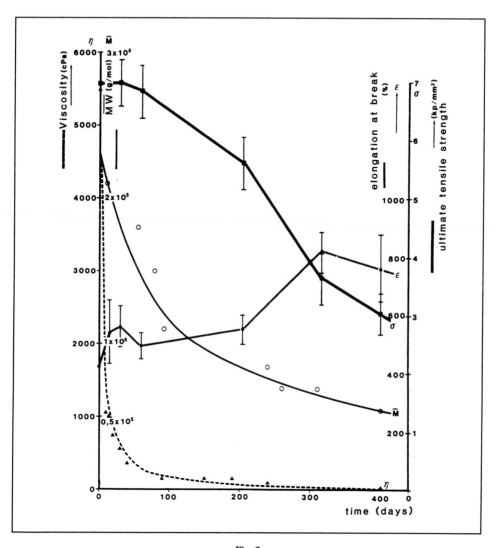

Fig. 3

Fig. 3

Dependence of ultimate mechanical properties of the decreasing average molecular weight of a thermally degraded polyurethane solution.

σ = *ultimate tensile strength*   M = *average molecular weight*
ϵ = *elongation at ultimate tensile strength* η = *viscosity of the polyurethane solution.*

Degradation (as demonstrated in *Fig. 3*) leads to

1. decreased ultimate tensile strength (σ)

2. decreased average molecular weight (M) in nearly the same way as σ

3. increased elongation at break (ϵ) because low molecular fragments in the material plasticize the molecular structure.

Surface changes.

The surface destruction caused by the attack of the biological environment can be visualized by any kind of microscopy, preferably by Scanning Electron Microscopy (SEM) which gives excellent three-dimensional images of the defects. Detailed test is given in chapter: *Surface Analysis*.

Ruptures and cracks of the surface of poly-urethane samples implanted sub-cutaneously for months into the backs of rats arise from defects or imperfections in the surface during manufacture or subsequent handling (*Lemm et al., 1981*).

In the presence of cytotoxic compounds (e.g. an organo-tin compound) the polyurethane surface seems to be protected for months (*Marchant et al., 1986*).

One of the current hypotheses suggests that the origin of such cracks and ruptures is a stress cracking of the soft polyether segments within a poly-etherurethane. The first step is the uptake of water which makes the polymer swell and weakens the hydrogen bonds of the hard urethane segments. The second step is a phase separation of two incompatible parts within the polymer: the amorphous and the crystalline segments. The hydrophillic polyether chains then move towards the surface where they are exposed to subsequent oxidation (*Weimer et al., 1984*).

Another hypothesis explains the crack formation as a simple leaching of low molecular parts of the polymer.

Whatever it be, the most realistic way to test all kinds of effects of biodegradation of medical polymers is by implantation of samples in their intended environment. However, these in vivo experiments as well as the subcutaneous implantation into the backs of the rats are very time consuming. The time of exposure should exceed

6 months. Furthermore, it is not clear if the outcome of such experiments can indicate the expected time of biofunctionality in contact with blood.

Therefore methods have been developed to simulate and accelerate the effect of biodegradation in vitro. The LINI Test which is generally accepted by the manufacurers of textile fibers is used to investigate the leaching of colours from textile fibers as well as the stability of synthetic fibers exposed to washing procedures (*LINI Test, 1965*).

STRATEGIES TO MINIMIZE THE DEGRADATION PROCESS

The most successful method to minimize the effect of biodegradation is the selection of resistent materials. Recently, new polyurethane types are described with no loss of mechanical properties or changes of the average molecular weight up to a period of 24 months of subcutaneous implantation or in contact with blood (*Braun et al., 1988*).

The fact that degradation preferentially starts at surface inhomogeneities demonstrates the necessity of an extremely careful preparation of blood contacting surfaces. Dust particles, bubbles and all similar imperfections must be avoided carefully.

In the case where the first step of degradation is a leaching of low molecular substances (oligomers or additives), an extraction of the raw material is recommended before the polymer solution is prepared. This extraction has to be adjusted individually to each polymer. The polymer solutions must be handled dustfree and stored under a nitrogen blanket, excluded from light or exposure to elevated temperatures and humidity.

When secondary bonds in the polymer provide increased mechanical strength, a thermal treatment (annealing) of a medical device made from e.g. polyurethane elastomer will improve the degree of crystallinity and consequently the stability of the polymer structure.

DEGRADABLE MATERIALS

In recent years there has been a growing interest for polymers intended to degrade slowly after implantation. The material will provide a mechanical or therapeutic support until the ingrowing natural tissue can take over its biological task. Surgical suture material is the best known application.

Prostheses for the replacement of sections of the oesophagus or the trachea is resorbed after 4-6 months. Within this period the natural tissue has formed a new tubular organ.

Vascular grafts are covered with a neointima and collagen fibers after a few months. A mechanical support is still considered to be necessary after this regeneration to avoid aneurism formation, because the normal structure is not achieved.

Subcutaneously implanted drug release systems are in clinical use today. They are slowly resorbed by their environment, providing a constant drug release.

Artificial skin for severely burned victims for temporary covering has been developed.

The mentioned examples reflect an important and rapidly expanding field of applied biomaterial research.

It is a difficult problem to design polymer compositions with degradation kinetics that are precisely adjusted to the healing process or regeneration of the natural organ.

REFERENCES

Braun, B., Grande, P.U., Lehnhardt, F-J, Jerusalem, C. & Hess, F. (1988) VASA, J. Vascular Diseases, Suppl. 22, 1-38.

Hennig, E. (1981) Proc. Europ. Soc. Artif. Organs, 8, 76-80.

Lemm, W., Krukenberg, G., Regier, R. & Gerlach, K. (1981) Proc. Europ. Soc. Artif. Organs, 8, 71-75.

Lemm, W. & Bücherl, E.S. (1986) in "Blood Compatible Materials and Their Testing" (Dawids, S. & Bantjes, A. eds.) pp.257-265, Martinus Nijhof Publishers, Dordrecht, Boston, Lancaster.

LINI Test (1965) Meilliand Textilberichte, 9, 1008-1009.

Marchant, R.E., Anderson, J.M. & Dillingham E.O. (1986) J. Biomed. Mat. Res., 20, 37-50.

Smith, R., Oliver, C. & Williams, D.F. (1987) J. Biomed. Mat. Res. 21, 991-1003.

Weimer, E. & Schaldach, M. (1984) Biomed. Technik, 29, 218-221.

Williams, D.F. & Mort, E. (1977) J. Bioeng., 1, 231-238.

Boiling Test for In Vitro Degradation of Biomaterials

Author: W. Lemm

Dept. of Experimental Surgery, Rudolf-Virchow-Clinic, Location Charlottenburg,
D - 1000 Berlin 19

INTRODUCTION

Name of method:

Boiling test for in vitro degradation of biomaterials.

Aim of method:

Screening test for accelerated degradation of polymers by boiling.

Physical background for the method:

Because in vivo studies of biodegradation are time consuming, simulation tests have been developed to get preliminary information on the relative stability of soluble polymers. One of these is the so-called "boiling test" which must be considered as an in vitro screening method to preselect biomaterials. The result of chemical manipulations to achieve a higher stability (e.g. cross-linking, stretching, annealing etc.) of the polymer structure can be identified within a short period by e.g. molecular weight determination with Gel Permeation Chromatography (GPC) (synonymous: Exclusion Chromotography = ECG) following a boiling test.

The experimental conditions of the boiling test are a slight modification of the established LINI Test which is carried out with a washing machine in which the samples are treated under reproducible conditions with different detergents and temperatures up to 135°C (*LINI Test, 1965*). A modification of the LINI Test being adjusted to the demands of biodegradation is the boiling test.

Polymer samples are boiled for up to 100 hours in distilled water either in the presence or under exclusion of oxygen. The indication for the progressing

89

S. Dawids (ed.), Test Procedures for the Blood Compatibility of Biomaterials, 89–96.
© 1993 *Kluwer Academic Publishers. Printed in the Netherlands.*

degradation process is the decreasing average molecular weight and the altered molecular weight distribution obtained by the GPC method. Dump-bell shaped samples are required to investigate the stress-strain-behaviour after the boiling (see chapter: *Detection of Mechanical Properties*).

In the presence of oxygen during boiling, both the hydrolytic and the oxidative breakdown of the polymer structure can be observed. Under exclusion of oxygen only the hydrolytical attack will occur.

Chemical bonds susceptible to hydrolysis may be analysed quantitatively by a chemical saponification as well: the samples are boiled for four hours in an alcoholic potassium hydroxide solution. Under these conditions e.g. ester-, urethane- and, if present, urea-groups are hydrolysed completely whereas ether-bonds remain stable. The saponification number represents the consumption of potassium hydroxide (mg KOH/g).

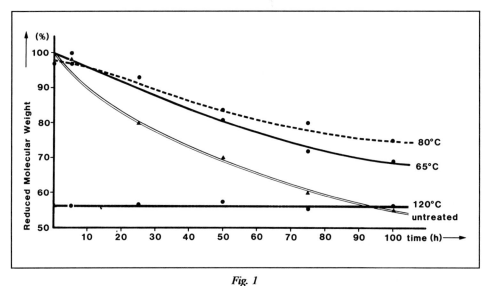

Fig. 1

Degradation in boiling water after a thermal treatment in polyurethane samples.

A thermal treatment of polyurethane elastomers may improve their stability. As an example: Polyurethane samples are annealed for seven days at different temperatures (65°C, 80°C and 120°C) protected by a nitrogen blanket. Finally the samples are boiled 100 hours in distilled water in the same way as described before. The reduced average molecular weight is analysed by GPC, and the results are compared to the untreated material (*Fig. 1*). A temperature of 120°C seriously degrades the material to a larger extent. An annealing temperature of 65°C does not very much improve the basic stability of the material against the conditions of the boiling test. The annealing process performed for seven days at 80°C yields the most resistant material.

The effect of manipulation of the polymer structure can be studied by this test method as well. Generally an amorphous polyurethane is less stable than the same material in its crystalline state.

Scope of method:

A general screening test to predict the resistance to degradation. The boiling procedure over 100 hours is followed by a number of analyses.

DETAILED DESCRIPTION OF METHOD

Equipment.

High Pressure Liquid Chromatography = HPLC-equipment for GPC with four separation columns in pore sizes of 10^{-4}, 10^{-5}, 10^{-6} and 10^{-7}m (Styragel®).

UV detector or differential refractometer with recorder. Several organic solvents in a purity for liquid chromatography. The choice for a particular solvent depends on the polymer (tetrahydrofuran, N,N-dimethyl-formamide, toluene, benzene, methylketones).

Two-neck glass flask (500 ml), reflux condenser, inlet-tube for nitrogen or oxygen.

Polymer reference standard: Polystyrene

Outline of method.

Samples are cut out of the polymer film with a thickness ≤ 0.2 mm and introduced in a two-necked flask filled with 300 ml of distilled water.

For polymers which cannot be dissolved in organic solvents, it is advisable in all cases to prepare dumb bell shape samples prior to boiling.

If the hydrolytical degradation is to be studied exclusively, the water has to be distilled previously in a nitrogen atmosphere or thoroughly boiled over nitrogen just prior to the procedure. During the boiling experiment, the access of oxygen has to be prevented. A continuous addition of nitrogen is recommended.

If a test of both the hydrolytic and the oxidative degradation of the polymer is intended, no particular care has to be taken of air exposure.

Description of procedure.

The polymer samples are boiled under reflux conditions up to 100 hours. After time intervals (0, 6, 12, 24, 48, 72 and 100 hours) specimen are removed and dried.

For soluble polymers 2% solution of the test material is then prepared in an adequate solvent (e.g. polyurethanes in N,N-dimethyl-formamide or tetrahydro-furan) and injected together with the internal standard (toluene) in the chromatograph where the identical solvent circulates in a constant flow rate of 1.5 ml/min percolating the separation columns each filled with gel particles of the same pore sizes (*Fig. 2*). A suggested 4 column system separates soluble polymers within a molecular weight range between 10^6 and 10^4 Dalton.

After the polymer has separated into the different molecular weight fractions they pass either an UV-detector or a differential refractometer connected to a recorder or computer unit providing a graphic display of the molecular weight distribution.

A polymer with cleaved covalent bonds will show a relatively lower average molecular weight and a different molecular weight distribution where oligomers or even monomers can be identified.

The GPC-method must be calibrated beforehand with a set of polymer standards of identical molecular weights which should cover the expected molecular weight range. The widely used and commercially available polymer standard is polystyrene.

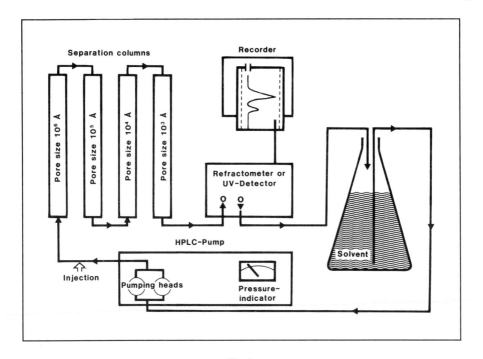

Fig. 2

Diagram of GPC equipment for the determination of the average molecular weight and molecular weight distribution.

Measurement of saponification.

The polymer film (1g) is placed in a 1 neck bottle (500 ml) with 300 ml ethanol 96% and potassium hydroxide 0.5 M. The sample boils for 4 hours. The degree of saponification is indicated by the hydrolysis and determined volumetric by titration with 0.5 M hydrochloric acid.

Safety aspects.

Apart from cautious handling of the boiling strong alcalic solution, no special precautions are necessary.

RESULTS

Soluble polymers.

To quantify the effect of degradation in the boiling test, a socalled half-life is defined, indicating the boiling time necessary before the tested material has lost 50% of its original average molecular weight. *Table I* compares the half-life of some selected materials.

Table I

The degradation of some biomaterials in boiling water.

Material	Chemical Composition	Half-life (Hours)
Pellethane 2363-80 AE	poly-ether-urethane	31
Pellethane 2363-55 D	poly-ether-urethane	21
Estane 5714	poly-ether-urethane	> 100
Lycra 420	poly-ether-urea-urethane	69
Biomer	poly-ether-urea-urethane	74
Plathuran 8300	poly-ester-urethane	75
Avcothane 51	poly-ether-urethane-PDMS-block-copolymer	> 100
Kraton 2104	synthetic rubber	> 100

DISCUSSION

Interpretation of results.

In some cases an increasing molecular weight with GPC analysis is observed after the exposure to boiling water. This is also the case in vivo. In spite of previous cleavages or cross-linkings an explanation is that oligomers are leached out of the polymer structure. The remaining solid part simulates a higher average molecular weight than before.

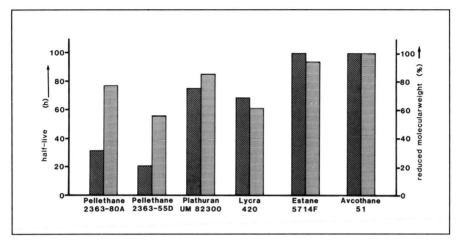

Fig. 3

Comparison of in vitro and in vivo degradation of some polyurethanes.
1st column: Half-life in vitro (h)
2nd column: Reduced average molecular weight after six months
of subcutaneous implantation (%).

Limitations of method.

Only soluble polymers can be tested by GPC-analysis. In cases of unsoluble materials the extent of degradation after the exposure to boiling water can only be identified by altered mechanical parameters.

Normally thermoplastics like polyethylene or polypropylene require high boiling hydrocarbon fractions to be dissolved at elevated temperatures. In such cases the whole GPC-analyser must be modified to be operated at this temperature where these solutions are stable. Therefore the method of the melt index is much more common in polymer technology to get an information on the molecular weight and molecular weight distribution.

All polymeric materials degrade to some extent under the different experimental conditions. The degradation in vitro correlates quite well with the results of the in vivo test as shown in *Fig. 3* where the half-life of several materials in the boiling test is compared to the results obtained from in vivo experiments with standard subcutaneous implantation in rats. It can be seen that the two test methods correlate reasonably well.

REFERENCE

LINI Test (1965) Meilliand Textilberichte, 9, 1008-1009.

Enzymatic In Vitro Degradation of Polymers

Author: W. Lemm

Dept. of Experimental Surgery, Rudolf-Virchow-Clinic, Location Charlottenburg,
D - 1000 Berlin 19

INTRODUCTION

Name of method:

Enzymatic in vitro degradation of polymers in the presence of enzymes.

Aim of method:

The following in vitro method should be understood as a suggestion how to manage individual studies on the interaction between polymers and enzymes to predict the expected biofunctionality of a biomaterial in-situ.

Physical background for the method:

Enzymatic degradation of polymers is a useful method to make individual studies on the interaction. This can provide some understanding of the expected biofunctionality of a biomaterial in vivo. A number of enzymes can be used for these studies as presented in *Table I*. These enzymes attack specific ester- and peptidebonds by hydrolysis.

Commercial enzymes vary a great deal in purity and specific activity. Furthermore concentration of enzymes in a physiological buffer will decrease with time due to normal degradation and bacterial attack. Thus an antibacterial additive is required (sodium azide). Gel-Permeation-Chromatography (GPC) is recommended as method to indicate the progression of the degradation process. Scissions of the polymers back-bone are recognized by a lower average molecular weight and a change of molecular weight distribution (peak shoulders, appearance of low molecular fractions etc.). In some rare cases no changes of the molecular weight distribution is observed. In these instances it should be realised that the split

97

S. Dawids (ed.), Test Procedures for the Blood Compatibility of Biomaterials, 97–103.
© 1993 *Kluwer Academic Publishers. Printed in the Netherlands.*

products may leach into the buffer solution and thus escape the GPC analysis. For GPC analysis the polymer should be dissolved in a suitable solvent. Some polymers for extrusion are generally more difficult to dissolve, but can in many cases be brought into solution in hot chlorobenzenes. GPC analysis can then be made at elevated temperatures (60° - 80°C) although this requires special adjustments of the equipment to work at elevated temperatures.

In solution enzymes loose their activity gradually. Therefore enzyme solutions must be replaced every 5-7 days during the testing.

Table I

Selection of hydrolases, their hydrolytic activity and polymers which might be attacked.

Enzyme	Hydrolytic Activity	Target Polymer	Company
Esterase	ester-bonds	polyester	Serva
Trypsin	peptides, amides	proteins	Serva
Trypsin	peptides, amides	polyamides	Serva
Trypsin	peptides, amides	polyurethanes	Serva
Papain	peptides, amides	polyurethanes	Serva
Chymotrypsin	peptides, amides	polyurethanes	Serva
Amylase	glycolysis	polysaccharides	Serva
Amylase	glycolysis	(polyether)	Serva
Cellulase	glycolysis	polysaccharides	Serva

Scope of method:

This method is suitable for experimental and screening purposes.

DETAILED DESCRIPTION OF METHOD

Equipment.

HPLC-equipment for GPC with four separation columns in pore sizes of 10^{-4}, 10^{-5}, 10^{-6} and 10^{-7}m (Styragel®)(*Fig. 1*).

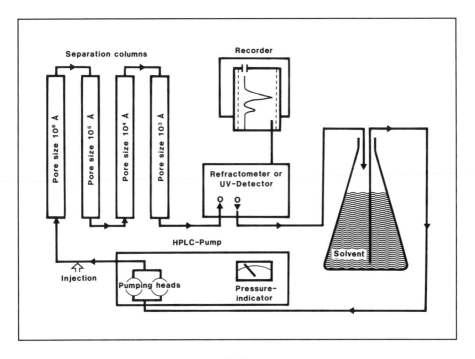

Fig 1

Diagram of GPC equipment for the determination of the average molecular weight and molecular weight distribution.

UV detector or differential refractometer with recorder. Several organic solvents in a purity for liquid chromatography. The choice for a particular solvent depends on the polymer (tetrahydrofuran, N,N-dimethyl-formamide, toluene, benzene, methylketones).

Glass flasks (100 ml) with wide neck.

Outline of method.

Samples are cut out of the polymer film with a thickness ≤ 0.2 mm and introduced in a flask filled with 30 ml of enzyme solution. The flask is sealed and left at room temperature in a dark place. Every day the content is mixed gently. Every 5 days the enzyme solution is replaced.

For polymers which cannot be dissolved in organic solvents, it is advisable in all cases to prepare dumb bell shape samples prior to testing.

Description of procedure.

The polymer samples are carefully washed in distilled water - in some cases with added detergent - and carefully rinsed and dried. The samples are preferably circular specimens (diameter: 10mm) cut out of films with a thickness of 0.1-0.2 mm.

The enzyme solution is prepared with the powdered enzyme in clean glass flasks. The concentration of each enzyme has to be adjusted to a final activity of 100 units/ml for all enzymes. A phosphate buffer solution should be used to adjust the enzyme solution to pH = 7.3. To each buffer solution sodium azide (NaN_3) is added in a concentration of 0.02% to protect the enzymatic proteins from bacterial attack.

For each material, up to 10 equal samples are prepared. One sample acts as control - i.e. pH and sodium azide added but no enzyme.

After intervals of 1 week, 2 weeks, 1 month, 2 months, 3 months, 4 months, 6 months, 12 months a sample is removed, carefully washed, dried and dissolved in a suitable solvent for the subsequent GPC analysis (detailed description, see: *Boiling test*).

Safety aspects.

No special precautions are necessary. One should avoid prolonged exposure of unprotected skin to the enzyme solutions due to their ability to protein hydrolysis. In rare cases allergic reactions may occur.

RESULTS

In *Fig.* 2 the degradation of a copoly-(poly-ethylene-oxide-ethylene-terephthalat) (60:40) is shown in the presence of different enzymes. This material is intended to degrade.

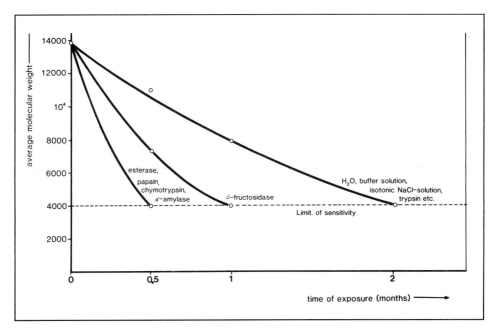

Fig. 2

Degradation of copoly-(polyethyleneoxide-ethylene-terephthalat) (60:40) in different enzyme solutions.

Figs. 3 and *4* show the decreased average molecular weights of a poly-etheruretha-ne in comparison to a poly-esterurethane after six months in several enzyme solutions as indicated in the figures. The polyester-type is obviously more degraded than the poly-etherurethane.

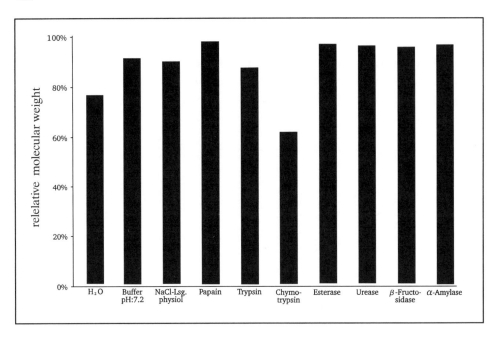

Fig. 3

Degradation of a poly-ether-urethane in different enzyme solutions.

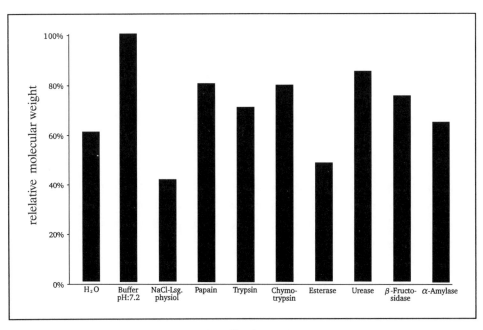

Fig. 4

Degradation of a poly-ester-urethane in different enzyme solutions.

DISCUSSION

Although the enzyme catalysed hydrolysis of polymers is a heterogenous reaction, it can be shown that some hydrolases are obviously responsible for biodegradation effects. The test results indicate that even stable polymers can be attacked by not very substrate specific enzymes.

Limitations of method.

The described method has to be considered only as a frame within which the investigation of enzymatic degradation of polymers can be carried out. Alternative materials require modified test conditions with another selection of enzymes individually adapted to the actual demand.

Some of the suggested enzymes are normally not present in the blood and in the majority of soft tissues. A potential degradation depends on the enzymes present in the surrounding tissue as well.

Tissue Implantation Test

Author: W. Lemm

Dept. of Experimental Surgery, Rudolf-Virchow-Clinic, Location Charlottenburg,
D - 1000 Berlin 19

INTRODUCTION

Name of method:

Tissue implantation test.

Aim of method:

In vivo testing of polymer degradation by subcutaneous implantation in the rat.

Physical background of method:

Although in vivo studies of biodegradation are extremely time consuming, there is no doubt that such experiments carried out in a relevant fashion provide the most realistic results concerning the biodegradation of synthetic materials. This is in spite of the lack of detailed insight in the complex degradation process. Even so, it remains controversial whether the results obtained by this tissue implantation test can used as indicant for the expected biofunctionality in blood.

The degradation mechanism of a polymer depends on whether it is in contact with blood or if it is surrounded by tissue. Degradation studies in blood environment are difficult to perform because additional events such as surface induced thrombus formation which often leads to occlusion or embolus formation limit the experiment. Therefore the technique of subcutaneous implantation into the back of animals is generally preferred.

The choice of the experimental animal will influence the results as well. For example, a polyurethane sample implanted for 30 months into the back of a dog

S. Dawids (ed.), Test Procedures for the Blood Compatibility of Biomaterials, 105–113.
© 1993 *Kluwer Academic Publishers. Printed in the Netherlands.*

will rarely show significant degradation, while the same material will degrade seriously within 9 months of implantation into the back of rats.

A comparison of results obtained from various in vivo biodegradation studies therefore requires a critical consideration of these facts before interpretation.

Scope of method:

A general purpose test for resistance to degradation in vivo.

DETAILED DESCRIPTION OF METHOD

Equipment.

Young rats of Wistar strain, each weighing 150 - 200 g (aged approximately 4 months).

Facilities for sterile surgery.

Outline of method.

The polymer samples are carefully washed in distilled water - in some cases with added detergent - carefully rinsed and dried and then sterilized. The samples should preferably be circular specimens (diameter: 10mm) cut out of films with a thickness of 0.1-0.2 mm.

For each material up to 10 equal samples are prepared. One sample acts as control.

After intervals of 1 week, 2 weeks, 1 month, 2 months, 3 months, 4 months, 6 months, 12 months samples are removed, carefully washed and dried. At least ten samples of each material should be tested to obtain a reasonable statistical evaluation. All samples should come from the same batch. For polymers which can be dissolved, this should be done in a suitable solvent for the subsequent GPC analysis described elsewhere (see: *Boiling test*).

For analysis by light- or electronmicroscopy circular specimens should be preferred.

For specimens aimed at subsequent mechanical testing (see chapter: *Detection of Mechanical Properties*), dumb-bell shaped samples should be cut out according to standard requirements (ISO 137).

Prior to implantation the samples must be sterilized. The most frequent sterilization method is ethylene-oxide sterilization. This method is very efficient but requires an aeration of the samples for approximately two weeks in well ventilated environment prior to implantation. An alternative is formalin sterilization which is also very lenient on polymers and only leads to minimum absorption of formalin. Decontamination with e.g. alcohol is sufficient if the samples have been handled in an aseptic fashion all the way through. Other types of sterilization may change or even destroy the material

A simple system for identification of the samples is necessary in order to locate them for recovery in the tissues. Implants often change their positions gradually due to movements of the animal. Only four specimens should be inserted into the back of each animal. The location should be over the paravertebral muscles on the fascia.

The implantation proper must be carried out under perfect sterile conditions to protect the animal and especially to avoid secondary local infection which is known to accelerate the degradation process through inflammatory cell reactions. Infections around a foreign material cannot normally be eliminated by the animal. The infection will either lead to incapsulation of the material with heavy scarring and inflammatory reactions (as well as accelerated attack by enzymes from the leucocytes) or lead to expulsion as a "normal" physiological reaction.

Description of procedure.

The animal is anaesthesized with diethyl-ether and fixed on a small table with small pieces of cardboard and rubber band. An area of approximately 5 x 10 cm on the back is shaved, taking care not to harm the skin. The surface is then disinfected with 0.5% iodine solution (higher concentrations will damage the skin). An incision of approximately 4 cm length is made perpendicular to the spine. The polymer specimen should not be bended during implantation as this will enhance degradation.

The wound should be closed with fine resorbable suture material. No wound dressing is necessary although fluid wound dressing (spray) can be used.

The animal should be returned to its cage before it awakens.

For practical purposes three persons should undertake the operation which can be carried out within 5-10 min. per animal.

For statistical purposes at least three animals should be used for each interval. In cases where several degradation tests (e.g. including mechanical parameters) are planned, at least five animals should be used for each time interval.

The specimens may remain in situ up to a period of 12 months, but at intervals as mentioned above samples are explanted using the same procedure. However, no sterile conditions are necessary during the explantation and the animal should be sacrificed (with an overdose of diethyl-ether).

The test samples are generally incapsulated after a period of two weeks and are difficult to discern in the tissue. The samples are removed together with the surrounding tissue and subsequently dissected and cleaned from the adhering tissue. It is then put into a proteolytic solution (trypsin) which eliminates adsorbed protein coatings. After three days in the solution at room temperatures, the samples are carefully washed in distilled water, dried and are now ready for the examinations.

In general the following tests are made:

- test for average molecular weight and molecular weight distribution using gel permeation chromatography (see: *Boiling Test*),

- test for stress-strain behaviour (see chapter: *Detection of Mechanical Properties*),

- test for surface analysis (see chapter: *Surface Analysis*).

Safety aspects.

The polymer specimens must be sterile and no remnant of sterilizing agents must remain on the surface (paraformaldehyde) or dissolved in the polymer (ethylene oxide).

Good health of the animals is essential to avoid reduced resistance to secondary infections. Rules of Good Laboratory Practice should be adhered to.

RESULTS

Although implantation tests only indicate the material properties in subcutaneous environment, it is nevertheless felt that it provides a guide for biofunctionality in other applications. Some typical results are given below.

The average molecular weight of all tested materials except Avcothane 51 (poly-etherurethane-poly-dimethylsiloxane block copolymer) and Estane 5714 generally decrease considerably upon six months of subcutaneous implantation as shown in *Fig. 1*.

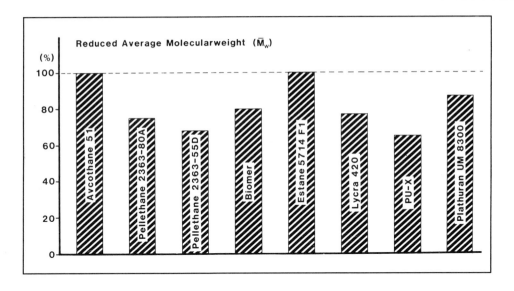

Fig. 1

Reduced average molecular weight of some polyurethanes after six months subcutaneous implantation.

In general the loss of average molecular weight does not have much influence on the stress-strain behaviour as shown in *Fig. 2* and *3*. The shown deviations are within statistical range. Synthetic caoutchouc undergoes considerable changes and may show an immense elongation at break after 6 months.

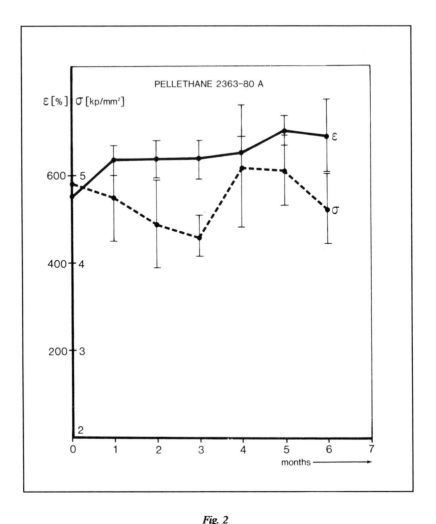

Fig. 2

Tensile strength and elongation at break of a poly-etherurethane after subcutaneous implantation.

Surface analysis by Scanning Electron Microscopy = SEM (see chapter: *Surface Analysis*) demonstrates the pattern of surface defects. The synthetic rubber will not show any surface degradation even after 9 months of testing although the mechanical strength is changed. On several poly-urethane surfaces severe changes may be observed with cracks and ruptures after approximately nine months of sub-cutaneous implantation (*Figs. 4 - 5*). The changes emanate from surface defects such as microbubbles close to the surface, inhomogeneities and spots where the samples are slightly damaged (e.g. where they are cut out of the films).

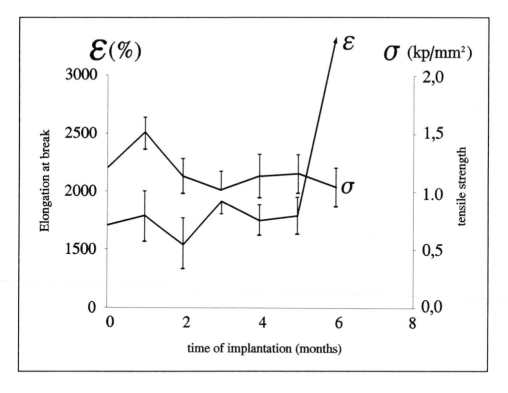

Fig. 3

Tensile strength and elongation at break of a synthetic rubber after subcutaneous implantation.

DISCUSSION

The in vivo test is in spite of its demand on resources considered as a reliable method to investigate the resistance to degradation. Results may differ widely from those obtained by the Boiling test which makes the Tissus Implantation test very essential in the preclinical evaluation of materials. It can also provide some information on the tissue reaction towards the material, but this is very difficult to interpret as there will be reactions from the surgical procedure. Other sources to inflammatory reactions are contamination due to bacteria, fungi, chemicals (sterilants) and glove powder. All these factors can result in a variable amount of infiltration of the interfase with e.g. leucocytes, macrophages etc. from the surrounding tissue together with build up of connective tissue, fibrosis and scarring.

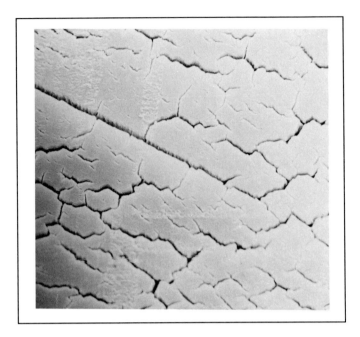

Fig. 4

Surface of Pellethane 2363-80AE after six months subcutaneous implantation (1:1000)

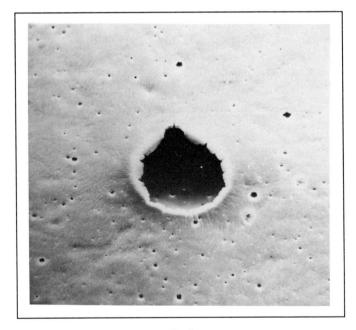

Fig. 5

Surface of Biomer® after nine months subcutaneous implantation (1:600)

Furthermore it should be reiterated that inflammatory reactions greatly enhance the chemical attack on the polymer and thus accelerate degradation. If an infection occurs in the vicinity of the sample, it is not possible for the animal to eliminate this and a chronic inflammatory reaction progresses.

A reduction of the average molecular weight does not initially seem to influence the fundamental mechanical properties of most materials.

As the test is very resource- and time-consuming it should be plannned carefully. The material should also have surface properties closely resembling those of the final items. One should carefully avoid even small damages and dents of the surface and during preparation of the samples, surface inhomogenities and defects should be careful avoided.

Limitations of method.

It is of importance to avoid inflammatory reactions in the tissue. As mentioned infections may disturb the test in two ways:

a)   a severe inflammatory reaction around the tissue cannot be separated from tissue reaction against the material per se,

b)   inflammatory reactions enhance the degradation processes greatly.

Presence of sterilisant remnants (e.g. dissolved ethylene oxide) may lead to local necrosis and subsequent inflammatory reaction. Presence of glove powder which adhere to the polymer surface by electrostatic forces may lead to acute inflammatory reactions and formation of foreign-body reactions. Thefore a "no touch" should be applied in the handling of the polymer samples.

The test does not fully reflect the performance of material under different conditions (e.g. vascular use) which may be a much harsher environment.

Chapter IV

TEST METHODS FOR SURFACE ANALYSIS

Chapter eds.: A. Baszkin[1] & R. Barbucci[2]

[1]   Physico-Chimie des Surface URA, CNRS 1218, Université Paris-Sud, F - 92296 Châtenay-Malabry

[2]   Dipartimento di Chimica, Universita di Siena, I - 53100 Siena

INTRODUCTION

The successful design of polymers intended for use as biomaterials demands careful characterization of the structure and chemical composition of their surfaces.

Processes such as protein adsorption, platelet and complement activation and fibrin polymerization are all surface induced phenomena and depend to a large extent on the composition of the outer few atomic layers of a polymer. The composition of this surface region of polymer can be significantly different from its bulk because polymer surfaces can adapt or restructure in response to the local environment. Enrichment or depletion of surface components may take place as a result of diffusion processes inside the polymer and reactions of components in the adjacent phase with the polymer surface.

For all these reasons, the techniques which can directly probe the polymer-solution interface are of extreme importance. Such a probe of the polymer-solution interface is necessary to directly measure the <u>in situ</u> interactions of ions, biopolymers (proteins and nucleic acids), low molecular-weight substances and cells with polymer surfaces and to learn how these interactions are affected by surface morphology or by specific chemical functional groups present near the polymer surface.

The analysis of polymer surfaces should provide information about their chemical composition before and after exposure to a biological fluid, as a function of the depth from the surface and of two directions normal to the depth. This analysis should provide data on: 1) elemental composition of the surface region; 2) surface

S. Dawids (ed.), Test Procedures for the Blood Compatibility of Biomaterials, 115–116.

functional groups present and extent of their preferential orientation; 3) elemental depth profiles for distances up to 100 Å into the sample; 4) surface crystalline order; 5) surface domain structure; 6) surface vertical and lateral heterogeneity; 7) morphology of the surface.

None of the existing analytic techniques can provide by itself all these data. The combination of many analytical techniques, each having its specificity, is therefore required to achieve this goal.

In this chapter, the description of the physico-chemical methods used to characterize polymers intended for medical applications is limited to the following test procedures:

    I.   ESCA (Electron Spectroscopy for Chemical Analysis).

    II.   SIMS (Secondary Ion Mass Spectroscopy).

    III.   FT-IR (Fourier Transform Infrared Spectroscopy). General Application.

         FT-IR (Fourier Transform Infrared Spectroscopy). Application to proteins adsorption.

    IV. Calcium-thiocyanate adsorption method for the in situ quantification of functional sites at the polymer-solution interface.

    V. Surface energetics of polymers through contact angle measurements.

    VI. SEM (Scanning Electron Microscopy).

The spectroscopy tests (I-III) are mainly required to obtain microscopic information on chemical composition, molecular structure and chemical bonds present in polymer surface region.

The advantage of the calcium-thiocyanate test consists in its unique ability to probe the polymer-water interface in situ and to quantitatively measure the number of acid or base type groups in the surface region of a polymer. - Contact angle measurements provide information relative to surface energetics and orientation of polymer surface groups in contact with water phase. - Finally, scanning electron microscopy is an extremely useful way to visualize in three-dimensional perspective morphological and structural aspects of polymer surfaces.

Electron Spectroscopy for Chemical Analysis (ESCA)

Authors: H. Bauser & G. Hellwig

Fraunhofer-Institut für Grenzflächen und Bioverfahrenstechnik, D - 7000 Stuttgart 80

INTRODUCTION

Name of method:

Electron spectroscopy for chemical analysis (ESCA), or X-ray photoelectron spectroscopy (XPS).

Aim of method:

Analysis of the chemical composition of surfaces including those of polymers.

Physical background of the method:

ESCA yields information about the chemical composition of the surface layer of a solid sample. This information is gathered from the kinetic energy spectrum of electrons which are emitted from the surface into a surrounding vacuum when the sample surface is irradiated with monochromatic X-rays. The electron emission is based on two different processes: the photo-effect and the auger effect.

X-ray photoelectrons and chemical shift.
The incident X-rays release inner core electrons from their parent atoms (*Fig. 1*) and furnish them with a kinetic energy high enough to leave the sample. Inelastic collisions of the released electrons with the electron sheath of other atoms reduce the kinetic energy. Electrons which originate from atoms very close to the surface have the highest chance of escaping without inelastic scattering. These unscattered electrons give rise to a line in the electron-energy spectrum (*Fig. 2, case A*).

Part of the inelastically scattered photoelectrons also leave the sample and contribute to the photoelectron spectrum in the form of a background (*Fig. 2, case B*) above which the distinct peaks of the unscattered electrons show up.

117

S. Dawids (ed.), Test Procedures for the Blood Compatibility of Biomaterials, 117–136.
© 1993 Kluwer Academic Publishers. Printed in the Netherlands.

The kinetic energy E_{kin} of the unscattered electrons is related to the binding energy E_b as follows:

$$E_{kin} = h\nu - E_b - \Phi \tag{1}$$

where $h\nu$ is the X-ray quantum energy and Φ is the electron workfunction of the spectrometer. E_b is referenced to the Fermi energy of the solid and is given a positive sign.

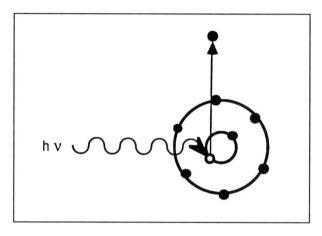

Fig. 1

The basic process of X-ray photoelectron spectroscopy:
an X-ray photon releases a core electron.

Since E_{kin} can be measured (see *Detailed description of method*), h depends on the instrumental set-up and Φ is determined by calibration, then E_b can be calculated. The core-level binding energies E_b are characteristic of the chemical elements and hence are used to identify the elemental composition. ESCA allows all chemical elements except H and He to be detected.

The position of the elemental peaks in an ESCA spectrum is influenced by the chemical environment of the atom. Depending on electronegativity differences with neighbouring atoms, the shielding of the positive nuclear charge by the outer electrons is reduced or enhanced, thus giving rise to an energy shift of the emitted electrons. Conversely from these "chemical shifts" information is gained about the molecular environment of the atoms or their oxidation state. In organic polymers, for example, the chemical shift of the C1s band allows discrimination between different binding partners (*Fig. 8, Table I*), the shift being highest for highly electronegative neighbour atoms like 0 and F.

In the customary "spectroscopic" notation for photoelectron emission lines, the initial level of the detached electron is designated by the atomic quantum number symbols (e.g. 1s, $2p_{1/2}$, $3d_{5/2}$).

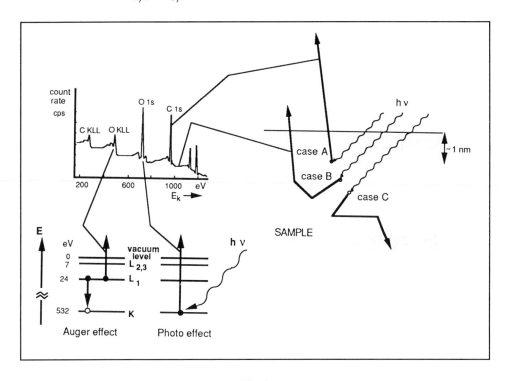

Fig. 2

A core electron can be detached either by photo emission or by Auger effect (bottom left). The detached electrons are either emitted from the surface (right) without inelastic scattering (case A) giving rise to a line (top left), or after inelastic scattering (case B, spectral background), or are not emitted (case C).

A less prominent feature of some ESCA spectra are shake up bands which occur with relatively low intensity when a valence-electron transition is coupled to the core-electron transition. The kinetic energy of the detached electron is thus decreased by the transition energy ("shake-up band") or ionisation energy ("shake-off band") of the valence electron. Shake-up type satellite bands can be found with aromatic C atoms and may be used to estimate the contributions of aromatic and aliphatic C-C bonding to the C1s band in a polymer (*Clark & Dilks, 1976*).

In Eq.1, an additional term should be included which takes account of the electrostatic surface charging of insulating or poorly conductive samples owing to electron emission (see *Description of procedure*).

Table I

Binding Energies of Oxygen-containing Structural Features (from Clark & Dilks, 1979)

Feature	Polymer	Binding energy (eV) C1s	O1s
-C-C-	vinyl-polymer backbone	285	
--C=O OH	polyacrylic acid	289,1	533.0
			534.3
-C=O OC	polyacrylates	288.9	532.8
		286.6	534.3
-C=O OC	polymethylacrylates	288.8	533.0
		286.7	534.4
-C-O-C-O-	polymethyleneoxide	287.8	533.6
C=O	polyacetyl-p-xylylene	287.6	533.6
-O			534.9
C=O	polycarbonates	290.6	533.0
O			534.9

Auger electrons.

After the detachment of a core electron, the vacancy in the electron shell is refilled by an electron of a higher state (*Fig. 2*). The energy released by this transition is either emitted as a photon (X-ray fluorescence) or is used up in the emission of another electron (*Fig. 2*). The latter process is known as the Auger effect, and it is the dominant deactivation process for the lighter elements ($Z < 20$). The accompanying Auger line in the electron spectrum is designated by the three X-ray shell symbols of the original and the final vacancies ("X-ray notation"), in the chosen example KLL (*Fig. 2*). The kinetic energy E_{kin} of Auger electrons is determined by the electron-energy level E_1 of the initial vacancy and the energies related to the final vacancies E_2 and E_3 (the latter being the binding energy of the auger electron emitted) according to the formula:

$$E^A_{kin} = E_1 - E_2 - E_3 - \Phi \tag{2}$$

In the chosen example Eq.2 reads $E^A_{kin} = E_K - E_L - E_L - \Phi$. Because the exciting energy is supplied by an intra-atom electronic transition, the kinetic energy of the Auger electron does not depend on $h\nu$[1]. However, in a binding-energy scale calculated according to Eq.1 the positions of the Auger peaks are "artificially" made to depend on the photon energy. The Auger peaks in the electron spectrum provide an additional tool for identifying the chemical states of the detected elements, since the chemical shifts of the Auger bands are different from those of the photoelectron bands.

Information depth and surface sensitivity.
For electrons in solids the mean free path of inelastic scattering λ is smaller by several orders of magnitude than the penetration depth of the X-rays in the usual energy range of 1 to 2 keV. Hence it is the escape depth of the unscattered electrons (which is used here synonymously for the mean free path) which determines the information depth. Escape depths for polymers from *Ashley (1980)* are plotted versus the kinetic energy of the electrons in *Fig. 3*. They are in the order of a few nanometers, although there is still some controversy about λ (*Clark et al., 1981*).

The number of loss-free electrons from a given atom type which is assumed to be uniformly distributed in the surface region obeys a Lambert-Beer formula:

$$I(t) = I_\infty \{ 1 - \exp(-t/\lambda \cos \Psi) \} \tag{3}$$

---

[1] *This is the basis for Auger Electron Spectroscopy (AES) and Scanning Auger Electron Microscopy (SAM) in which the excitation energy is provided by an electron beam. Owing to the high current density of the electron beam, polymer surfaces are altered in their chemical constitution during measurement. Hence, AES/SAM is not a suitable method for polymeric biomaterials. Another problem is the considerable negative electrostatic charging which must be compensated by flooding the exposed area with positive ions.*

where I(t) is the count rate for emitted electrons for a sample (or a top layer) of thickness t, I_∞ is the count rate for a sample of a thickness t>>λ, and Ψ is the escape angle (Ψ being zero for emission normal to the surface). A similar formula is derived for electron emission from underneath a top layer.

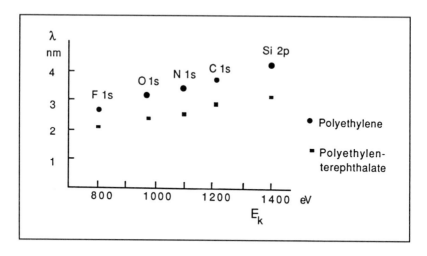

Fig. 3

Escape depth of electrons from two polymers (data from Ashley, 1980).

From Eq.3 it can easily be inferred that 95% of the possible total electron intensity originates from a surface layer of thickness 3λ if Ψ = 0, i.e. if only electrons leaving the sample normal to the surface are considered. Hence, the information depth for polymeric solids extends up to about 10 nm (Ψ = 0). Under such conditions, with a detection limit of roughly 0.1 %, the surface sensitivity corresponds to about one tenth of an adsorbed monolayer. The thickness of the 95%-layer would drop to 0.5 λ if the electrons were collected under an escape angle of 80°. Therefore ESCA exhibits a potential for depth profiling down to about 1 nm (or even less, *Paynter, Ratner & Thomas, 1984*) provided the surface is flat enough and the equipment suited to exact angular analysis. Under such favourable conditions the surface sensitivity is increased to about 1% of a monolayer.

Depth profiling according to Eq.3 requires a laterally homogeneous sample. Insular deposits lead to a different angular dependence (*Andrade, 1985*). Without angle-dependent measurements, additional information is required in order to discriminate between uniform and insular deposits.

Since the information depth of nearly 10 nm comprises approximately 10 - 20 molecular layers, ESCA can be applied to "real" biomaterials which usually - even if handled under clean conditions - have an adsorption layer acquired from the environment (air, packaging, moulds etc.).

Scope of the method.

As a chemical surface analysis method ESCA is useful for control of the production, preparation (e.g. cleaning, sterilization) and surface treatment of polymeric biomaterials, but also for studies relating chemical surface composition to biocompatibility. Degradation effects can also be investigated. Surface accumulation of migrating components such as plasticizers as well as surface contamination from the environment (e.g. from moulds) can be detected, and information can be gained about thin layers (their chemical composition, thickness and insular structure). Deposits on explanted biomaterial and the chemical effects of friction and wear can be studied with ESCA.

DETAILED DESCRIPTION OF METHOD

Equipment and detailed testing conditions.

The process of photoemission and electron spectroscopy requires an X-ray source, an electron-energy analyzer and an electron detector which are mounted in a vacuum chamber (*Fig. 4*).

X-ray source.
For excitation a sufficiently monochromatic X-ray source having its emission energy range of one to several keV is required. The characteristic $K\alpha_{1,2}$ emission of Mg or Al X-ray anodes meets this requirement, and these anodes are usually used in commercial instruments. The Mg $K\alpha_{1,2}$ emission has a photon energy of 1253.6 eV with a halfwidth of 0.7 eV, Al $K\alpha_{1,2}$ has 1486.6 eV and a halfwith of 0.85 eV. In addition to the $K\alpha_{1,2}$ emission which contains about 1/2 of the total emitted X-ray intensity, both anodes emit a continuous bremsstrahlung spectrum extending over several keV and some minor characteristic lines at higher energies, e.g. Mg $K\alpha_3$ at 1262 eV with 8% of the $K\alpha_{1,2}$ intensity, or Al $K\alpha_3$ at 1496.4 eV with 6.4% of $K\alpha_{1,2}$ intensity. Such lines have to be taken into account in trace analysis because they

cause small satellite lines characteristically spaced at lower binding energies from each major photoelectron peak.

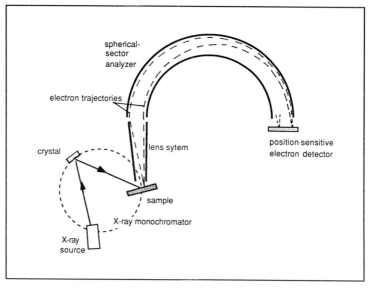

Fig. 4

Schematic sketch of hemispherical analyzer with X-ray gun, monochromator and position sensitive detector.

Higher resolution can be gained with an X-ray monochromator, at the expense of intensity, i.e. the time necessary for data acquisition increases greatly. A monochromator suppresses X-ray satellite lines and ghost lines (which originate in the X-ray tube from metals other than the anode) and eliminates the background component due to bremsstrahlung. Monochromators make use of lattice diffraction at a reflecting crystal which is mounted in a Rowland-circle arrangement. They may reduce the halfwidth of the Mg or Al $K\alpha_{1,2}$ emission to about 0.4 eV.

Energy analyzer.

The electron-energy analyzer receives the emitted electrons under a fixed or variable stepwise aperture angle and deflects their trajectories by electrostatic fields between two curved electrodes. In this way electrons with an energy in the interval $E_p \pm 1/2 \, \Delta E_p$ are focussed on the output aperture. E_p is the pass energy and ΔE_p the energy spread of the electron passing the analyzer.

The most common analyzer used in ESCA is the concentric hemispherical analyzer with a retardation grid and focussing lenses placed in front of the input aperture

(*Fig. 4*). The electron trajectories are deflected approximately in semicircles along equipotentials.

The energy resolution $\Delta E_p/E_p$ is a function of the geometrical design and the aperture width. In order to get the same energy spread ΔE_p = const throughout the energy range under study, the analyzer is set at a constant pass energy and the spectrum is swept by varying the retardation voltage. This enables electrons of the respective energy intervals to be successively slowed down to the fixed pass energy. In the alternative mode, $\Delta E/E$ = const, retardation and analyzer potential are changed simultaneously with a fixed retardation ratio, i.e. the pass energy is kept at a fixed fraction of the energy loss in the retardation field. Since the resolution of the analyzer is related to its sensitivity, spectra in the latter mode exhibit relatively higher intensities in the high E_k region.

The total halfwidth ΔE of the ESCA lines contains two additional terms as follows:

$$(\Delta E_{tot})^2 = (\Delta E_{nat})^2 + (\Delta E_{X\text{-ray}})^2 + (\Delta E_{analyzer})^2 \tag{4}$$

The natural linewidth of the ligher elements is of the order of 0.1 to 1 eV.

Detector system.
The detector collects the electrons which pass the analyzer output aperture. As the intensity of the incident photoelectrons is low, single electron counting is necessary with an electron multiplier or a channel electron multiplier (channeltron) of high gain (typically 10^8).

Position sensitive detectors have recently been developed. They offer two advantages. In a conventional hemispherical analyzer electrons of different energies have their focal spots at different positions in the aperture or detector plane. With a multichannel analyzer simultaneous counting at the different positions is possible. Simultaneous data acquisition for different energy ranges thus enhances the efficiency and may make up part of the sensitivity loss associated with a monochromator.

Another application for position-sensitive detectors is in small-spot ESCA. By careful focussing and using coupled input and output apertures, energy-selected electrons from a small area of about 100 μm or less can be focussed into a small spot on the detector. It is thus possible to scan small surface areas with a resolution of some ten micrometers.

In addition to multichannel analyzers with independent counting chains, resistive film-type detectors are used which, however, do not allow simultaneous counting at different locations, but on the other hand make a continuous small-spot possible.

Vacuum system.
The vacuum chamber is fitted for ultrahigh vacuum conditions (UHV) in order to avoid changes due to adsorption or oxidation in the uppermost molecular layers during measurement.

For good spectrometer performance a vacuum of at least 10^{-4} to 10^{-5} mbar is essential in order to avoid electron scattering losses and glow discharge effects in the components. Samples with high vapour pressure may contaminate the vacuum system. The analysis chamber may therefore be heated to 250° - 400°C.

In order to reach good UHV in the analysis chamber, pumps which do not cause oil contamination are necessary (turbomolecular, ion getter, titanium sublimation or cryogenic pumps).

The vacuum chamber contains several other components according to the degree of sophistication of the system (vacuum-controls, ion gun for ion etching, quadrupole mass spectrometer for residual gas analysis, microscope and laser for sample adjustment etc.).

Sample, sample holder, loading device.
The area analysed is normally about 10 mm^2, but in the more modern instruments this area can be reduced to a spot of 200 μm diameter or even less below. The sample size in spectrometers with fast loading systems is usually restricted to an area of about 1 cm^2 and a thickness of a few millimeters; however, special options enable the use of larger samples (up to about 20 cm diameter and 2 cm thickness).

For sample insertion, special loading devices and lock systems have been designed to maintain a good vacuum during sample loading. Experimental chambers can be used in which different types of sample treatment are possible; the chemical effects of surface treatment can thus be investigated in situ.

Samples with high vapour-pressure components (e.g. wet samples) can be cooled down to liquid nitrogen temperature with the aid of special sample holders and loading devices. This technique has been applied in the study of proteins in frozen solutions.

Computerization and automation.

Commercial suppliers offer hardware and software which provides high precision measurement and automation by microprocessor controlled component performance, computerised data collection and data processing. Smoothing, spectral deconvolution and satellite subtraction programmes are typical examples of data processing.

Suppliers of ESCA instruments.

- Finnigan, San José, Ca. 95134, USA
- Kratos Ltd., Urmston, Manchester M31 2LD, UK
- Perkin Elmer Corporation, Physical Electronics Division, Eden Prairie, Mn 55344, USA
- Riber Division, 92500 Rueil-Malmaison, France
- Surface Science Instruments, Mountain View, Ca. 94043, USA
- VG Scientific Ltd., East Grinstead, West Sussex, UK

Outline of method.

Electrons leaving the sample surface upon monochromatic X-irradiation travel through an energy analyser and are counted. Their count rate is recorded as a function of kinetic energy. From these energy spectra, the ratio of the chemical elements in their various binding states is quantitatively evaluated in an approximately 10 nm thick surface layer. For samples with a very flat surface, depth profiling is possible by varying the electron escape angle.

Description of procedure.

The sample is prepared as described in *Safety: General remarks and necessary precautions*, mounted in the sample holder and inserted into the analysis chamber by means of a load-lock system. The parameters for a survey spectrum are then chosen, e.g. (figures in parentheses refer to *Fig.5*) anode/excitation energy (Mg/1253.6 eV), excitation voltage (14 kV), emission current (20 mA), analyser mode (ΔE = const), pass energy (202 eV), detector voltage (2.1 kV), energy range (0 to 1200 eV), intensity range (0 to 7 x 10^5 cps), scan time (15 min.), number of scans (30), step time (10 ms), step energy (400 meV). From the survey spectrum one decides which lines or line groups are to be measured under high resolution. For each high resolution spectrum the parameters have to be chosen appropriately. The data processing may be accomplished by the computer; in this case, however, it is advisable to be aware of the prerequisites of the computer programmes. Data processing comprises several main steps: smoothing of the spectrum, background subtraction, spectrum deconvolution, calibration of the energy scale, converting E_k to E_b, integrating deconvoluted lines (peak areas), converting peak areas to relative concentrations, identification of stoichiometric relations (assigning to compounds, functional groups, segments etc.). Depending on the degree of sophistication of the particular ESCA system, part of the data processing is done while the spectrum is observed on the screen, part after printing by the plotter.

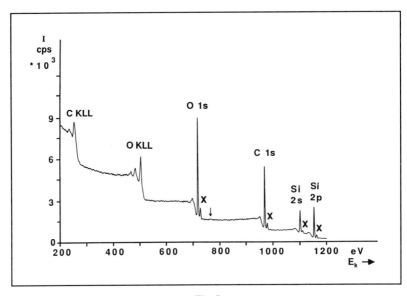

Fig. 5

Survey spectrum of polydimethylsiloxane. (X: X-ray satellites, arrow: position of Sn 3d lines).

The line positions have to be calibrated each time since there is an unknown energy shift caused by electrostatic charging. The dominant C1s line is usually chosen as a reference line. This line is attributed to the C-C feature of either polymer backbone or adsorption layers and is positioned at 285 eV. This is usually safer than a reference taken from a thin layer of evaporated gold, as the electrical potential of metallic islands might differ from that of the polymer surface. If there is line broadening caused by unhomogeneous charging, charge neutralization by a flood gun may be necessary. For qualitative analysis the corrected line positions are compared with tabulated line positions for photoelectron and Auger transitions (*Wagner et al., 1979*). The prominent lines (C, O, N) are first identified together with the associated Auger and satellite lines. The remaining lines may be regarded as the main lines of trace elements. Interferences with coinciding lines must be kept in mind. Weak lines of suspected trace elements are sometimes hard to discriminate from X-ray satellite lines or weak Auger lines of other elements present. In these cases it often helps to compare the spectra gained with different anode materials. In other cases calibration should be done with model samples of known chemical composition.

For quantitative analysis, i.e. the determination of the relative concentrations of the components, the peak areas have to be standardized by means of atomic sensitivity factors S which for a given spectrometer link the intensities I in terms of peak areas to the relative concentrations of the various elements as follows:

$$I_i = S_i \, n_i \text{ and hence } n_1/n_2 = (I_1 S_2)/(I_2 S_1) \tag{5}$$

where $S_i \simeq f \, \delta_i \, \theta \, TA \, \lambda_i$ and f is the X-ray photon flux incident on the sample, δ_i the cross section for photoelectron generation of the line under investigation, θ accounts for the angular dependence of the photoelectron yield, T is the transmission of the spectrometer, A is the analysed surface area of the sample and λ_i is the escape depth. For homogeneous samples the S_i values - or at least their ratios - are virtually independent of the sample matrix, in which case the use of tabulated values supplied by the instrument manufacturer gives an error of 10 - 20%. Calibration procedures and/or calculations are required in other cases.

Safety aspects: General remarks and necessary precautions.

ESCA involves a vacuum, high voltage, X-rays and occasionally the heating of the vacuum chamber up to 400°C. Although good safety standards are assured by modern ESCA systems, it is well to be aware of these conditions. Maintenance should only be done by a specialist. If the sample cooling to liquid nitrogen temperature is adopted, then the appropriate safety regulations for handling liquid nitrogen must be observed.

The recommended sample preparation methods depend on the purpose of the ESCA analysis. In any case precautions have to be taken so as not to contaminate the surface when cutting off a sample. In many cases samples are analysed as received, since ESCA is often used to characterize surface composition of a "real" sample, ready for implantation for example. In such cases any cleaning procedure (except those used before implantation) might distort the information sought. In other cases, in particular if clean surfaces are to be studied, standard cleaning methods like those described in the test of contact angle measurements can be applied. Volatile material can be removed by keeping the sample in a clean vacuum for a long time before insertion into the ESCA chamber.

ESCA instruments are usually equipped with facilities for argon ion etching for depth profiling, but this is not recommended for polymeric materials because chemical bonds are destroyed and there may be selective sputtering. Both effects would distort the information.

RESULTS

Normal values for reference materials.

Fig. 5 displays a survey spectrum of a polydimethyl siloxane sample supplied by Mercor Inc. This material is recommended by NHLBI as a primary reference material. The prominent lines are the photoelectron peaks of the main elements C, O and Si, and the Auger peaks of C and O. X-ray satellites accompanying the photoelectron peaks on the high (kinetic) energy side are illustrated in this spectrum which is taken without a monochromator.

For quantitative analysis and the detection of trace elements high resolution spectra of the individual peaks have to be obtained as described in *Description of procedure*. Examples of high resolution spectra are given in *Alternative typical experimental values for testing of well-known materials* since other polymers give a better illustration of chemical shifts. Quantitative evaluation of the high resolution spectra confirms within limits of errors the ESCA analysis results quoted by the supplier:

$$C : O : Si : Sn \quad : Cl$$
$$2 : \ 1 : 1 \ : 0.001 : 0.006$$

The spectra of polyethylene which is also a recommended NHLBI primary reference material have not been included since they do not illustrate additional features.

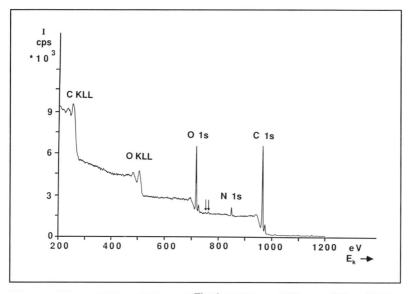

Fig. 6

Survey spectrum of a poly(etherurethane).

Alternative typical experimental values for testing of well-known materials.

A survey spectrum of a poly(etherurethane), Pellethane, cast from 4,4'-dimethylformamide is shown in *Fig. 6*. With respect to *Fig. 5* there is an additional photoelectron line peak N1s, and two tiny humps at approximately 1100 and 1150 eV attributed to Si (this assignment is confirmed by high resolution spectra). Two minute humps near 765 eV indicate the presence of tin. This corresponds to approximately 490 eV on a binding energy scale (cf. the doublet in *Fig. 9*).

A high resolution spectrum of the C1s band which can be resolved into four lines is displayed in *Fig. 7*. Unlike the survey spectra, high resolution spectra are plotted here accordingly to binding energy. Other examples of chemical shifts in the C1s band are shown in *Fig. 8* where the deconvoluted high resolution spectrum of polyurethane is compared with the C1s lines of polydimethylsiloxane, polymethylmethacrylate and polycarbonate.

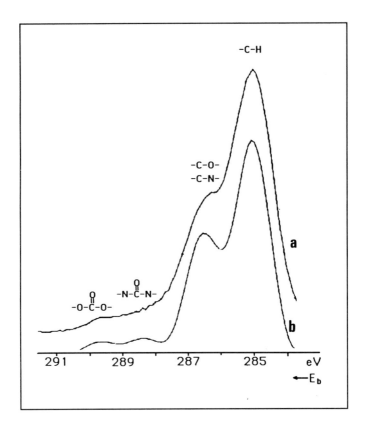

Fig. 7

*High resolution ESCA spectrum of C1s region of poly(etherurethane)
a) as measured; b) deconvoluted*

In *Fig. 9* the Sn $3d_{3/2}$ and Sn $3d_{5/2}$ lines which are barely discernable in the survey spectrum (*Fig. 6, arrows*) are shown as an example of the presence of the trace element.

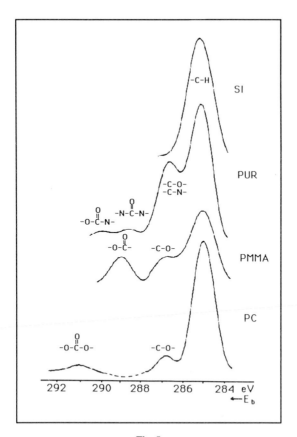

Fig. 8

High resolution ESCA spectra in the C1s region of different polymers.

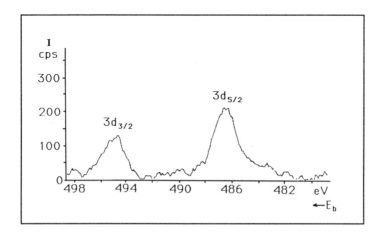

Fig. 9

High resolution ESCA spectrum in the Sn region of a poly(etherurethane) sample.

DISCUSSION

Interpretation of results.

The surface of the reference material, polydimethylsiloxane, has the same "theoretical" and stoichiometric composition as the bulk. The same amounts of trace elements (Sn, Cl) have also been observed by other laboratories[2]. This indicates that a high degree of reproducibility in the preparation and packaging of this material has been achieved.

In the poly(etherurethane) the Cls spectrum (286 eV line) contains the contributions of -C-N- and -C-O. Since the latter is dominant, this line is indicative of the polyether segment. The carbamate line at 289.5 eV accounts for the hard segment; however, as it is a minor component of the C1s band, errors due to incorrect background subtraction and deconvolution are larger. The surface composition of segmented polyurethanes depends on the preparation process and often deviates from the bulk composition (*Lelah et. al., 1983*). The relative peak areas of the C1s spectrum have been used to correlate the biocompatibility of poly(etherurethanes) to the ratio of their soft and hard segments at the surface (*Lelah et al., 1983, Merril et al., 1982, Paynter et al., 1984*). For very smooth surfaces an angular dependent change in this ratio gives the depth profiling of the chemical composition of the sample (*Paynter et. al., 1984*).

The Sn 3d line positions in *Fig. 9* ($3d_{5/2}$ at 486.5 eV) correspond to SnO_2. However, as there is coincidence with the line positions of some other tin compounds, the peaks cannot be definitely assigned. The tin lines appear only on the glass-dried (not the air-dried) side of the film. Since the glass surface is often coated with tin containing compounds (e.g. tin dioxide) used to fill the microcracks of the surface and enhance its hardness and tensile strength, traces of this coating may be transferred to the polymer.

Limitations of the method.

Although ESCA is basically a non-destructive method, prolonged exposure to X-rays may chemically change the surface of polymeric samples. Prolonged exposure

---

[2] *NHLBI-DTB Primary Reference Materials, Polymethylsiloxane sheet, Handling instructions and characterization results, Mercor Inc. Berkeley, Calif.*

may become necessary when high resolution is required for the detection of trace elements, and when depth-profiling by angular variation is performed. Identification of the binding type of the elements by analysis of chemical shifts has its limitations. Peak overlap (e.g. C-O and C-N in *Fig. 7*) is a consequence of low energy resolution (0.1 to 1 eV). In the case of trace elements, the error in peak positions may become sizeable. Energy referencing may become difficult with hetorogeneous surfaces which do not charge uniformly (see *Description of procedure*). The surface sensitivity of about 10% of a monolayer may not be sufficient in some cases.

The lateral resolution of several square millimeters or about $0.01mm^2$ with more modern instruments is a decisive limitation if the heterogeneity of the surfaces is considerable, or when insular deposits are to be investigated.

Rough surfaces do not allow the full possibilities of the method to be exploited; in particular, they do not allow angular dependent depth profiling (the dimensions of surface irregularities must not be larger than the mean free path of inelastic scattering; the latter is in the order of nanometer).

The "slowness" of ESCA (several hours per sample), in terms of possible radiation damage and also in terms of associated expenses, is its major disadvantage.

REFERENCES

Andrade, J.D (1985) in "Surface and Interfacial Aspects of Biomedical Polymers" (Andrade, J.D. ed.), vol.1, pp.105-195, Plenum Press, New York.

Ashley, J.C. (1980) IEEE Trans. Nucl. Sci. NS-27, 1454-1458.

Clark, D.T. & Dilks, A. (1976) J. Polym. Sci., Polym.Chem. 14, 533-542.

Clark, D.T. & Dilks, A. (1979) J. Polym. Sci., Polym. Chem. 17, 957-976.

Clark, D.T., Fok, Y.C.T. & Roberts, G.G. (1981) J. Electron. Spectrosc. 22, 173-185.

Lelah, M.D., Lambrecht, L.K., Young, B.R. & Cooper, S.L. (1983) J. Biomed. Mat. Res. 17, 1-5.

Merril, E.W., Sa Da Costa, V., Salzman, E.W., Brier-Russel, D., Kuchner, L., Waugh, F., Trudel, G., Stopper, S. & Vitale, V. (1982) in "Biomaterials: Interfacial Phenomena and Applications", pp. 95-107, Am. Chem. Soc. Washington, DC.

Paynter, R.W., Ratner, B.D. & Thomas, H.R. (1984) in "Polymers as Biomaterials" (Shalaby, S.W., Hoffman, A.S., Ratner, B.D. & Horbett, T.A. eds.), pp.121-133, Plenum Press, New York.

Wagner, C.D., Riggs, W.M., Davis, L.E., Moulder, J.F. & Muilenberg, G.E. (1979) Handbook of X-ray Photoelectron Spectroscopy, Perkin-Elmer Corporation, Eden Prairie.

Secondary Ion Mass Spectroscopy (SIMS)

Authors: H. Bauser & G. Hellwig

Fraunhofer-Institut für Grenzflächen und Bioverfahrenstechnik, D - 7000 Stuttgart 80

INTRODUCTION

Name of method:

Secondary ion mass spectroscopy (SIMS).

Aim of method:

Analysis of the chemical composition of biomaterial surface.

Physical background of the method:

Secondary ions which are ejected from the surface upon the impact of a beam of (primary) ions analyzed in a mass spectrometer.

The primary ions used in the SIMS are typically in the 100 eV - 30 keV energy range. When such ions impinge on a surface, they penetrate into the sample creating an atomic collision cascade during which some atoms, molecules or fragments of molecules gain enough energy and momentum to be ejected from the surface (*Fig. 1*). This process is called sputtering. Most of the ejected species are neutral, but some carry a positive or negative charge (secondary ions), and these can be detected and analyzed in a mass spectrometer. The escape depth of secondary ions from polymer surfaces has not been determined yet. For inorganic solids typical reported values are of some tenth of a nanometer for the escape depth, and approximately ten nanometers for the penetration depth (*Benninghoven et al., 1987*).

A great part of the incident ions remains embedded in the sample surface and changes its chemical composition. An enrichment in primary ions up to a few percent is possible. By the same token the structure of the exposed surface is being

S. Dawids (ed.), Test Procedures for the Blood Compatibility of Biomaterials, 137–150.
© 1993 *Kluwer Academic Publishers. Printed in the Netherlands.*

changed (the thickness of the altered layer is about twice the penetration depth of the primary ions).

The number of emitted secondary ions of species i per unit of time N_{Si} can be defined as follows:

$$N_{Si} = N_p \, y_i \, \alpha_i \, c_i \, \partial_i \, \eta_i$$

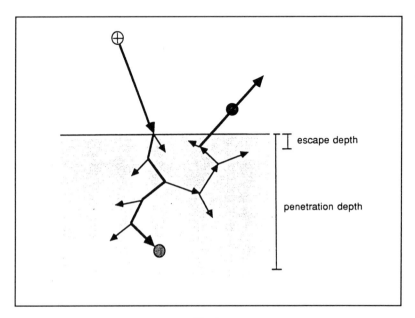

Fig. 1

The basic process of SIMS. A primary ion penetrated into the solid and, after dissipating its energy in a collision cascade, is possibly implanted. Some recoil atoms at the surface are ejected.

N_p designates the number of incident primary ions i per unit of time; y_i the sputtering yield (mean number of emitted atoms per incident particle); α_i the ionisation probability; c_i the atomic concentration; ∂_i the relative abundance of the isotope; η_i the efficiency of secondary ion measurement (this entity contains equipment parameters such as transmission of mass spectrometer and other).

The sputtering yield y_i depends on the target material (i.e. on the type of the emitted particles and their local binding energy in the matrix), on the ion beam (energy, particle, current density) and on geometrical factors. Whereas sputtering yields can be predicted in many cases of single element materials, not much is known about the yields of molecular fragments of a polymeric material. Uncertainties about sputtering yields and about ionization probabilities render the SIMS method poor in quantitative exactness (though high in sensitivity).

Details on physical and technical aspects of SIMS are treated in *Benninghoven et al., 1987*.

Scope of method.

High sensitivity qualitative information may be gained on the nature of chemical species in the outermost molecular layer of polymer surfaces. Comparison with the known fingerprint spectra (see *Interpretation of results*) furnish structural information of polymer surfaces and enable detection of trace elements and mapping of the lateral distribution of components. This depth profiles of trace components may be established and the adsorption of contamination films can be investigated. SIMS detects all elements (including hydrogen) and molecular fragments discriminating isotopes, and - with the scanning instrumentation - provides information on lateral distribution of elements. These features make of the SIMS a valuable tool to control manufacturing procedures of polymers in regard to the cleanliness and reproducibility. Physical basis for correlating of chemical surface properties to the biocompatibility of materials can thus be obtained.

DETAILED DESCRIPTION OF METHOD

Equipment and testing conditions.

The terms SIMS in its general meaning comprises a variety of instruments with different functions and operating in different modes. Similar to other surface analysis systems engaging charged particle beams (i.e. SEM/EDX, AES/SAM). Depending on the primary function, on main components and on operating modes employed, a variety of different SIMS systems is in use.

The basic components of a SIMS instrument are the ion source with the primary ion mass filter and the focussing optics, the sample holder and the sample load system, the transfer (or extraction) optics, the secondary mass analyzer, and the ion detector (*Fig. 2*). These components are mounted in an ultrahigh vacuum chamber.

The necessary electronic equipment, computer system, recorder or plotter can be attached to a SIMS equipment.

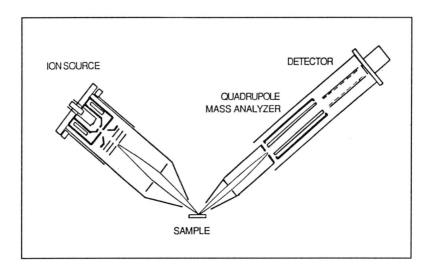

Fig. 2
Schematic view of a SIMS spectrometer

Additional equipments are: a scanning equipment for the ion gun (synchronously coupled to the scanning electronics of a cathode-ray tube), an electron flood gun for charge-neutralization on surface of insulating samples, a sample holder allowing controlled lateral adjustment, a preparation chamber etc.

Ion source.
The ion source is usually either a plasma (i.e. gas discharge) source for ions like Ar^+, N_2^+, O_2^+, O^-, Xe^+, or a liquid-metal ion source for Cs^+, In^+, or Ga^+. In the standard duoplasmatron ion source, for example, positive or negative ions are extracted from a plasma arc which is confined to a small volume by an intermediate electrode between cathode and anode and by a strong magnetic field. A liquid metal source comprises for example a tungsten frit through the pores of which the heated metal pervaporates. The metal vapour atoms become ionized by interaction with the hot tungsten surface and are extracted through a pierced electrode.

A number of other types of ion sources such as cold cathode sources, electron impact sources, electrohydrodynamic sources are available. The appropriate choice of an ion source depends on the intended purpose. For example, high detection of electropositive elements require the use of negative ions such as oxygen. Thus the duoplasmatron oxygen sources are used to obtain a high sensitivity detection of electronegative elements, and liquid gallium ion sources in microfocussing guns of imaging systems.

A primary ion mass filter is usually installed in order to remove impurity ions and neutral particles from the ion sources. It is often either of the magnetic sector-field type or a Wien filter.

The charging of insulating samples by ions can be avoided by converting the primary ions into atoms in a charge exchange cell. The use of primary neutrals - known as fast atom bombardment mass spectrometry (FABMS) - reduces radiation damage and is recommended for analysis of polymeric materials (*Briggs et al., 1983*).

Sample.

Most of the instruments can handle small samples (about one square centimeter) with the thickness not exceeding few millimeters. This means that an appropriate sample has to be cut out from a material under investigation. Modern instruments enable the use of samples with diameters up to 200 mm and with a thickness up to 20 mm. The samples are usually mounted on metal supports. A precision sample holder allows exact (light) microscope-controlled positioning of the sample ($\pm$ 1 μm) and exact choice of the incident-beam angle.

Mass analyzer.

The ion extraction system collects a maximum part of emitted ions and focusses them on the entrance aperture of the mass spectrometer. The mass analyzers differ according to their resolving power and transmission (including energy acceptance). In the quadrupole mass spectrometers the ions travel along the central region between four parallel rods in a sinusoidially modulated transversal electric field. For a given geometry and characteristics of the electric field components and the frequency only ions of a particular Ze/m (charge to mass ratio) can pass and reach the detector.

The mass spectrum can be recorded, e.g. by simultaneous variation of ac and dc field amplitudes at a fixed frequency, while the line width Δm can be varied by change of the ratio of the dc to ac field amplitudes. Quadrupole mass spectrometers have a relatively low mass resolution which is partially caused by the energy spread of the secondary ions. The mass resolution can be improved by an energy preselector. This decreases the energy acceptance (by some eV), i.e. a great part of the ions entering the mass spectrometer is being lost. Quandrupole mass spectrometers are sensitive to the changes in charging of insulating samples. The maximum mass resolution $m/\Delta m$ is often below 1,000 but may reach 2,000 in modern instruments. In the $m/\Delta m$ = const mode the transmission is virtually independent of m. In the constant line width mode (Δm = const) which is more frequently used, a relatively high transmittance is obtained at lower masses at the expense of resolution. The choice of Δm = 0.5 amu for example, gives a mass resolution of $m/\Delta m$ = 60 at m = 30 amu (for comparison, separation of the isobars N_2 and CO requires $m/\Delta m$ = 2,500, of CO and C_2H_4 770 and of N_2 and C_2H_4 1,113). The transmission of quadrupole mass analyzers is in the order of 1% and decreases towards higher masses. Advantages of these analyzers are their flexibility in regard to fast scanning and to pole reversal, and their relatively low cost.

Double focussing mass spectrometers engage an electrostatic energy analyzer and a magnetic sector field. They have a much higher resolution ($m/\Delta m$ is in the range of 25,000) and a high energy acceptance of nearly 200 eV, leading to a transmission of roughly 10%. By virtue of the higher sensitivity, the double focussing mass analyzers can be used in the imaging systems.

A more recent development of SIMS instrumentation involves application of a time-of-flight mass spectrometer. It employs a pulsed primary ion current of a pulse width of about ten nanoseconds or below at a repeat frequency of the order of kilohertz. The transient time for the secondary ion pulses is a function of m/ze. Time-of-flight mass analyzers reach mass resolutions of above 5,000. The range of mass numbers can exceed 10,000 amu and is considered to be virtually unlimited. Energy focussing is achieved in a toroidal condenser. With an energy band pass of several hundred electron volts, time-of-flight mass spectrometers have a very high transmission (beyond 50%) and an extremely high sensitivity. Due to the small primary ion currents (typically $1nA/cm^2$ on time average, with 10 keV ions) and the high mass range the TOF-SIMS are particularly suitable for polymer surface analysis.

Detectors.
Sensitive detectors are open dynode type electron multipliers or channeltron multipliers which are more resistant to deterioration by venting. Multipliers have a gain (i.e. electrons per incident ions) in the range of 10^4 - 10^8. Faraday cup

detectors are less sensitive by several orders of magnitude. Scintillator plates are used either as direct ion targets or with ion-electron converter electrodes; the emitted photons are detected with photo multipliers, sometimes linked to the scintillator screens by light guides. Channel plates combine ion-electron conversion in a two-dimensional array of channeltron type tubelets with a scintillator screen. Such two-dimensional detectors can be used in the imaging systems.

Scanning SIMS.

In the scanning instruments the primary ion beam can be rastered synchronously with a cathode ray oscilloscope beam and thereby display the lateral intensity pattern of a chosen mass line on the screen. A display of atomic or molecular species (simultaneous and in different colors) allows chemical mapping of main components of the surface. The primary ion beam can be focussed down to a 50 nm diameter. Line scans and three-dimensional pictures (if the depth profiling information can be included) are other features of the scanning systems.

Vacuum.

Measurements are performed under ultrahigh vacuum conditions in order to keep the background spectrum low to avoid surface contamination of the sample and scattering of ion beams.

Computerization.

A high degree of computerization is necessary for modern SIMS instruments. Commercial suppliers offer the hardware and software systems for the operation, spectra acquisition and data processing. Also mass spectra libraries are becoming available.

Modes of operation.

SIMS can be operated in two different modes which are called the static and the dynamic modes.

In the static mode the erosion of the surface due to sputtering is so slow that the surface composition is only negligibly altered during a run. At a primary ion current density of the order of 10^{-9} A/cm^2 the "lifetime" of the uppermost surface layer is in the order of hours (*Benninghoven, 1985*). Beam scanning or pulsed operation ease the application of low overall current densities.

In the dynamic mode with a high current density of 1 A/cm^2 or more a monolayer of the surface is etched off in the order of milliseconds. The dynamic mode is applied for depth profiling with high sensitivity. The ppb range can be reached for many elements on surfaces of inorganic materials, and dynamic detection ranges up to six orders of magnitude for some elements can be obtained. Depth profiling of insulating surfaces requires charge neutralization, i.e. by a low energy-electron flood gun and extraction field adjustment. The dynamic mode appears not to be appropriate for polymeric materials at the present state of the art since chemical damage and selective sputtering are expected to distort the information.

Secondary neutrals mass spectrometry.
Since most of the secondary particles are neutrals, their detection is better suited for a quantitative analysis than ordinary SIMS spectra. Post-ionization of neutrals can be achieved with a low energy electron beam or by interaction with a low pressure gas discharge plasma. Both methods do not seem to be suitable for polymers at present, since the ejected molecular fragments (being isolated from an energy accepting matrix) can be easily destroyed by these ionization methods. In this respect the post-ionization by a laser beam seems to be more promising.

Suppliers of SIMS instruments or of SIMS components.
- AEI instruments, Manchester, UK
- Applied Research Laboratories, Sunland, Ca., USA
- Atomica Technische Physik GmbH, 8000 München, FRG
- Balzers AG, Balzers, Liechtenstein
- Cameca SA, Courbevoie, France
- Finnegan, San José, Ca. 95134, USA
- Hitachi Company Ltd., Instruments Division, Chiyolaku, Tokyo 1000, Japan
- Kratos Ltd., Manchester M31 2LD, UK
- Perkin-Elmer Corporation, Physical Electronics Division, Eden Prairie, Mn 55344, USA
- Riber Instruments SA, 92503 Rueil-Malmaison, France
- Surface Science Instruments, Mountain View, Ca. 94043, USA
- VG Ionex Ltd., East Grinstead, West Sussex RH19 1UB, UK

Outline of the method.

The secondary ions released from the sample surface upon irradiation with a (primary) ion beam are analyzed in a mass spectrometer, and their count rate is recorded as a function of their mass-to-charge ratio m/ze. Thus the secondary ions give rise to spectral lines. The calibrated positions of the lines correspond to the charged molecular fragments.

Description of procedure.

For the majority of existing SIMS systems the basic procedure may be described (with the parameters chosen for *Figs. 3 & 4* in parenthesis). The sample preparation is similar to that used for ESCA . One should, however, be aware that SIMS is more sensitive to contaminations, even to those which are in quantities less than a monolayer. The sample is mounted onto the sample holder and inserted into the analysis chamber by a load-lock system. The type of the primary ion (Ar^+) and of the ion source, if there are more than one available, has to be chosen. The primary ion energy (5 keV) and the emission current (1 nA; primary ion current to the sample approx. 1 nA) are then adjusted. The beam has to be aligned through the primary ion mass filter. In the static SIMS conditions, the primary ion beam is either defocussed or, as in the scanning mode, focussed on the sample. For polymer analysis, it is ideally delivering 1 nA/cm^2 to the sample surface (*Briggs, 1988*). For the secondary ions, the extraction potential and the mass analyzer parameters are to be adjusted (e.g. for a quadrupol mass spectrometer the mean voltage and the line width). An electron flood gun is necessary for polymer sample analysis. When the qualitative analysis is performed, the lines belonging to the molecular fragments of the polymer should be identified first. The assignment of lines to the inorganic surface contaminants may be ascertained by their isotope patterns and by comparing between the positive and negative ion spectra. To identify the contamination due to the organic compounds, comparison with model compounds may be necessary (*Briggs, 1988*). Polymer spectra usually exhibit a great wealth of lines which makes a complete line interpretation cumbersome if not impossible. Useful information can be drawn by comparing different surfaces of the same sample or of different samples of the (nominally) same material. The surface concentration of different polymer segments may depend on the manufacturing procedure and may be different from the bulk concentration. In this sense SIMS provides structural information about various surfaces of samples of the same material (*Graham & Hercules, 1981*).

Safety aspects, general remarks and necessary precautions.

High voltage, X-rays, vacuum and occasionally heating of the vacuum chamber up to 400°C are associated with SIMS. Although modern SIMS systems have good safety standards, one has to be aware of these conditions. The maintenance work should be restricted to the specialist.

The recommended sample-preparation methods depend on the purpose of the SIMS analysis. Necessary precautions have to be made to avoid contamination of the surface when a sample is cut off from a polymer film. As the depth of the SIMS analysis is only of about one molecular layer of the material, even thin adsorption films may mask the surface of the sample. If the latter is to be studied, gentle cleaning methods such as those described in the contact angle measurement test can be applied. Ion beam etching may distort the information due to selective sputtering.

RESULTS

Normal values for reference materials.

SIMS spectra of the NHBLI-recommended reference polydimethyl siloxane and polyethylene are shown in *Figs. 3 & 4.* They were obtained with 5 keV Ar$^+$ primary ions (see *Description of procedure*) and with a neutralizing electron flood gun.

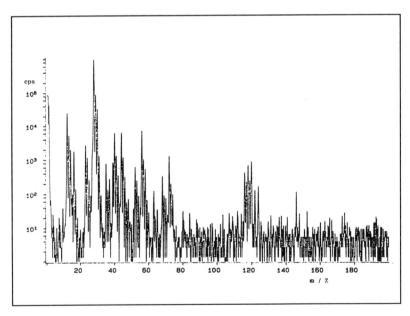

Fig. 3

Positive ions spectrum of polydimethylsiloxane

Fig. 3 shows a positive ion spectrum of the polydimethyl siloxane. The spectrum contains intensive lines of small molecular fragments of the polymer, including atomic ions such as ^{12}C, ^{28}Si and isotopes. It contains also many other lines in the whole mass range recorded. The lines corresponding to the repeat groups (74 and 128 amu) are smaller than the lines which may be assigned to the doubly deprotonated repeat groups. A noticeable feature is the Sn$^+$ mass spectrum which consists of mainly 7 isotope lines between 116 amu and 124 amu and, hence, can unambiguously be identified as a trace element at the surface. Traces of Na and Cl suggest NaCl contamination (identification as in *Fig. 4*).

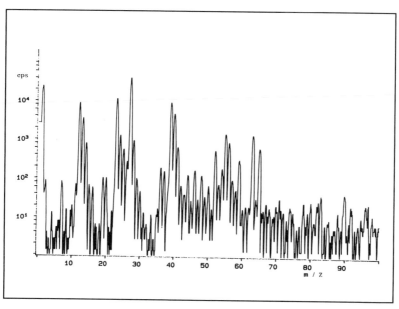

Fig. 4

SIMS spectra of polyethylene: positive ions.

Figs. 4 & 5 compare a positive with a negtive ion spectrum of polyethylene. The molecular fragments in the negative spectrum display a higher degree of deprotonation than in the positive spectrum. Electropositive trace elements (^{23}Na$^+$) appear predominantly in the positive spectrum while the electronegative ones (^{35}Cl$^-$ and ^{37}Cl$^-$) are present in the negative spectrum.

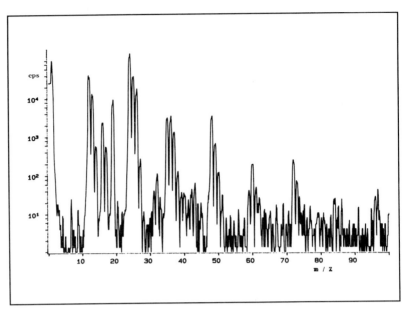

Fig. 5

SIMS spectra of polyethylene: negative ions.

Alternative typical experimental values for the test on well-known materials.

In the recent years SIMS spectra of several solid polymers (*Briggs et al., 1983; Briggs, 1988; Gardella & Hercules, 1980; Graham & Hercules, 1981*) as well as of liquid phase polymers (*Briggs, 1988*) and of polymers deposited on metallic substrates (*Benninghoven, 1985; Bletsos et al., 1987*) have been published. Spectra of similar polymers such as polyethylene and polypropylene for example, or of poly(etherurethanes) with different segment lengths of polypropyleneglycol, exhibit noticeable differences (*Briggs, 1988*). The use of low primary ion current densities and electron flooding provides mass spectra with lines up to 200 amu and beyond. A predominance of hard or soft segments at the surface of different poly(etherurethanes) has been detected. Also a trace impurity of the bis-ethylene stearamide which has not been identified with the ESCA has been demonstrated on a commercial biomedical polyurethane surface (*Briggs, 1988*). SIMS spectra enable to distinguish between structural differences of mold facing and air facing surfaces of solution cast polymers (*Graham & Hercules, 1981*).

DISCUSSION

Interpretation of results.

SIMS spectra of polymer surfaces contain a host of different lines. Spectra of solid polymer surfaces exhibit fragments of the polymer, but obviously not the un-fragmented molecules. Although the molecular fragments giving rise to lines are indicative for each polymer, it appears that presently the exact chemical constitu-tion cannot be yet deduced merely from the line pattern. Such spectra have to be used rather as fingerprint spectra to allow identification by comparison with the standard spectra accumulated under comparable experimental conditions. With the improved resolution (as in the case of TOF-SIMS) ambiguities in interpetation will be removed, and the information to be drawn from the SIMS spectra will certainly increase. Furthermore, SIMS analysis is very helpful to increase the understanding of chemical reactions generated by the surface treatment of polymers.

SIMS has a high potential to detect tiny traces of impurities on the surfaces. This is illustrated in *Figs. 3 & 4*. These trace elements (Na, Cl, Sn) can be detected with a much higher sensitivity than with the ESCA. The minute amounts of organic impurities can be identified by comparison with model compounds (*Briggs, 1988*).

Limitations of method.

SIMS spectra, at the up-today knowledge, are of a fingerprint quality. A basic limitation on their complete quantitative evaluation and in the unambiguous assignment to the unknown polymer surfaces are the uncertainties of the local sputter yield and the ionization probabilities (see *Physical Background of the method*). For polymer materials, the mutual intertanglement of the chains in the surface region increase these uncertainties by influencing the degree of fragmenta-tion and apparently preventing the ejection of unfragmented molecules. The advantage of the high surface sensitivity restricts information to the uppermost surface layer. The information which can be obtained on deeper layers of a polymer by depth profiling in the dynamic mode is at least very questionable. The applicable ion beam intensity is restricted since polymer surfaces are highly amenable to chemical and structural distortion by impinging ions.

Related relevant testing methods.

SIMS is not directly related to, but in several respects complementary to ESCA as a highly surface sensitive method for chemical analysis.

REFERENCES

Benninghoven, A. (1985) J. Vac. Sci. Technol. A3, 451-460.

Benninghoven, A., Rüdenauer, F.G. & Werner, H.W. (1987) Secondary Ion Mass Spectrometry, J. Wiley, N.Y.

Bletsos, I.V., Hercules, D.M., van Leyen, D. & Benninghoven, A. (1987) Macromolecules 20, 407-413 .

Briggs, D., Brown, A., van den Berg, J.A. & Vickerman, J.C. (1983) in Ion Formation from Organic Solids, (Benninghoven, A. ed.) 162-166, Springer-Verlag, Berlin.

Briggs, D. (1988) Surface & Interface Analysis Series, SIA 12: ECASIA 1987, Proc.Eur.Conf. on Applications of Surface and Interface Analysis, Fellbach 1987, Wiley, 391-404.

Gardella, J.A. & Hercules, D.M. (1980) Anal. Chem. 52, 226-232.

Graham, S.W. & Hercules, D.M. (1981) J. Biomed. Mat.Res. 15, 465-477.

Fourier Transform Attenuated Total Reflection Infrared Spectroscopy (ATR/FT-IR)

Authors: M. Nocentini & R. Barbucci

Dipartimento di Chimica, Università di Siena, I - 53100 Siena

INTRODUCTION

Name of method:

Fourier Transform Infrared Spectroscopy (FT-IR) coupled with Attenuated Total Reflection Spectroscopy (ATR).

Aim of method:

Infrared Spectroscopy gives information about chemical groups and molecular structure. FT-IR combined with ATR can be used to study molecular and surface morphology of polymer surfaces. Important information may be obtained about blood-surface interactions by monitoring protein adsorption at polymer surfaces in contact with flowing blood (*Gendreau & Jakobsen, 1976. See also "Fourier Transform Attenuated Total Reflection Infrared Spectroscopy (ATR/FT-IR): Application to Proteins Adsorption Studies" by Magnani & Barbucci*).

Background of the method (*Griffths & Haseth, 1986; Ferraro & Basile, 1979; Harrick, 1979*):

Surface infrared spectroscopy couples two powerful techniques: FT-IR and ATR which enable the surface of a sample for a finite depth into the bulk to be examined. The electromagnetic radiation is totally internally reflected at the interface of an optically transparent material and the sample only when the index of refraction of the optically transparent material is greater than the index of refraction of the sample.

S. Dawids (ed.), Test Procedures for the Blood Compatibility of Biomaterials, 151–170.
© 1993 *Kluwer Academic Publishers. Printed in the Netherlands.*

The spectrum obtained by coupling infrared spectroscopy with the internal reflection technique is characteristic of the sample surface for a finite depth into the bulk. The surface depth ranges from about 0.5 to 3 μm for polymers.

FT-IR.

FT-IR uses an interferometer to collect the spectral data instead of the monochromator used in conventional dispersive infrared spectrometers. In this way the radiation is not dispersed. The interferometer produces a plot of light intensity in the time domain (interferogram) which is then Fourier transformed into a more useful light intensity versus frequency spectrum.

A simple scheme of the Michelson interferometer is shown in *Fig. 1*. It consists of two mutually perpendicular plane mirrors, one of which is fixed and the other moves along an axis perpendicular to its plane. A beamsplitter placed between two mirrors partially reflects the incident beam from an external source to the fixed mirror and partially transmits it to the movable one. When the two beams are recombined at the beamsplitter, a path difference has been introduced and the interference condition created. The two beams interfere at the beamsplitter and are again partially reflected and partially transmitted. The beam that returns to the source is not of interest, and only the output beam travelling in the direction perpendicular to the input beam is measured.

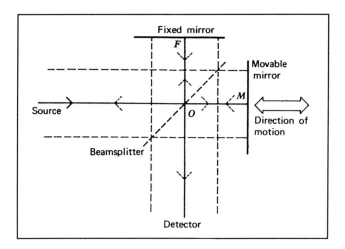

Fig. 1

Schematic representation of a Michelson interferometer.

Let us consider an idealized situation: a source of monochromatic radiation produces an infinitely narrow, perfectly collimated beam and the beamsplitter is a non-absorbing film reflectance and transmittance of which are both 50%.
Then if λ is the wavelength of the radiation and $\bar{\nu}$ the wavenumber they are related by the equation $\bar{\nu} = 1/\lambda$.

The path difference is equal to 2 (OM-OF) and is called the retardation. It is usually given by the symbol δ (*Fig. 1*). When $\delta = 0$, the two beams are in phase and interfere constructively, producing a maximum (*Fig. 2*). When $\delta = 1/2$, the beams are out of phase and interfere destructively; all the light returns to the source and nothing reaches the detector (*Fig. 2b*).

As the mirror is moved at constant velocity, the signal at the detector varies sinusoidally and a maximum is registered each time the retardation is an integral multiple of λ ($\delta = n\lambda$).

The intensity of the beam at the detector measured as a function of δ is given by:

$$I(\delta) = 0.5 \; I(\bar{\nu}) \; \{1 + \cos 2\pi\bar{\nu}\delta\}$$

where $I(\bar{\nu})$ is the intensity at any point where $\delta = \lambda n$
and is equal to the intensity of the source. Only the component:

$$I(\delta) = 0.5 \; I(\bar{\nu}) \; \cos 2\pi\bar{\nu}\delta$$

is important and this is generally referred to as the interferogram. $I(\delta)$ is also proportional to beamsplitter efficiency, detector response and amplifier characteristics; all these values are constant for a given apparatus.

So the equation may be modified by a simple correction factor: $H(\bar{\nu})$ to give:

$$I(\delta) = 0.5 \; H(\bar{\nu}) \; I(\bar{\nu}) \; \cos 2\pi\bar{\nu}\delta$$

if $0.5 \; H(\bar{\nu}) \; I(\bar{\nu}) = B(\bar{\nu})$

$$I(\delta) = B(\bar{\nu}) \; \cos 2\pi\bar{\nu}\delta$$

Mathematically, $I(\delta)$ is said to be the cosine Fourier Transform of $B(\bar{\nu})$.

The spectrum is calculated from the interferogram by computing the cosine Fourier Transform of I(δ) which explains the name given to this spectroscopic technique.

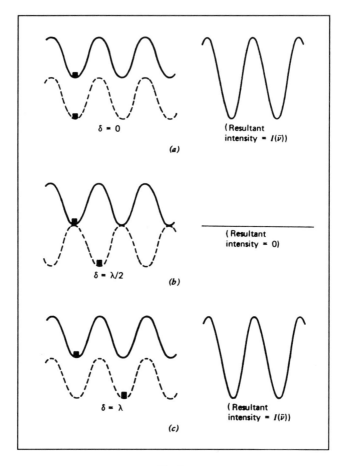

Fig. 2

Schematic representation of the phase of the electromagnetic waves from the fixed mirror (solid lines) and movable mirror (broken lines) at different values of the optical retardation δ.

If the source is not monochromatic but is a continuum, the interferogram is represented by the integral:

$$I(\delta) = \int B(\bar{\nu}) \cos 2\pi\bar{\nu}\delta \, d\bar{\nu} \qquad (1)$$

which is one-half of a cosine Fourier Transform, the other being

$$B(\bar{\nu}) = \int I(\delta) \cos 2\pi\bar{\nu}\delta \, d\delta \qquad (2)$$

It should be noted that $I(\delta)$ is an even function, so that *Eq. 2* may be rewritten

$$B(\bar{\nu}) = 2\int I(\delta) \cos 2\pi\bar{\nu}\delta \, d\delta \qquad (3)$$

From *Eqs. 1 & 3* it is evident that in order to have a complete spectrum from 0 to $+\infty$ at infinitely high resolution, we would have to scan the moving mirror an infinitely long distance with δ varying from 0 to $+\infty$. If we measure the signal over a limited value, the spectrum will have finite resolution.

The advantages of the Fourier Infrared Spectrometer over dispersive instruments for surface studies can be summarized as follows (*Hirschfeld, 1983 and Hirschfeld, 1984*):

- The multiplex advantage due to the fact that the detector views all the frequencies simultaneously during mirror movement. This determines an improvement in the signal to noise ratio (SNR) per unit time.

- The throughput advantage. The total source output can be continuously passed through the sample. This results in a gain in energy from 80 to 200 times greater than for dispersive instruments.

- The speed advantage. It is possible to obtain interferograms on a msec time scale.

- Frequency precision. The helium-neon laser monitor of mirror position provides greater frequency precision in the final spectrum. The interferogram contains all the spectral information, but it is not readily interpreted in the form obtained (*Fig. 3*). The conversion of the interferogram to an infrared spectrum requires the use of a computer. One of the most important uses of the computer is the co-adding of successive interferograms which increases the signal to noise ratio.

Since the signal increases as the number of scans and the noise as the square root of the number of scans, the signal to noise ratio should increase as the square root of the number of scans. This is very useful for low intensity spectra and also when it is necessary to demonstrate the existence of small differences between spectra.

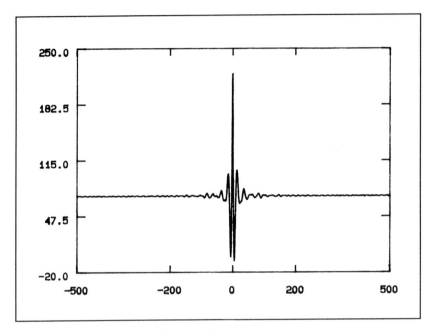

Fig. 3

Interferogram

ATR.

At an interface which separates two media of different refractive indices, the electromagnetic waves are either totally or partially reflected back into the original medium or refracted into the second medium depending on the indices of refraction and the propagation parameters of a wave. Total internal reflection is a special case of the general phenomenon of reflection and refraction of incident electromagnetic radiation at an interface. If the index of refraction of the initial medium is greater than that of the second medium ($n_c > n_s$) and the incident angle (θ) is greater than the critical angle, $\theta > \sin^{-1}(n_s/n_c)$, then the incident electromagnetic waves are completely reflected (*Fig. 4*).

In an internal reflectance element, a standing wave (the evanescent wave) is established near the surface of the crystal. The amplitude of the electrical field decreases exponentially with the distance from the surface of the IRE (Internal Reflection Element). The distance at which the amplitude of the electric field reaches 1/e of its initial magnitude is defined as the penetration depth dp. The magnitude of dp is a function of the wavelength of the radiation λ, the refractive index of the prism n_c, the refractive index of the sample n_s, and the angle of incidence θ of the beam at the surface of the IRE:

$$dp = \frac{\lambda}{2\pi n_c \, [\sin^2 \theta - (n_s/n_c)^2]^{1/2}} \qquad (4)$$

where $n_s < n_c$, for most materials dp is close or equal to 0.1. The smallest values of dp are for materials of high refractive index and for high angles of incidence.

By changing θ and n_c it is possible to vary dp. Maintaining constant all the other parameters, a depth profile can be obtained by varying the angle of incidence of the IR beam.

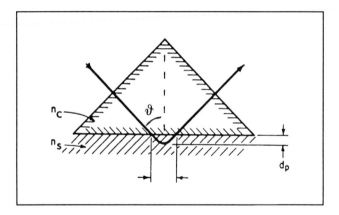

Fig. 4

Representation of total internal reflection. The radiation penetrates fractions of a wavelenghth (dp) into the material of refractive index n_s. The crystal has a refractive index n_c.

Scope of the method.

Analytical infrared spectroscopy methods are designed to characterize polymers in the virgin state (fundamental chemical bonds) and following exposure to biological media. Information can also be obtained about the orientation of a polymer surface.

DETAILED DESCRIPTION OF METHOD.

Equipment.

FT-IR.
A schematic representation of a commercial interferometer for the mid-infrared region is illustrated in *Fig. 5*. For the mid-infrared region the most commonly used light source is a silicon-carbide Globar.

The interferometer has one movable and one fixed mirror. The velocity of the movable mirror may be varied, typically between 0.5 and 60 mm sec^{-1}. Usually a He-Ne laser beam is used to set mirror velocity. A uniform speed is necessary in order to avoid increasing the noise level of the spectrum and distorting the instrument lineshape.

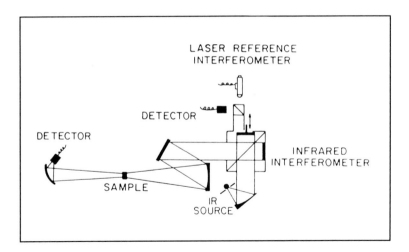

Fig. 5

Schematic representation of Fourier Transform spectrometer.

One of the most important components governing the performance of the michelson interferometer for infrared spectrometry is the beamsplitter. The beamsplitters are generally thin films. Their reflectance is determined by the refractive index of the material, the thickness of the film, the angle of incidence of the beam and the wavenumber of the radiation. Germanium or silicon films are generally used for mid-infrared spectrometry. The plate on which the film is deposited may be KBr or CsI or CsBr.

Infrared detectors can be divided into two types: thermal detectors and quantum detectors. Thermal detectors operate by sensing the change in temperature of an adsorbing material. They can be used over a wide range of the wavelengths, but as the temperature of the adsorbing material must change, they are usually slow (0.01 to 0.1 sec.). The alternative method of detecting infrared radiation relies on the interaction of radiation with the electrons of a solid. The excitation of electrons to a higher energy state thus created depends on the quantum nature of the radiation. The most commonly used quantum detector is of mercury cadmium telluride (MCT). When a MCT detector is used in conjunction with a rapid scanning interferometer, the mirror velocity should be set as high as possible.

ATR.

A schematic diagram of a particularly simple ATR accessory that can be used with several FT-IR spectrometers is shown in *Fig. 6.*

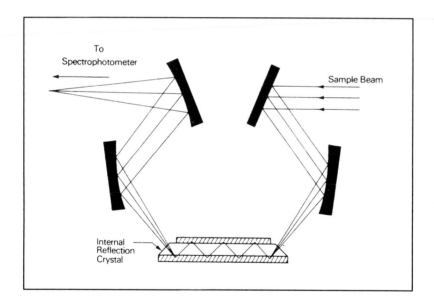

To Spectrophotometer

Sample Beam

Internal Reflection Crystal

Fig. 6

Optical layout of the multireflection ATR attachment.

In some commercially available ATR accessories it is possible to change the incident light angle usually between 35°-60°. In this way the depth of penetration dp can be changed. The cross section of the internal reflection element of most ATR accessories is rectangular. However, special purpose crystals for different kinds of samples with square circular cross sections have been designed for FT-IR spectrometers and are commercially available. Crystalline materials used for internal reflection elements are listed in *Table I* together with the index of refraction and the transmission region.

Table I

Materials and optical properties of internal reflection elements.

Crystalline materials	Index of refraction (n_c)	Transmission region (cm^{-1})
Germanium (Ge)	4.00	4000 - 900
Silicon	3.42	4000 - 1500
Zinc selenide (ZnSe)	2.42	4000 - 700
Thallium bromide- Thallium iodide (KRS5)	2.35	4000 - 400

The infrared bands associated with some internal reflection elements may be intense and can mask the weaker bands of the polymer spectra. Although the refractive index of polymers ranges from 1.38 to 1.63, an average value of 1.5 can be assumed for the following evaluation of sampling depth. The effect of the dependence of sampling depth on wavelength is much more pronounced in the case of ATR spectra than in the case of transmission spectra obtained with the same sample under the same conditions.

In *Fig. 7* the sampling depths of the most commonly used IRE materials are plotted as a function of the wavenumbers.

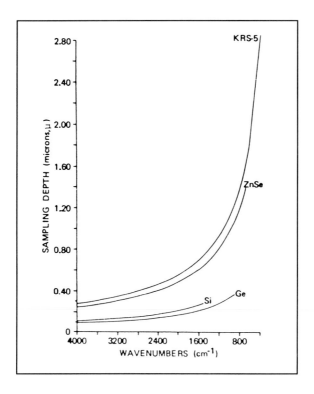

Fig. 7

Sampling depth versus wavenumbers for several internal reflection elements.

Spectral substraction (*Koening, 1975*).

Computerized spectral substraction permits the removal of an interferent from the sample spectrum without a reference of the same optical thickness. In our case the interferent may be a substance adsorbed at a polymer surface.

Usually two single beam spectra are collected: one of the sample surface and one of the reference. Different effective pathlengths can be compensated for by multiplying the reference spectrum by a factor K and subtracting this spectrum from the sample spectrum.

The scaling factor K is calculated to fulfil the condition (A - KB) = 0, where A is the absorbance of the sample spectrum and B the absorbance of the reference spectrum. The best method to perform such a spectral subtraction is to isolate a single band characteristic of the contaminant spectrum and to calculate K. This factor should then be used to subtract the entire spectrum of the contaminant and to obtain the spectrum of the polymer surface.

Description of procedure.

Sample preparation.
Unlike conventional spectroscopy in which sample preparation is an important factor, in internal reflection spectroscopy one should be particularly careful to ensure good contact between sample and crystal. For the solid-like polymer films the contact area is determined by their surface roughness. Thus for the study of thin films, the contact between the sample and the internal reflection element is a critical factor. Usually a good contact and a good quality spectra are obtained when polymer films can be directly deposited from the polymer solutions. Another critical factor to obtain good spectra is the cleanness and the transparence of the crystal. A good way to obtain clean crystals is to wash them in an ultrasonic bath.

Sample registration.
The background spectrum of the crystal should be recorded first. The instrument has to be purged with N_2 or with dry air to avoid the bands due to CO_2 and H_2O. Then the sample spectra should be recorded and the background substracted. The last step is computer manipulation to substract the background spectrum, and baseline correction or smoothing to reduce the noise level when necessary.

General remarks and necessary precautions.

For the practical application of internal reflection spectroscopy it is necessary to avoid distortion and shift due to dispersion. It is important to stay well away from the critical angle. If the angle of incidence cannot easily be changed, distortion can be eliminated keeping the angle of incidence fixed and employing an internal reflection element of higher refractive index.

Two methods of correcting for distortion are shown in *Fig. 8*. The internal reflection spectrum of polypropylene shown in *Fig. 8a* taken with a KRS5 crystal at $\theta = 38°$ is highly distorted. By increasing the angle of incidence, an undistorted spectrum is obtained (*Fig. 8b*). Distortion can be eliminated even at a low angle of incidence by replacing the KRS5 with a Ge IRE (*Fig. 8c*). Spectrum distortion may also be caused by bad contact between the sample and the crystal.

It is necessary to be careful in the spectral subtraction operation. Scale absorbance subtraction is not possible when absorbance is too high. In other words, to be sure that Beer's law is obeyed, it is important that the spectra under study have an absorbance of less than 0.5 absorbance units. Some errors in quantitative analysis are inherent to FT-IR spectrometry itself. One special effect due to FT-IR spectrometer performance is the stability of the interferometer. Instability of the interferometer may result in frequency shifts caused by the incoherence of co-added interferograms.

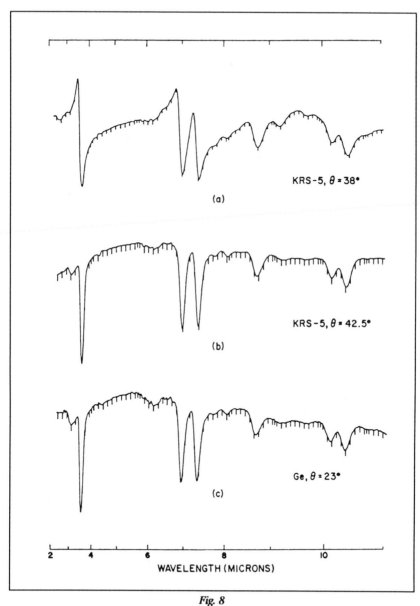

Fig. 8

Typical Spectrum of polypropylene n_s ≈ 1.5: (a) KRS5 at θ = 38°; (b) KRS at θ = 42.5°; (c) Ge at θ = 23°.

RESULTS

Normal values for reference materials.

The spectrum of polyurethane (Pellethane 2363-80E) shown in *Fig. 9* may be regarded as a reference spectrum. This ATR spectrum is obtained with a KRS5 crystal internal reflection element and an incident angle of 45°. For our purposes there are two important regions (i) 3400-3500 cm^{-1} relative to hydrogen bonded and non hydrogen bonded NH streching and (ii) 1800-1600 cm^{-1} which is assigned to C=O stretching.

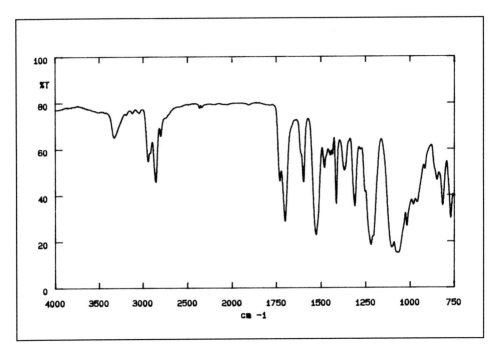

Fig. 9
ATR spectrum of Pellethane

Alternative typical experimental values for the testing of materials.

The morphology of block copolymers such as polyurethanes and polyurethanes-ureas have been extensively studied by FT-IR techniques in the past few years *(Wang & Cooper, 1983)*.

The unusual range of physical and chemical properties associated with block polymers results from the unique morphological separation between hard and soft domains. The primary driving force for domain formation is the strong inter-molecular interaction between the urethane units which are capable of forming interurethane hydrogen bonds. It has generally been found that the fraction of bonded urethane carbonyls increases with increasing hard segment content which suggests an increase in hard domain ordering. Other factors that control the degree of microphase separation include copolymer composition, block length, crystal-linity of other segments and the method of sample preparation.

As an example of application of the ATR/FT-IR technique to understand the chemical structure of materials, we report here the spectra of a new material PUPA synthetized by CRISMA (Siena, Italy) which contains poly(amidoamine) chains and a polyurethane as a matrix (*Barbucci et al., 1989*). Its structure is shown in *Fig. 10* and its chemical composition can be changed by increasing the quantity of polyurethane as shown in *Table II*.

Table II
Chemical composition of PUPA

Material	Composition		
	PU	Hexam	N_2LL
	g	ml	g
PUPA	1.00	0.42	0.40
PUPA bis	2.00	0.42	0.40
PUPA penta	5.00	0.42	0.40

PU: polyurethane. Hexam: hexamethylene-diisocyanate. N_2LL: poly(amidoamine).

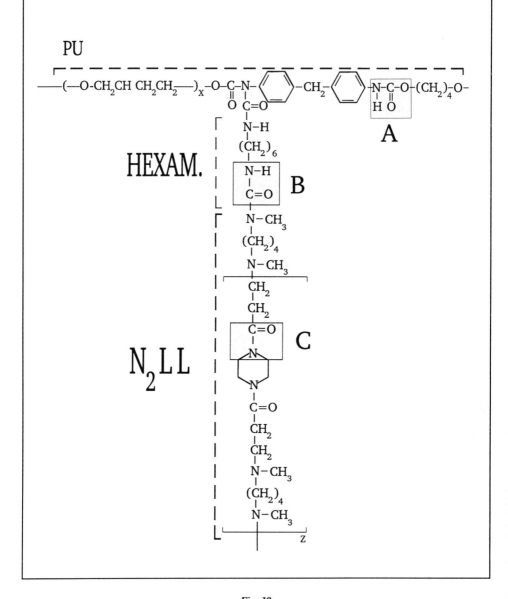

Fig. 10

PUPA structure

The spectrum of PUPA using a KRS5 internal reflection element at $\theta = 45°$ is shown in *Fig. 11*.

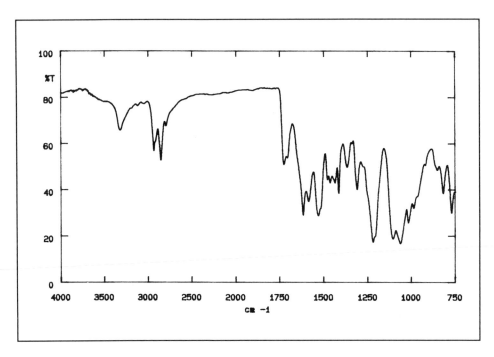

Fig. 11

PUPA spectrum between 4000 and 750 cm^{-1}

DISCUSSION

Interpretation of the results.

The spectra of the different PUPAs exhibit a large variation in the region between 1800 and 1500 cm^{-1} (*Fig. 12*). The assignment of the frequencies is reported in *Table III*.

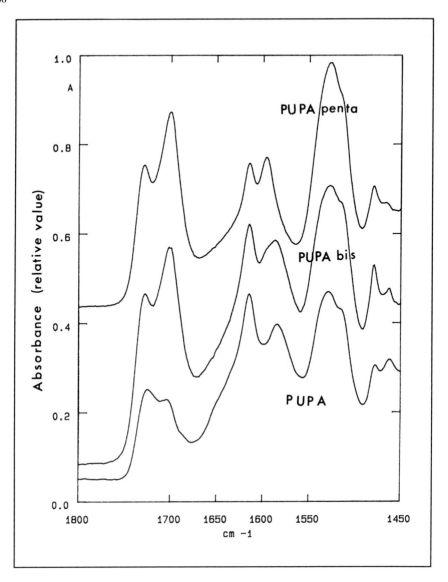

Fig. 12

PUPA, PUPA bis, PUPA penta spectra between 1800 and 1500 cm⁻¹

The band at 1615 and the shoulder at 1585 cm^{-1} assigned to the amide I and amide II respectively may be correlated with the presence of the urea carbonyl group (B) (*Fig. 10*). The band at 1595 cm^{-1} assigned to the C=C stretching frequency of the aromatic ring is indicative of the presence of Pellethane and increases in intensity from PUPA to PUPA penta (*Table II*).

Table III

Assignment of some frequencies in PUPA spectra.

Band assignment	Frequencies (cm^{-1})
NH stretch. of H-bonded NH group	3320
C=O stretch. of non H-bonded urethane group (A)	1730
C=O stretch. of H-bonded urethane group (A)	1700
C=O stretch. of N$_2$LL amidic group (C)	1650
Amide I (B)	1615
C=C stretch. of the PU aromatic ring	1595
Amide II (B)	1585

The broad shoulder at about 1680 cm^{-1} can be correlated with the C=O stretching of the amidic group (C) (*Fig. 10*) of polyamidoamine chains. In this case a useful correlation between polymer composition and the infrared spectra was found.

The spectra also provide information about the change in morphology of the samples. The band at 3320 cm^{-1} would indicate hydrogen bonded NH groups. As there are no bands at frequencies beyond 3320 cm^{-1} which would indicate non hydrogen bonded NH groups, it is evident that all NH groups in all samples are involved in hydrogen bonds.

Changes in the relative intensity of the bands at 1730 and 1700 cm^{-1} are indicative of a change in the quantity of hydrogen bonded urethane carbonyl groups (A) (*Fig. 10*). As for other polyurethaneureas, there is a correlation between the quantity of hydrogen bonds and phase separation.

Limitations of the method.

The most important limitation of the method is the penetration depth of radiation into the sample which is expressed by *Eq. 4*. The spectra obtained are not only spectra of the polymer surface, but of the adjacent polymer bulk too.

In many cases it has been demonstrated that changes in the angle of incidence or refractive index lead to spectral changes due to different penetration depths. This shows that the bulk spectrum is different from that of the surface and that this technique can give useful information about the polymer surface, even if the spectra obtained are not exactly surface spectra.

REFERENCES

Barbucci, R., Benvenuti, M., dal Maso, G., Nocentini, M., Tempesti, F., Losi, M., Russo, R. (1989) Biomaterials, 10, 299-308.

Ferraro, J.R. & Basile, L.J. (1979) "Fourier Transform Infrared Spectroscopy Applications to Chemical System", Vol. 1, 2 & 3, Academic Press.

Gendreau; R.M. & Jakobsen, R.J. (1976) J. Biomed. Mat. Res. 13, 893-906.

Griffths, P.R. & Haseth, J.A. (1986) "Fourier Transform Infrared Spectroscopy, Chemical Analysis", Vol. 83, Wiley Intersci.

Harrick, N.J. (1979) "Internal Reflection Spectroscopy", Harrick Scientific Co.

Hirschfeld, T. (1983) ESN Eur. Spectroscopy News 51, 13-18.

Hirschfeld, T. (1984) ESN Eur. Spectroscopy News 55, 15-21.

Koening, J.L. (1975) Appl. Spectroscopy 29, 293-308.

Wang, C.B. & Cooper, S.L. (1983) Macromolecules 16, 755-786.

Fourier Transform Attenuated Total Reflection Infrared Spectroscopy (ATR/FT-IR): Application to Proteins Adsorption Studies

Authors: A. Magnani & R. Barbucci

Dipartimento di Chimica, Università di Siena, I - 53100 Siena

INTRODUCTION

Name of method:

Fourier Transform Infrared Spectroscopy (FT-IR) coupled with Attenuated Total Reflection Spectroscopy (ATR).

Aim of method:

This method can be applied for in vitro and ex vivo experimental procedures. FT-IR can be used for monitoring biological processes (*Fink & Chittur, 1986*). When FT-IR is coupled with ATR technique a powerful tool is created for studying processes at interface. Information can be obtained on blood compatibility of polymeric materials by monitoring the plasma protein adsorption at a polymer surface in contact with flowing plasma or blood (*Gendreau & Jakobsen, 1979*).

Background of the method:

(see: *"Fourier Transform Attenuated Total Reflection Infrared Spectroscopy"* by *Nocentini & Barbucci* in this chapter).

Scope of method:

The method is designed for "on-line" analysis of protein adsorption to surfaces under physiological conditions. Three kinds of information may be derived from this investigation:

S. Dawids (ed.), Test Procedures for the Blood Compatibility of Biomaterials, 171–184.
© 1993 Kluwer Academic Publishers. Printed in the Netherlands.

1. the types of proteins adsorbed onto the polymer surface (*Gendreau, 1986*),
2. the rate and the amounts of adsorption of proteins (quantitative aspect of the process) (*Fink et al., 1987*),
3. the conformational changes in the adsorbing species (structural aspect of the process) (*Lenk et al., 1989*).

The technique is mainly designed for experimental purposes using small amounts of polymer coated on the surface of a Germanium crystal.

DETAILED DESCRIPTION OF METHOD

Equipment.

FT-IR spectrometers.
Perkin Elmer Corp., Analytical Instruments, Norwalk, CT 06856, USA
BIO-RAD, Digilab Div., Cambridge, MA 02139, USA
Nicolet Analytical Instruments, Madison, WI 53711, USA

ATR apparatus.
ATR optics and flow cell: Harrick Scientific Inc., Ossining, N.Y., USA
Plumbing setup: Disco S.R.L. - International Ismatec, Milan, Italy.

A schematic diagram of a particular ATR accessory designed for a FT-IR PERKIN-ELMER M 1800 spectrometer is in principle shown in *Fig. 1*.

In order to maximize the energy throughout to the detector, the geometry of the ATR accessory may be slightly varied according to the different geometry of the sample compartment of the FT-IR spectrometer.

Flow cell design.
The flow cell used for protein adsorption studies is shown in *Fig. 2*. It is narrow-gap flow cell, and the particular design represents a compromise between optimal spectral performance and reasonable hydrodynamic conditions within the cell.

The height of the flow chamber is roughly 0.8 mm which allows shear rates in the range of 500-700 sec^{-1} for volumetric flow rates of about 75 ml/min. The cell is designed to mount on a pair of guide bars to slide back and forth inside the focusing accessory. This provides for experiments in a dual channel configuration in which one channel can be used as a blank for experiments performed in the other channel, or two separate experiments may be conducted simultaneously under the same flowing conditions.

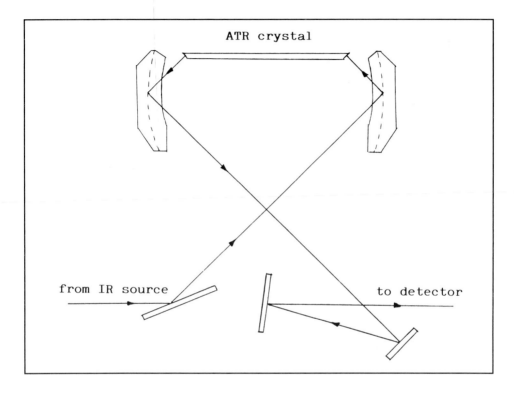

Fig. 1

Schematic diagram of the ATR accessory for flow experiments designed for Perkin-Elmer M 1800.

The crystals used are 45° Germanium ATR crystals and are designed to provide approximately 50 reflections while still allowing an energy output in the range of 10-15% of the input. They are 105 x 10 x 2 mm, and only one of the two available surfaces is utilized.

Due to the high refractive index of Ge (n_c = 4.00), the beam slightly penetrates into the bulk of the crystal surface system allowing the study of the process at the interface between solution and crystal surface.

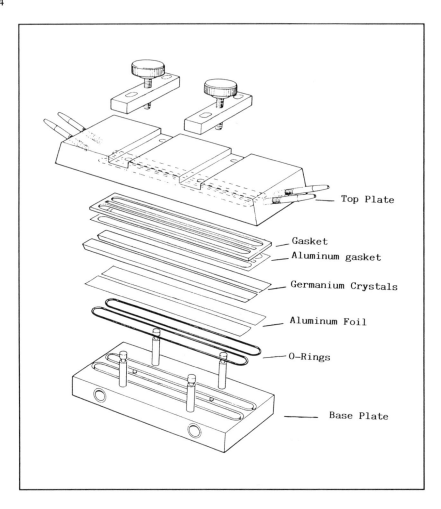

Fig. 2

Exploded view of the dual channel ATR flow cell.

Spectral subtraction.

One of the major advantages of FT-IR over previous infrared techniques is the capability of studying samples under physiological parameters. The improved SNRs (including the stability and reproducibility) of the newer FT-IR spectrometers contribute to the possibility of digitally compensating for the absorption due to the water in aqueous protein solution. There is a difficulty with digital water subtraction in knowing exactly how much to subtract from the spectrum of protein solution, because of underlying absorbance of the protein Amide I vibration in the water absorption region (1670-1610 cm^{-1}).

Generally, when dealing with protein, two subtraction criteria are mainly followed (*Gendreau, 1986*):

1. Subtraction to a fixed band ratio.

    The intensity of the Amide I band (in the 1670-1620 cm^{-1} region) is affected by the water band subtraction while the intensity of the Amide II band (in the 1570-1530 cm^{-1} region) is not affected. Thus studies in the dry state and with D$_2$O have provided guidelines as to the expected 1640:1550 (Amide I:II) band ratio, and subtractions are scaled to provide this band ratio.

2. Subtraction to achieve a flat baseline.

    Due to the interaction between water band and Amide I vibration a residual dispersion curve appears at roughly 1700 cm^{-1} if one oversubtracts (i.e. subtracts more water than is present and begins to subtract the Amide I band). Thus one subtraction technique quite successful is to attempt to maintain a flat baseline at frequencies above the Amide I absorption band.

Outline of method.

During multiple internal reflections inside the internal reflection element the infrared radiation penetrates into the sample for a finite depth, enabling the molecular vibrations within the superficial layers of the sample to be studied.

The ending radiation, the intensity of which is attenuated for the absorption of the sample, is directed to the detector and processed by the Optical Unit Processor. A fast Fourier Transform (FFT) is used by the processor to transform the time-based interferograms into the frequency related infrared spectral data. Spectra are monitored visually. They can be stored and recorded graphically or processed by the computer.

Because of the magnitude of sampling dedth (less than 1 μm using Germanium as internal reflection element), the spectral data obtained are mainly due to the polymer-solution interface.

Description of procedure.

Schematic representation of the flow cell and plumbing set-up for the protein adsorption experiments is shown in *Fig. 3*.

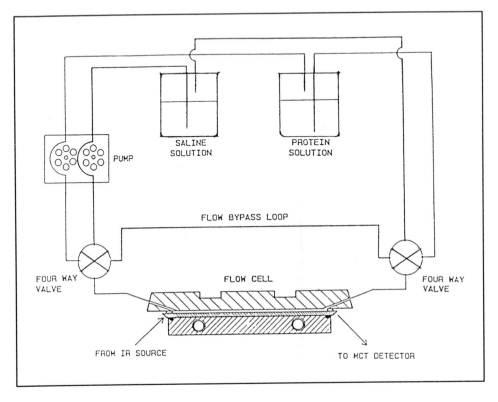

Fig. 3

Schematic representation of the apparatus for protein adsorption experiments.

Sample preparation.

The ATR crystal can be precoated with materials of interest to study surface interaction. The coating has to be very uniform and thin (about 10-20 nm) much thinner than the depth of penetration of the IR evanescent wave (about 400-600 nm over the region of interest (1800-1200) cm^{-1}) to allow protein adsorption to be followed directly on the polymer surfaces.

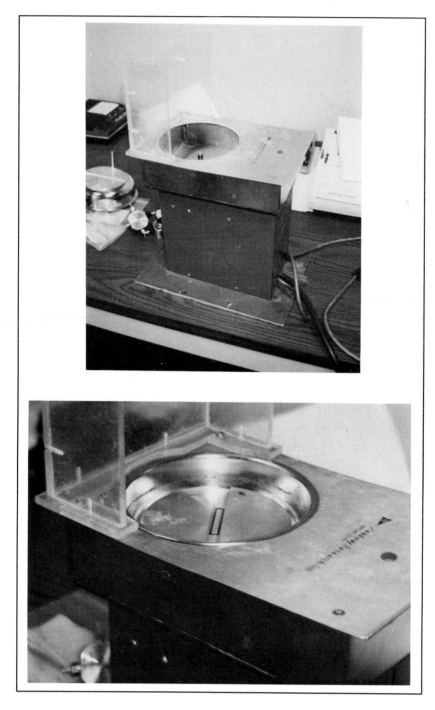

Fig. 4

*Spin-coater machine (**above**). Particular of the spin-coater: the sample holder (**below**).*

The coating is performed by spreading a known volume of a dilute solution (of known concentration) of the polymer evenly over the surface of the ATR crystal and spinning in a specially designed vacuum holder: the spin coater machine (*Fig. 4*).

Polymer solution is used cool within 24 hours of preparation to obtain coatings which are clear and uniform in appearence. The polymer films over the ATR crystal surface obtained by this procedure are about 10-20 nm thick. The thickness can be determined by ellipsometry (*Cuypers et al., 1986*).

Sample registration.
Preliminary procedure:

1. The Attenuated Total Reflection (ATR) flow cell which has been assembled and tested for leakage, is aligned inside the focussing accessory.
2. The instrument is then purged by N_2 to remove all the atmospheric water vapor.
3. The reference background value of the cell with the crystal coated with a polymer film is read and collected.
4. A flow of saline solution is circulated through the flow cell with a roller pump. The protein solution is circulated into the bypass circuit (see *Fig. 3*) prior to starting the adsorption experiment (air bubbles should be avoided). The polymer surface is equilibrated with saline for 1 hour.
5. The spectrum of the cell filled with saline solution is collected and stored for later use.

Experimental procedure.
Adsorption experiment. The two four-way valves are turned to introduce the protein solution into the flow cell and data collection starts. Spectra are collected at predetermined short time intervals (adsorption experiment). All the spectra are collected in the single beam (SB) mode. Absorbance spectra are automatically calculated by the following ratios:

$$\frac{\text{Spectrum of (Protein + Saline + Polymercoated crystal)}_{SB}}{\text{Spectrum of (Polymer-coated crystal)}_{SB}} = A_{protein} + A_{saline} \quad (1)$$

$$\frac{\text{Spectrum of (Saline + Polymer-coated crystal)}_{\text{SB}}}{\text{Spectrum of (Polymer-coated crystal)}_{\text{SB}}} = A_{\text{saline}} \qquad (2)$$

Subtraction of Eq. 2 from Eq. 1 gives the spectrum of the protein.

When protein adsorption takes place from a protein mixture or from whole plasma or blood (in vitro and ex vivo experiments), quantitative and structural information is obtained of the process with the use of commercially available computer programs for multicomponent analysis which are based on Matrix algebra (*Brown et al., 1982; Chittur et al., 1986*).

Washing experiment. A similar scanning sequence is used when the bulk protein is removed from the flow cell by the solvent wash.

Safety aspects and general remarks.

No special precautions are needed. - Proteins have to be used as pure as possible (not less than 98%). All protein solutions have to be prepared freshly in buffer (pH = 7.4). Experiments should be conducted at 37°C. At the end of each experiment the entire apparatus has to be rinsed carefully with sodium dodecyl sulphate (SDS) or other detergents to remove the proteins definitively.

RESULTS

When an ATR experiment is carried out in flow cell, the surface of the polymer coated is continuously monitored and can be measured frequently as protein adsorbs. Quantitative changes in the amount of the adsorbed proteins can be detected as well as changes in the conformation of the adsorbed species, but only when working with a single protein. The type of the adsorbed species can be determined by comparison with reference protein spectra. The given measurements are given on "naked" Germanium crystal.

Fig. 5 shows the adsorption kinetics from a single protein solution (30 mg/ml albumin in saline solution)(period A-B) followed by displacement of the protein solution by saline (period C). The amount of protein seen in each spectrum obtained is directly related to the absorbance at 1550 cm^{-1} (Amide II band) (*Fink et al., 1984*).

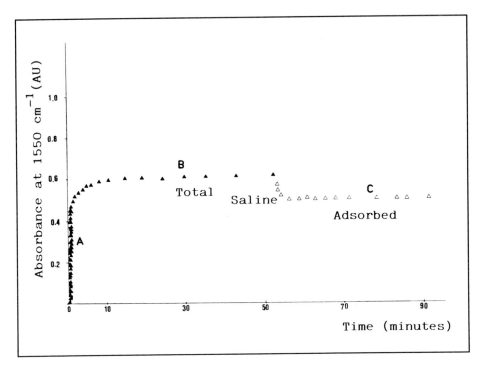

Fig. 5

Adsorption kinetics for albumin solution (30 mg/ml) over Germanium at 60 ml/min flow rate.

DISCUSSION

Interpretation of results.

When a protein solution is introduced into the ATR flow cell it gradually displaces the saline solution. Two phases can be identified: diffusion and adsorption of protein. In the first phase (period A) the protein solution displaces the saline in a laminar flow fashion leaving an evanescent boundary saline layer. The protein diffuses through this layer with increasing speed and immediately begins to adsorb to the surface. This phase is characterized by a rapid rise of the IR absorbance intensity due to the combination of the protein molecules in solution and adsorbed.

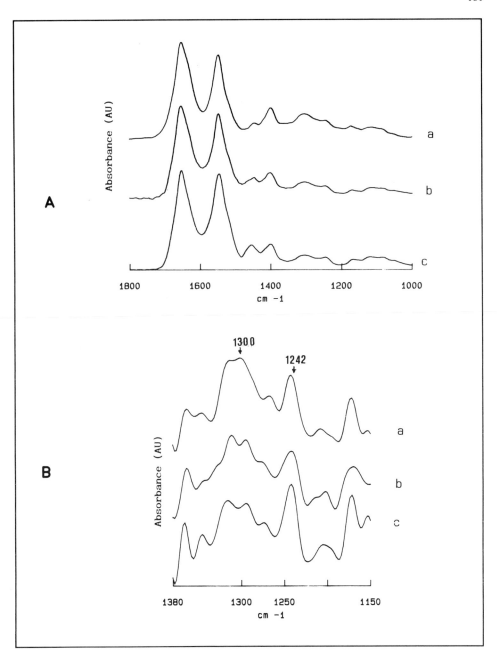

Fig. 6

A: *Representative spectra of a)* albumin in solution, *b)* albumin adsorbed on Ge (adsorption
time: 2 min.) *c)* albumin adsorbed on Ge (adsorption time: 50 min.).

B: *Deconvolved Amide III region of the spectra a), b) and c).*

This absorbance gradually rises until the concentration of the soluble protein in the boundary field reaches that of the protein solution. The second phase (period B) then begins where only spectral changes are observed due to adsorption/desorption and/or rearrangement of the protein layer.

When saline is introduced into the cell displacing the protein solution, the third phase (period C) starts. The spectrum of the adsorbed protein is then seen where there is no contribution of protein in the solution. The difference between the spectrum at the end of the period B and that at the end of the period C is thus the spectrum of the protein in the bulk. Successive subtractions of the derived spectrum from the absorbance spectra obtained in the period B give the spectra of the growing adsorbate film at the time intervals desired. The correction for the spectra collected in the period A is complicated because of the rapidly changing composition of the boundary layer.

As it can be seen in *Fig. 5* two important points can be emphasized:

1. the signal of the protein in solution relative to the bulk signal of protein (i.e. spectrum at the end of B minus spectrum at the end of C) is a significant fraction of the total signal,
2. the adsorption is seen as occurring in two distinct stages: an initial rapid adsorption occurring during the laminar filling of the measuring cell (period A) and a later stage of continuous slower completion of the adsorption (period B).

In *Fig. 6* are shown representative spectra of albumin adsorbed on the naked germanium crystal at two different times of adsorption together with the spectrum of albumin in solution. These spectra can be considered as reference graphs as almost all types of ATR equipment exclusively depend upon germanium crystals.

The calculated secondary structure assignments are given in *Table I*.

Comparison of the Amide III region of the spectrum (a) and (b) shows that there are similar, but not identical, 1300 cm^{-1}/1242 cm^{-1} intensity ratios between the two spectra (see *Table II*). Thus, after two minutes of adsorption the helix structure is still maintained to some extent, even if it clearly begins to change. This relates to the 1300 cm^{-1}/1242 cm^{-1} ratio which is slightly higher for the dissolved compared with the adsorbed albumin. As the adsorption time progresses, the structure of the adsorbed film changes and after about 1 hour the adsorbed film has achieved a stable structure. Thus the 1300 cm^{-1}/1242 cm^{-1} ratio decreases with increasing adsorption time and reaches a stable value after 50 minutes. This indicates the conformational changes of the adsorbed protein molecules.

Table I

Observed Amide I and Amide III frequencies (aqueous solution) for Albumin (3%wt)

Frequencies[a] related to conformation (cm^{-1})			
AMIDE I		**AMIDE III**	
α-Helix	Random coil	α-Helix	Random coil
1655s	1655s, 1678sh	1320-1300m,br	1265w, 1242w

s = strong, **m** = medium, **w** = weaker, **sh** = shoulder, **br** = broad

[a] the frequencies are those of deconvoluted spectra.

Table II

Values of the 1300 cm^{-1}/1242 cm^{-1} band intensity ratios.

Albumin in solution	1.23 ± 0.03
Albumin adsorbed on Ge (adsorption time: 2 min.)	1.15 ± 0.02
Albumin adsorbed on Ge (adsorption time: 50 min.)	0.80 ± 0.02

Limitations of method.

The limitation of the method is related to the inherent complexity of the system to be studied. Moreover, a high number of steps is necessary to sample and manipulate the data. The technical problems which make the analysis of the process difficult, especially from a quantitative point of view, involve:

- quantitative subtraction of the water absorption from the protein solution spectra,
- the complexity of the ATR spectra which include contributions from both adsorbed and soluble protein in the boundary layer adjacent to the ATR surface,
- the general similarity of the protein IR spectra which requires protein spectral recognition on the basis of weaker secondary bands or on slight variations in band intensities and/or frequencies,
- the sensitivity of the IR signals of proteins to microenvironmental factors such as pH and ionic strength.

REFERENCES

Brown, C.W., Lynch, P.F., Obremski, R.J. & Lavery, D.S. (1982) Analytical Chemistry 54, 1472-1476.

Chittur, K.K., Fink, D.J., Leininger, R.I. & Hutson, T.B. (1986) J. Colloid Interface Sci. 111, 419-433.

Cuypers, P.A., Hemker, H.C. & Hermens, W.Th. (1986) in "Blood Compatible Materials and Their Testing" (Dawids, S. & Bantjes, A. eds.) pp. 45-56, Martinus Nijhoff Publ.

Fink, D.J. & Chittur, K.K. (1986) Enzyme Microb. Technol. 9, 568-572.

Fink, D.J., Hutson, T.B., Chittur, K.K. & Gendreau, R.M. (1987) Analytical Biochemistry 165, 147-154.

Gendreau, R.M. & Jakobsen, R.J. (1979) J. Biomed. Mater. Res. 13, 893-906.

Gendreau, R.M. (1986) Spectroscopy in the Biomedical Science, CRC Press, Inc.

Lenk, T.J., Ratner, B.D., Gendreau, R.M. & Chittur, K.K. (1989) J. Biomed. Mater. Res. 23, 549-569.

Calcium-Thiocyanate Method for the Quantification of Sites on Polymer Surfaces

Author: A. Baszkin

Physico-Chimie des Surfaces, URA CNRS 1218, Université Paris-Sud,
F - 92296 Châtenay-Malabry

INTRODUCTION

Name of method:

Calcium-Thiocyanate adsorption method for the quantification of functional sites at the polymer-water interface.

Aim of method:

Quantification of the number of polar groups present in the surface zone of a polymer (outer few atomic layers of a sample), accessible to the adsorbing Ca^{2+} or SCN^- ions at the polymer-water interface.

Background of the method:

The principle of the method is based upon the use of radioactive isotopes emitting soft ß-radiation (^{14}C, ^{45}Ca). When a polymer film bearing functional sites acid or basic type on its surface is placed in contact with an aqueous solution containing $^{45}Ca^{2+}$ or $S^{14}CN^-$ ions, then the adsorption of these ions at the polymer-solution interface makes it possible to quantify the polymer functional groups under in situ conditions. The radioactivity measured above the solution polymer interface comes from the molecules adsorbed in excess at the interface and from the radioactive ions in a very thin layer of the solution adjacent to the polymer. As the mean free path of this radiation in aqueous solution is equal to 300 μm for ^{14}C and 650 μm for ^{45}Ca, all radiation originating from the solution below this depth is attenuated.

S. Dawids (ed.), Test Procedures for the Blood Compatibility of Biomaterials, 185–196.

Scope of method.

The technique is especially suited to monitor in situ the adsorption of either calcium or thiocyanate ions on flat polymer films bearing functional sites which can thus be quantified.

Most existing spectroscopic methods do not satisfy these requirements not only because of the extremely low absolute number of these groups present in the surface zone of a polymer (10^{15} - 10^{17} sites/cm^2) but also because they are not used in the in situ conditions.

The technique is recommended for the general screening of polymer surface functionality after reactions employed to create new or to modify existing chemical groups at the surface (chlorination, oxidation, grafting of polar monomers onto hydrophobic polymers etc.).

The surface of a polymer, as defined by this method, is that part of a polymer which is accessible to reagents that are soluble in water, but insoluble in the polymer.

DETAILED DESCRIPTION OF METHOD

Equipment.

The in situ $^{45}Ca^{2+}/S^{14}CN^-$ adsorption measuring apparatus is schematically represented in *Fig. 1*.

The radiation originating from the solution/polymer interface region is detected by the gas flow tube, measured by the electronic counting device and continuously recorded by a potentiometric Y = f(time) recorder.

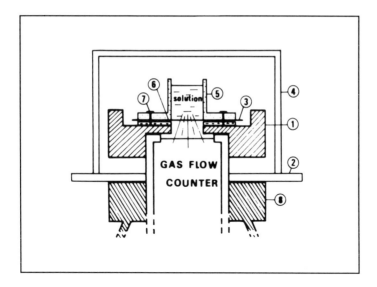

Fig. 1

(1), (2), (8) Supports ensuring reproducibility of geometry; (3) polymer film; (4) cover; (5) glass container; (6) "O" ring; (7) cell assembling screws.

The four principal units of the radioactivity measuring device are:

a) The flow gas detector "2Pi" with preamplifier incorporated manufactured by Berthold, Elancourt, France is in form of a 50 mm diameter hemispherical steel tube. Inside the tube are a pair of easily removable anodes. The screw ring at the bottom of the tube serves to fix a Mylar foil (6 μm) which forms a window facing the cell containing the radioactive solution. The radioactivity detector is continuously flushed with methane gas from a cylinder. The input of methane is controlled by a gas bubble meter attached to the tube.

b) Dual scaler with clock incorporated in a removable power supply bin. The dual scaler and the bin are manufactured by Enertec, Schlumberger, Montrouge, France. The dual scaler includes two independent scales and can be connected to a printer. The main characteristics are: 8 digit display, 10^8 capacity with overflow lamp, 20 MHz counting frequency, link to the recorder.

c) Linear ratemeter which provides an analog DC voltage output proportional to the average input count rate. It incorporates a digital display for direct reading of the count rate in counts per second. It includes three outputs: two recorder outputs, one for a voltage recorder, one for a current recorder and one gate input for the scaler.

d)   Potentiometric $Y = f(time)$ recorder, Servotrace 250 mm, Sefram, Velizy, France.

The measuring cell shown in *Fig. 1* consists of two dismantling parts: a circular glass open bottomed container with a flat ground collar and a polystyrene ring with a window of the same diameter as the glass cell. The glass container can be made by a laboratory glass maker. Molten paraffin is spread onto the flat ground part of the open glass cell. A polymer film, cut with a circular punch, is inserted between the two parts of the measuring cell and tightly sealed with screws by means of a Viton "O" ring incorporated in a slot in the polystyrene ring. The thickness of the film should preferably not exceed 50 μm.

The cell is filled with radioactive solution (c.a. 3ml) and placed in a special support above the flow counter. The counter measures the radioactivity and displays it on a recorder as a function of time.

Materials and reagents.

Radioactive calcium chloride (^{45}Ca) and potassium thiocyanate (^{14}C) may be purchased from any radioactive material supplier under conditions stipulated by law.

The radioactive material is isotopically diluted with unlabelled material to obtain specific activities suitable for counting. A convenient counting range for a gas flow counter is 2.10^2 - 10^5 counts/min. The lower value (2.10^2 counts/min) should be at least twice as big as that of the background count of the device. The upper value is limited by the dead time of the counting device.

Water used in these experiments must be triple distilled out of glass or quartz. Its surface tension is normally 72.1 mN/m. Water purified by ion-exchange or other column methods cannot be used.

Analytical grade products should be used for the preparation of buffers, when adsorption of either Ca^{2+} or SCN^- ions is performed at a given pH.

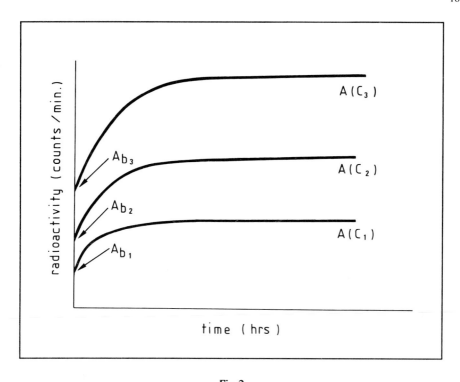

Fig. 2

Typical Ca²⁺/SCN⁻ adsorption kinetics curves at the polymer bearing surface functional sites/solution interface. Three different Ca²⁺/SCN⁻ solution concentrations.

Outline of the method.

A series of experiments is performed in which radioactivity is measured as a function of adsorption time at a given calcium or thiocyanate solution concentration. *Fig. 2* shows typical adsorption kinetics curves for three different Ca^{2+}/SCN^- solution concentrations.

The intercepts of these curves with t = 0 axis yield the A_b values corresponding to the radioactivities originating only from the solution at each concentration of Ca^{2+}/SCN^-. The A_b varies linearly with the ion solution concentration.

The total radioactivities (A_t) measured at equilibrium, (when the plateau values are attained), originate from the excess of Ca^{2+}/SCN^- ions adsorbed at the solution/-polymer interface (A_{ad}) and that of the ions present in the solution bulk (A_b). The A_b and A_t values are then plotted against solution concentration of Ca^{2+}/SCN^- ions (*Fig. 3*).

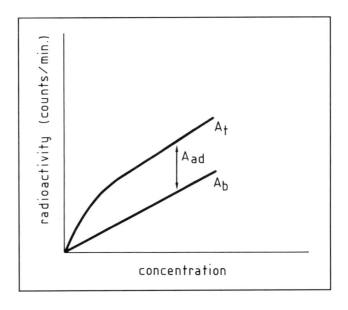

Fig. 3

*Schematic representation of asorption measuring technique. A, radioactivity measured through
a polymer sample. A_b radioactivity originating from the solution only (plotted from Fig. 2 at
$t = 0$). A_{ad} radioactivity corresponding to the amount adsorbed at the solid/liquid interface.*

We therefore have $A_{ad} = A_t - A_b$. When a polymer surface is devoid of functional
sites capable of adsorbing small inorganic ions from solution, the $A_{ad} = 0$ and $A_t = A_b$.

The A_{ad} values are then recalculated in surface densities of polymer functional
sites (δ) by means of a calibration graph (*Fig. 4*).

The calibration graph is obtained by depositing, drying and counting on polymer
window surfaces known amounts of $^{45}Ca^{2+}$ or $S^{14}CN^-$ ions. Counting should be
performed under the same geometrical conditions as in the adsorption kinetics
experiments (*Fig. 1*) and with a polymer film of the same thickness as in the
adsorption measurements.

The plateau value of each $\delta = f(c)$ isotherm corresponds to a number of functional
sites in a polymer zone available for adsorption in given in situ conditions.

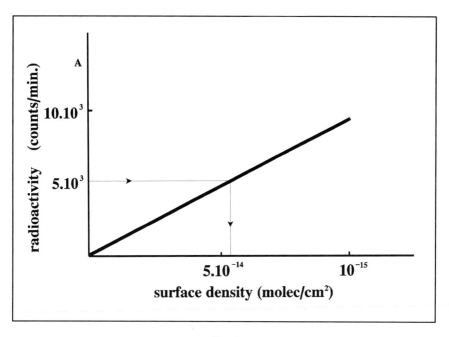

Fig. 4

Calibration graph. Radioactivity vs. amount of deposited Ca^{2+}/SCN^{-} molecules at the solid surface.

Description of procedure.

Solution preparation.

The first step is the preparation of aqueous solutions containing $^{45}Ca^{2+}$ or $S^{14}CN^{-}$ ions. Radioactive calcium chloride (^{45}Ca) or potassium thiocyanate (^{14}C) are normally furnished in small volumes (1-2 ml) in vials with rubber stoppers. The radioactive product is transferred to a 25 ml or 50 ml volumetric flask by means of a Pasteur pipette. The flask is adjusted with distilled water or an appropriate buffer. If the specific activity of this solution is too high, giving counts which are too high, then by isotopic dilution with a "cold", non radioactive calcium chloride or potassium thiocyanate aqueous solution, the reduction in the specific activity may be obtained. The isotopic dilution should provide a solution of an approximate specific activity and concentration (10^{-2} - $5.10^{-2}M$). The vial should be washed several times with small amounts of water (1-2 ml) and all rinses transferred to the flask, so that no radioactivity is lost.

A series of solutions is then prepared from the concentrated (mother) solution by dilution with water or an appropriate buffer. Each solution of a given concentration is prepared in a separate volumetric flask (100 ml).

To establish an isotherm of Ca^{2+} or SCN^- adsorption on a given polymer surface, 6-8 solutions of increasing concentration in the 10^{-6} - 10^{-4} M range are normally required.

Adsorption measurement.
Once a polymer film is fixed between the glass and the polystyrene part of a cell (*Fig. 1*) a radioactive solution (about 3 ml) is poured into the cell with a pipette. It is not necessary to measure this volume with high accuracy since the radioactivity which originates from the solution is greatly attenuated by the solution itself. The cell is then covered to keep out impurities and dust and reduce evaporation.

The adsorption experiment is carried out at room temperature. The adsorption course is automatically recorded as a function of time until a plateau value is reached. At the end of the adsorption experiment, the radioactive solution is pipetted out of the cell, the cell is washed several times by pouring in and pipetting out small amounts of water and then dismantled. The time necessary to reach equilibrium depends on the surface density and accessibility of polymer sites to the adsorbing ions. The recommended procedure is to leave the adsorption experiment overnight as the adsorption process in most cases takes about 15-18 hours before an equilibrium is attained.

Calibration graph.
The calibration graph is established by depositing small droplets of radioactive solution of a known concentration on the polymer films (inner side of the windows). By means of a calibrated microliter pipette or syringe, 5 μl, 10 μl, 15 μl and 20 μl are deposited in small droplets on four polymer windows. When dried, the surfaces are counted. The liquid droplets should be deposited uniformly over the polymer surface, and they should not touch each other. The abcissa of the graph is given by:

$$\frac{vol\ (\mu l).10^{-6} \times conc.\ (M) \times 6.023 \times 10^{23}}{surf.\ (cm^2)}$$

Linear relationships should be obtained between the radioactivity (counts/min) and the deposited quality of radioactive molecules molec/cm^2 (*Fig. 4*).

Adsorption isotherms (A_{ad} in counts/min) vs. concentration are then replotted in (A_{ad} in molec/cm^2) vs. concentration by means of a calibration graph. For each A_{ad} an equivalent in molec/cm^2 is found from *Fig. 4.*

Safety aspects of the method.

Although the hazards involved in the use of radioactive isotopes emitting soft ß-radiation are considerable less than for other radioisotopes, all the precautions for the manipulation of radioactive compounds should be respected. A special radioactive room has to be set up in the laboratory according to the local legislation rules. The use of the radioactive fume cupboard, protective glasses, gloves and plastic laminated filter paper is highly recommended. Solid and liquid radioactive wastes should be stored in the barrels supplied for the purpose by specialized companies dealing with the disposal of these wastes.

The decontamination of glass-ware used in the adsorption experiments has to be done with care. Furnishers of radioactive isotopes should be consulted as special solutions and products are available for the purpose (example: French product, Decon 90, supplied by Prolabo, Rhône Poulenc). In the absence of such products, contaminated glass and plastic ware can be immersed in a 2 N HCl for 24 hours and rinsed in abundant water. This has been found to be efficient for $^{45}Ca^{2+}$ and $S^{14}CN^-$ contaminated ware.

RESULTS

To our knowledge, the described method exists only in the author's laboratory. Furthermore, reference materials with a well known surface density of functional sites do not exist . Accordingly, only the results obtained in this laboratory may be presented here.

The calcium adsorption isotherms on surface oxidized polyethylene are given by way of example in *Fig. 5.*

Oxidation of low density polyethylene films (Cryovac C, Grace Company, France, thickness 19 μm) was performed in oxidizing mixtures prepared from sulfuric acid (sp.gr.1.84 pure grade) and potassium chlorate (reagent grade),

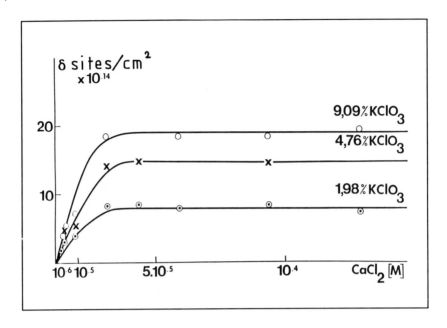

Fig. 5

Surface density of Ca^{2+} ions vs. $CaCl_2$ concentration in solution. Three oxidizing mixtures (wt%); time of immersion in oxidizing mixture - 30 sec.

As shown in *Fig. 5* the number of polar sites created at the polyethylene surface varies with the concentration of potassium chlorate in the oxidizing mixture.

Other typical values were obtained in the laboratory by the described method on poly(maleic acid) grafted polyethylene films. Such films may be transformed through a number of chemical reactions into polyethylene bearing quaternized polyamine sites. While the calcium adsorption isotherms on poly(maleic acid) grafted polyethylene surfaces yielded the quantity of dissociated COOH groups, thiocyanate adsorption isotherms were used to determine the number of quaternized polyamine groups on the polyethylene surfaces. Depending on the grafting conditions, pH and ionic strength of the aqueous adjacent phase, the surface density of functional groups on these polymers varied in the range 10^{15} - 10^{17} sites/cm^2. Further details of this method may be found in *Baszkin & Ter-Minassian-Saraga, (1971), Baszkin & Deyme (1974) and Baszkin et al. (1976).*

DISCUSSION

Interpretation of the results.

The isotherms shown in *Fig. 5* are of the Langmuir type I which implies that one Ca^{2+} ion adsorbs on one anionic site.

The adsorption equilibrium constant K may be calculated from the equation δ/c $(N-\delta) = K$, where c is the concentration, N the number of sites accessible for adsorption of Ca^{2+} ions and δ the number of calcium ions adsorbed at the surface. Plotting c/δ vs. c, one obtains a straight line relationship of slope $1/N$ and intersection with the ordinate for $c = 0$ at $1/KN$. Thus the number of sites and the constant K can easily be found.

The equilibrium constant K is independent of the degree of oxidation of the surface and is equal to $K = 1.66 \times 10^5$ litres/mole.

Limitations of the method.

The major limitations in the application of this method are:

- the method can only be applied to polymer film
- the method can only be applied to the polymer which do not absorb calcium or thiocyanate ions
- the thickness of polymer films for the experiment should not exceed 50-70 μm. For films of greater thickness, the absorption of ß-radiation by the film itself is too high
- the limited choice of soft ß-radiation emitting ions having a sufficiently long half life

Related testing methods and their comparative value.

The method described by *J.R. Rasmussen, E.R. Stedronsky & G.M. Whitesides: "Introduction, Modification & Characterization of Functional Groups on the Surface of Low Density Polyethylene Film" (J. Am. Chem.Soc., 99, 4736-4745, 1977)* was used for the quantification of carboxylic and ketone or aldehyde groups generated on the density polyethylene by treatment with concentrated chromic acid solution at temperatures between 25° and 75°C. Oxidized polyethylene films were submitted to a number of chemical reactions enabling the coupling of a fluorescent probe and the performance of fluorimetric assays on the subsequently uncoupled fluorescent molecules in solution.

The numerical values of surface densities of functional groups obtained with this method are in the same range as those obtained by the calcium/thiocyanate method. However, the conditions of treatment were different from those employed in the experiments represented in *Fig. 5.*

REFERENCES

Baszkin, A., & Ter-Minassian-Saraga, L. (1971) J. Polymer Sci. Part C. 34, 243-252.

Baszkin, A. & Deyme, M. (1974) C.R. Acad. Sci. Paris, Série C, 278, 1365-1367

Baszkin, A., Deyme, M., Nishino, M. & Ter-Minassian-Saraga, L. (1976) Prog. Colloid & Polymer Sci. 61, 97-108.

Rasmussen, J.R., Stedronsky, E.R. & Whitesides G.M. (1977) J. Am. Chem. Soc. 99, 4736-4745.

Contact Angles and Surface Free Energies of Solids

Author: A. Baszkin

Physico-Chimie des Surfaces, URA CNRS 1218, Université Paris-Sud,
F - 92296 Châtenay-Malabry

INTRODUCTION

Name of method:

Contact angle measurements on solid surfaces.

Aim of method:

Determination of thermodynamic quantities characterizing the polymer surface and polymer-water interface.

Background of method:

Contact angle geometry is based on the three phase equilibrium which occurs at the contact point of the solid-liquid-vapor or solid-liquid-liquid interfaces.

Basic wetting equations.
The work of adhesion for a solid and a liquid is given by:

$$W_{(SL)} = \gamma_S + \gamma_L - \gamma_{SL} \tag{1}$$

where γ_S, γ_L and γ_{SL} are the free energies per square centimeter of the solid, liquid and solid-liquid interfaces. They are reported in dyn/cm, mN/m or mJ/m^2. The γ_S is the free energy of the solid in high vacuum in the absence of the film of adsorbed liquid vapor.

Since there are presently no direct methods of measuring γ_S and γ_{SL} values, we resort to Young's equation (*Fig. 1*):

197

S. Dawids (ed.), Test Procedures for the Blood Compatibility of Biomaterials, 197–209.
© 1993 *Kluwer Academic Publishers. Printed in the Netherlands.*

$$\gamma_{SV} = \gamma_{SL} + \gamma_L \cdot \cos \theta \tag{2}$$

where θ is the equilibrium contact angle at the three-phase junction, γ_{LV} the surface tension of the liquid in equilibrium with the vapor, γ_{SL} the solid-liquid interfacial free energy and γ_{SV} the free energy of the solid in equilibrium with the vapor of the liquid.

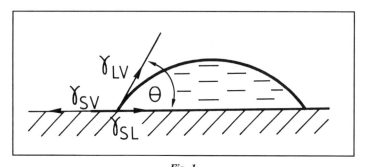

Fig. 1

Equilibrium force vectors at the contact point between a liquid drop and a solid surface.

The surface energy of the solid will favor spreading of the liquid drop. The solid-liquid interfacial energy and the vector of the surface energy of the liquid (liquid surface tension) in the plane of the solid surface will hinder the spreading of the liquid on the solid.

The surface pressure of the adsorbed film of the vapor of the liquid on the solid Π_S is:

$$\Pi_S = \gamma_S - \gamma_{SV} \tag{3}$$

From Eqs. 1, 2 and 3 one can deduce the work of adhesion to the film-covered solid surface $W_{(SLV)}$:

$$W_{(SLV)} = \gamma_S - \Pi_S + \gamma_L - \gamma_{SL} = \gamma_L (1 + \cos \theta) \tag{4}$$

Several kinds of intermolecular interaction occur between two phases. They include long range interactions (London - van der Waals), but may also include interactions of other kinds (acid-base, electrostatic). The solid-liquid work of adhesion is thus the sum of these interfacial interactions (*Fowkes, 1987*):

$$W_{(SLV)} = W^d + Ip \tag{5}$$

The polar component of the polymer-water work of adhesion may be obtained from Eq. 6 with the octane-under-water contact angle (θ_o) on a given polymer surface (*Andrade et al., 1979*):

$$Ip = 50.6 \ (1 - \cos \theta_o) \tag{6}$$

If the γ_{SV} or γ_{SL} values are required to characterize a polymer surface, then a second function in addition to the Young equation, relating γ_{SV}, γ_{SL} and γ_{LV} is necessary. This equation was developed by *Neuman et al. (1974)* by means of macroscopic observations and with well-documented assumptions. The equation known as the equation of state enables γ_{SV} and γ_{SL} to be calculated from a single contact angle value.

$$\gamma_{SL} = \frac{(\sqrt{\gamma_{SV}} - \sqrt{\gamma_{LV}})^2}{1 - 0.015 \sqrt{\gamma_{SV} \ \gamma_{LV}}} \tag{7}$$

$$\cos\theta = \frac{(0.015 \ \gamma_{SV} - 2.00) \ \sqrt{\gamma_{SV}\gamma_{LV}} + \gamma_{LV}}{\gamma_{LV} \ (0.015\sqrt{\gamma_{SV}\gamma_{LV}} - 1)} \tag{8}$$

For these calculations the contact angle of a liquid drop advancing slowly over a polymer surface (advancing contact angle - θ_A) is used. This angle is always larger than the receding angle (θ_r) obtained when a liquid is being withdrawn from the surface of the solid. The difference between θ_A and θ_r is called the hysteresis of the contact angle.

Contact angle hysteresis.
Intrinsic hysteresis: The hysteresis of contact angle caused by surface roughness is the intrinsic hysteresis of the surface and is also known as thermodynamic hysteresis. Roughness of less than 0.5 μm does not lead to significant hysteresis (about 10°). With increasing roughness, θ_A increases and θ_r generally decreases.

Heterogeneity of the composition of the surface as a result of the presence of low energy (hydrophobic) and high energy (hydrophilic) fractions gives rise to a large intrinsic hysteresis (up to 110° in some cases). Heterogeneity for patches less than 0.1 μm does not lead to significant hysteresis.

Environmental hysteresis: Polymer surfaces may reorientate or restructure in response to their local environment. The ability of polymer surface functional groups to move to and from the surface in contact with their adjacent phase is generally a time-temperature dependent phenomenon. The reorientation of polymer mobile segments may cause a considerable variation in contact angle values.

The adsorption of different solution constituents at the solution-polymer interface may also give rise to large contact angle hysteresis.

Deformation and swelling of polymers can considerably change contact angle values and be an additional cause of environmental hysteresis.

Scope of the method.

The techniques and the described method are recommended to calculate the free surface energies of the polymer and the polar/apolar components of the polymer-water work of adhesion. These quantities are closely related to protein adsorption and cell and tissue adhesion which, in turn, are important factors governing the blood compatibility of polymers.

DETAILED DESCRIPTION OF METHOD

Equipment and detailed testing conditions.

The present method of surface energy analysis of polymers by contact angle measurements is a combination of the drop-on-plate technique and the air and octane-under-water contact angle measuring technique (*Hamilton, 1972*).

While advancing contact angles (θ_A) are representative of dry surface energetics, the receding angles (θ_r) characterize wet surfaces. When the interactions of blood compatible polymers with aqueous media are of interest the θ_A are measured with water by the drop-on-plate method and the θ_r by the air-under-water technique. Octane-under-water contact angle data are used to calculate the polar component of the polymer-water work of adhesion.

Drop-on-plate method. The parts of the thermostatically heated apparatus shown in *Photograph 1* are listed below:

1) A double walled dark room chamber, thermostatted by means of an external liquid circuit. The chamber has two opposite windows and is installed on a plate which may be moved in two horizontal directions at right angles. The sample holder inside the chamber is fixed above a small dish filled with liquid. This allows saturation of the air phase with the vapour and thus avoids evaporation of the liquid drop during measurement.

2) An exterior screw permits rotation of the sample holder and measurement of the contact angle at various sites on the sample surface.

3) A microsyringe mounted at the top of the chamber by means of which a liquid drop (5μl) may be placed on the surface of a tablet. The microsyringe may be displaced in two opposite directions so that the drops can be deposited in a line on the surface.

4) A cathetometer with x 15 magnification, equipped with a goniometer that measures the contact angles.

A camera may be attached to the cathetometer. Although a temperature controlled chamber is preferable, thermostatting is not necessary for measurements taken at room temperature. An open sample stage which can be raised or lowered can be used especially with liquids that boil above 90°C.

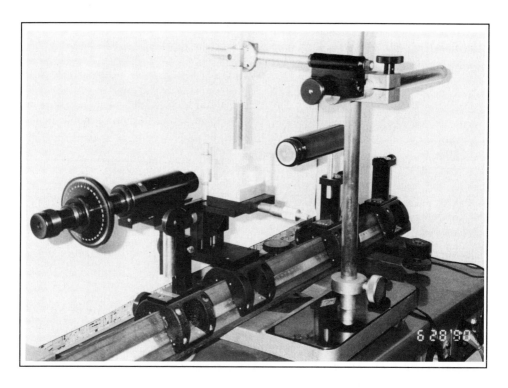

Air-under-water method. *Photograph 2* shows a typical device for receding contact angle measurements on hydrated surfaces. It consists of a travelling goniometer with x 15 eyepiece, a variable intensity light source and a micrometer-adjustable X-

Y stage vertically mounted on an optical bench. The stage contains a plexiglass container in which a teflon plate is suspended. The polymer sample is held on the underside of the teflon plate by means of small teflon clips. The container is then filled with triple distilled water and the plate with sample lowered into the container until the sample is completely immersed.

A bubble of air with a volume of about 0.5 μl is then formed at the tip of the microsyringe, below the surface, detached and allowed to rise to the polymer-water interface.

Octane-under-water contact angles. To measure contact angles of octane drops on hydrated polymer surfaces, the same optical device as for air-under-water contact angles is used.

Materials and reagents.
Hamilton precision syringes (10 μl) from Micromesure B.V., The Hague (Holland) are used for the air or octane-under-water contact angle measurements. Two separate syringes are necessary, one for the formation of air bubbles and the second for octane drop deposition.

Liquids used in contact angle measurements should be of high purity. Analytical grade products are generally satisfactory for the purpose. n-Octane from Merck (reference substance for gas chromatography) is recommended. It has a surface tension of 21.6 mN/m.

Water used for surface measurements must be triple distilled out of glass or quartz. Its surface tension in normally equal to 72.1 mN/m. Water purified by ion-exchange or other column methods cannot be used.

Outline of the method.

The geometry of sessile drops (drop-on-plate method) and of air or octane-under-water bubbles or drops is shown in *Fig. 2*.

The contact angle of sessile drops is always measured inside the drop (*Fig. 2a*). The contact angle of air bubbles is always measured outside the bubble (*Fig. 2b*). The contact angle of octane drops under water is measured inside the octane drop (*Fig.2c*).

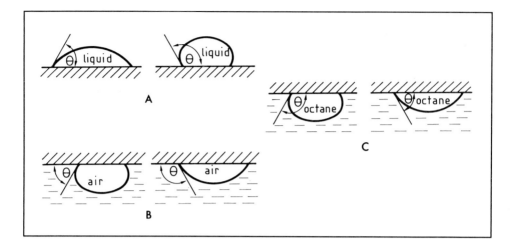

Fig. 2

Contact angle profiles (a) sessile drops (b) air bubbles under water (c) octane drops under water.

Description of procedure.

The contact angle may either be determined directly using a goniometer eyepiece with a crosshair or at leisure from a photograph. It can also be determined from a single trigonometric relation $\mathrm{tg}\,\theta/2 = 2h/d$ where d is the diameter of a segment of a circle (a spherical cap viewed in a direction parallel to the drop base) and h is the height.

To measure a contact angle directly the eyepiece crosshair is focused on and aligned with the base of the drop and its central point is placed where the drop terminates on the polymer surface. Turning the crosshair and placing it tangentially to the edge of the drop provides the contact angle.

The mean value of at least ten contact angles measured on drops or bubbles at different points on a sample is normally considered to be representative of a given surface. Angle measurements may be made to the left and right sides of the drops or bubbles.

It is important to correctly locate the crosshair center at the edge of the drop or bubble and then line it up parallel to the tangent to a drop arc. An error of 5° may easily be made if the operator is not aware of this necessary precaution.

The reproducibility of contact angle readings is normally ± 2°.

Safety aspects, general remarks and necessary precautions.

Different suppliers of optical equipment make contact angle measuring devices. When selecting such equipemnt one should bear in mind that it is preferable to use an enclosed chamber in which the drop can be saturated with the liquid vapour and is protected from contamination by atmosphere dust. It is useful to be able to regulate light source intensity and rotate the solid sample holder.

All glassware used in contact angle measurements (glass cells, pipettes, syringes etc.) has to be cleaned with a freshly prepared sulfochromic mixture and abundantly rinsed with triple distilled water. Protective glasses and gloves should be used when cleaning the glassware with the sulfochromic mixture.

A standard and extremely accurate cleaning procedure of polymer surfaces is necessary prior to contact angle measurements. Such a procedure may include Soxhlet extraction of samples in an appropriate solvent (not reacting with the polymer), ultrasonic bath cleaning, cleaning of polymer surfaces with soaps and cotton balls. Abundant and repeated rinsing of samples with triple distilled water is necessary whatever the cleaning procedure used. In this way the different polymer additives are leached out of the material. The last water used should be checked to see that its surface tension corresponds to the literature value.

Careful drying of samples under vacuum in desiccators is recommended prior to drop-on-plate contact angle measurements. Storage of samples in triple distilled water to ensure their full hydration is recommended prior to under-water contact angle measurements. The water should be changed several times.

A liquid surface tension measuring device (Wilhelmy plate method or Du Nouy ring method) is necessary standard equipment in any laboratory involved in contact angle measurements. Water and other liquids used for contact angle measurements should have surface tension values corresponding to those generally found in literature.

RESULTS

Normal values for reference materials.

To calibrate a contact angle measuring apparatus, reference polymer surfaces are recommended. However, such reference materials are rare. Polymer surface properties are very often considerably different for materials of identical bulk composition. The National Institute of Health (NIH) has prepared only two standard polymers: low density polyethylene and poly(dimethylsiloxane).

An air-under-water contact angle (θ_r) of 88.4 ± 3.6° on flat sheets of NIH standard polyethylene was recently reported in *Lelah et al. (1985)*.

Alternative typical experimental values for the test on well-known materials.

Smooth highly hydrophobic surfaces of polymers which do not contain functional groups in their structure may serve as reference materials in the absence of standard polymers. Typical values of water advancing contact angle (θ_A) measured by the drop-on-plate method, taken from *Dann (1970)* are:

Polyethylene: 94°; paraffin: 110°; Polytetrafluoroethylene: 112°.

DISCUSSION

Interpretation of results.

Polymer surface free energy (γ_{SV}), polymer-water interfacial free energy (γ_{SL}), polymer-water work of adhesion (W_{SLV}), its polar (Ip) and apolar (W^d) components and their ratio (Ip/W^d) are necessary to fully characterize polymer surface energetics and obtain information on the type of interactions across the polymer-water interface. Contact angle hysteresis is an additional source of information when investigating for composition-biological interaction correlations of biomaterials.

The present methods enable the determination of the following surface energetic quantities:

a) the γ_{SV} and γ_{SL} values may be calculated from *Eqs. 7 & 8* with the advancing contact angles obtained by the drop-on-plate technique,

b) contact angle hysteresis is the difference between θ_A and θ_r,

c) polymer-water work of adhesion is obtained from *Eq. 4* with either advancing or receding contact angle values,

d) the hysteresis of the polymer-water work of adhesion is calculated from:

$$H = \gamma_L (\cos \theta_r - \cos \theta_A),$$

e) polar components of polymer-water work of adhesion, Ip, are calculated using octane-under-water contact angles (θ_0) from *Eq. 6*,

f) dispersion components of the polymer-water work of adhesion W^d are obtained from *Eqs. 4 & 5*:

$$W^d = W_{(SLV)} - Ip$$

The application of this method in the calculation of the energetic parameters (a - f) of polymer-water interfaces is given in *Barbucci et al. (1987)*.

Limitations of the method

A major limitation of the method is that it can normally only be used on flat polymer surfaces. However, a special device has recently been proposed (*Lelah et al., 1985*) for measurement of air or octane captive bubble contact angles in water on polymer tubing. The contact angle is given by an equation relating several bubble dimensions which are directly measured on a bubble using a goniometer with attached vertical or horizontal micrometers.

One has also to be aware of the tedious aspects of the sessile drop (drop-on-plate) and captive bubble methods which require a great number of contact angle readings at different points on a polymer surface. The heterogeneity of polymer surfaces may be the cause of considerable scatter of contact angle values. An average contact angle value characterizing a given polymer surface must therefore be correctly established.

Related testing methods.

Other methods for contact angle measurements on flat polymer surfaces include:

a)   the Wilhelmy gravitational plate method
b)   the capillary rise at a vertical plate method
c)   the tilting plate method

All these methods are described in *Good (1979)*. In addition, method (a) is very well described in *Neuman et al. (1975)* and *Andrade (1985)*.

With the Wilhelmy gravitational method, advancing and receding contact angles are assessed directly measuring the force exerted on a rigid polymer plate suspended from an electro-balance.

In the capillary rise at a vertical plate method, instead of measuring the capillary pull on the plate, the capillary rise of the liquid at the vertical plate surface is measured optically.

The kinetic hysteresis of contact angles obtained with (a) and (b) corresponds to the effect of swelling and penetration, polymer surface mobility and functional group reorientation to and from the surface.

The main drawback of these methods is that the polymer plate must have a constant perimeter. Also the composition and morphology of the polymer plate at all surfaces (front, back and both edges) have to be the same.

Polymer samples for contact angle measurements using methods (a) and (b) have to be rigid. If one wants to measure these angles on polymer films deposited on glass or metal substrates (plasma polymerization, evaporation from organic solution etc.), it may be difficult to ensure the required uniformity of polymer surfaces.

In the tilting plate method, a rigid polymer plate is held in an adjustable holder capable of tilting a plate to any angle about the axis of rotation. The plate is tilted to a position at which the liquid surface remains undistorted up to the line of contact with the solid. The angle between the plate and the horizontal is the contact angle. When the tilting plate is immersed in the liquid, the advancing angle is measured. Raising the plate of the liquid provides the receding angle.

The main difficulty with this method is that the contact angle readings do not correspond exactly to the advancing or receding angles, but tend to lie somewhere between them.

Comparison of the described method with the contact angle methods (a, b, c) shows that the latter do not offer the possibility of determining the polar component of the polymer-water work of adhesion (Ip).

Since the Ip/W^d ratio correlates in many cases with protein adsorption or platelet adhesion to the polymer surface, the Ip component is an important energetic parameter of the polymer-water interface. Thus the combination of the drop-on-plate with the captive air and octane bubble technique seems to give a more meaningful surface energetic characterization of polymer surfaces.

REFERENCES

Andrade, J.D., King, R.N. Gregonis, D.E. & Coleman, D.L. (1979) J. Polymer Sci. Polymer Symp. 66, 313-336.

Andrade, J.D., Smith, L.M. & Gregonis, D.E. (1985) in Surface Chemistry & Physics, (Andrade, J.D. ed.), vol.1, pp.249-291, Plenum, New York.

Barbucci, R., Baszkin, A., Benvenuti, M., Costa, M.L. & Ferruti, P. (1987) J. Biomed. Mater. Res. 21, 443-457.

Dann, J.R. (1970) J. Colloid & Interface Sci. 32, 302-320.

Fowkes, F.M. (1987) The Journal of Adhesion Science and Tech., 1, 7-27.

Good, R.J. (1979) in Surface and Colloid Science (Good, R.J. & Stromberg R.R. eds.), vol. 11, Plenum, New York.

Hamilton, W.C. (1972) J. Colloid & Interface Sci. 40, 219-22.

Lelah, M.D., Grasel, T.G., Pierce, J.A. & Cooper, S.L. (1985) J. Biomed. Mater. Res. 19, 1011-1015.

Neumann, A.W., Good, R.J., Hope, C.J. & Separ, M. (1974) J. Colloid & Interface Sci. 49, 291-304.

Scanning Electron Microscopy (SEM)

Authors: S. Cimmino, E. Martuscelli & C. Silvestre

Istituto di Richerche su Tecnologia del Polimeri e Reologia, CNR, I - 80072 Arco Felice

INTRODUCTION

Name of method:

Scanning Electron Microscopy (SEM).

Aim of method:

Analysis of morphological and structural properties of polymer surfaces and biopolymer interface with living tissues.

Background of method:

Scanning electron microscopy relies on the interactions between an electron beam and a material. A comprehensive description is given in several text books (*Sawer & Grubb, 1987, Hoff et al., 1974, Goldstein & Yakowitz, 1975*). Its application to polymer and biological materials is review by *White & Thomas (1984), Echlin (1980) and Hayat (1977).*

When an electron beam impinges on a material, the many interactions it undergoes can be divided into two groups:

(i)   elastic scattering with a large angular deflection in the path of the incident electron and essentially no energy transfer,

(ii)   inelastic scattering with energy loss due to the electron-electron interactions.

This phenomenon occurs when the atomic electrons are excited to higher energy states leaving holes which are subsequently filled by electronic transition. Several

S. Dawids (ed.), Test Procedures for the Blood Compatibility of Biomaterials, 211–228.

phenomena related to the filling of these holes make SEM imaging possible. Radiation from the specimen or any type of the interaction between the specimen and the beam is used to form an image. The appearance of the image will depend on the interaction involved and on the detection and signal processing used.

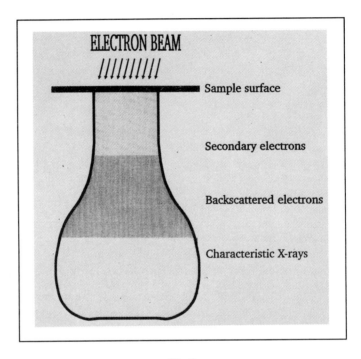

Fig. 1

The incident electron beam induces secondary electron release from the material subsurface. Some electrons are classically backscattered. The interactions generate characteristic X-rays. The 3 types of radiation are used for topographical, chemical and material analysis.

Fig. 1 is a scheme of the signals produced by the impact of an electron beam with the sample volume (interaction volume). The interaction volume increases with accelerating voltage and for specimens of low atomic number.

The most important signals are backscattered electrons, X-rays and secondary electrons. Backscattered electrons (BE) are primary beam electrons which have been classically scattered by nuclei of the sample and escape from the surface. They have high energy and may emerge from a depth of 1 μm or more depending on the specimen. They leave the surface over a wide area and provide a resolution of 1 μm. The number of BE depends on the angle of incidence of the electron beam and on the position of the detector and can give information about local chemical surface composition and topography.

Any sufficiently energetic electron beam incident on a solid specimen will generate X-rays. The X-rays have intensities and wavelengths that are characteristic of the quantity of individual elements composing the specimen. It is convenient to use X-radiation to determine variations in composition within a specimen. This may be done by measuring the characteristics of the emitted standard samples under similar excitation conditions. However, as the optimum excitations for X-ray emission do not coincide with those of high resolution electron images before such analysis is attempted, the excitation conditions have to be adjusted.

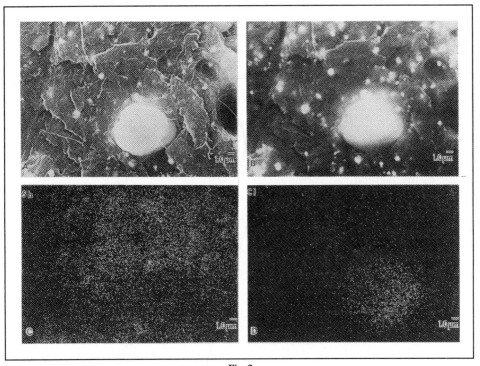

Fig. 2

Micrographs of a multiphase polymer material obtained using SEI (a), BEI (b) and X-ray microanalysis (c & d). These micrographs show that the dispersed phase particles have a higher atomic number than the matrix (a & b), and that the small particles contain antimony (c), whereas the larger particles contain chlorine. (From L.C. Sawyer & D.T. Grubb "Polymer Microscopy", Chapman and Hall, 1987, by permission).

Backscattered Electrons Imaging (BEI) and X-ray microanalysis are useful for the study of multiphase structures when one phase contains elements not contained by the others.

Secondary Electrons (SE) are emitted with low energy, less than 50 eV, and they come from the uppermost few nanometers of the material. If the beam falls on a tilted surface or onto an edge, more secondary electrons will escape from the specimen. If a beam falls onto a valley or a pit, fewer SE escape because there is less interaction volume near the surface. Secondary electrons yield high resolution topographic images.

Secondary Electron Imaging (SEI) is widely applied in SEM because of the ease of image formation and the efficiency of signal formation and collection.

The combination of X-ray, BEI, SEI and microanalysis enable the identification of the chemical composition and the distribution of elements within the sample. An example is given in *Fig. 2.*

The application of SEI to the study of the morphology of biopolymer surfaces is described in this text.

Scope of the method.

Scanning electron microscopy is essentially adopted to study the morphological properties of polymer surfaces (roughness, texture, porosity). The interaction of these surfaces with proteins (total amount of adsorbed proteins, kinetics of protein adsorption, their special arrangement and polymer tissue interactions) can be analysed indirectly.

Compositional or structural heterogeneity in the place of the surface can be revealed by the reaction of different surface domains to chemical treatment of the polymer.

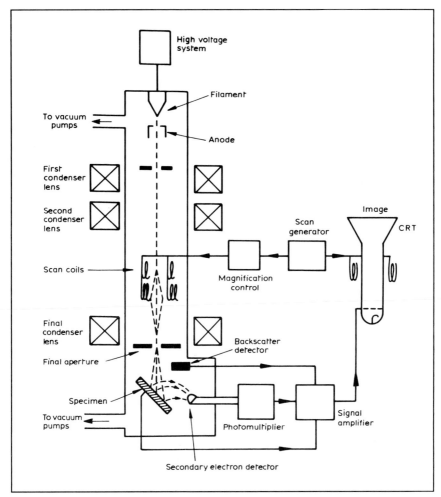

Fig. 3

Schematic diagram of a scanning electron microscope. (From L.C. Sawyer & D.T. Grubb "Polymer Microscopy" Chapman & Hall, 1987, by permission)

DETAILED DESCRIPTION OF THE METHOD

Equipment.

In Fig. 3 a schematic diagram is shown of a scanning electron microscope. The electron source commonly used is a thermoionic tungsten filament in an electron gun. The electrons are emitted by the cathode of the gun and accelerated by a field produced by an anode having a positive potential of about 1 KV relative to the cathode. A third electrode is placed between the anode and the cathode and is negative relative to the cathode. This third electrode is generally described as a Wehnelt electrode.

Electrons leaving different sources along the filament have a Maxwellian distribution of initial velocities which cause a crossover at a point beyond the anode. The crossover can be regarded as a small electron source and the demagnified image of this crossover forms the electron probe impinging on the specimen. This image is formed by the time sequence of a single picture point, produced by the sweep of the electron probe over the specimen surface. The detected signal is displayed as a TV type image. The electron beam in the display tube moves in synchrony with the probe on the specimen.

A modern scanning microscope can resolve[1] 4-6 nm. The magnification[2] on the display screen can range between 20 and 1×10^5.

The microscope can be connected with several devices and recording systems (cameras, video tapes, computers). In order to record photographically the images a polaroid camera is usually applied to the SEM equipment providing good and instant photographs.

Outline of the method.

An electron beam scans the specimen surface. The induced radiations produced by the electron beam-specimen interactions are used to form the SEM image by modulating the raster of a cathode ray tube (CRT) which is scanned in synchrony with the scan of the electron beam on the sample. Each point on the CRT tube corresponds to a point on the specimen surface. The SEM images can be observed visually on a monitor screen, recorded photographically, video taped or processed by computer. The images have a great depth of field and a natural and almost three dimensional appearance. The SEM technique can contribute to the study of

---

[1] *The resolution is the minimum distance between two points at which they can still be distinguished.*

[2] *The magnification in a scanning electron microscope is given as the ratio between the size of the image and the size of the region scanned on the specimen.*

the surface morphology of a specimen when the size of the details to be resolved is greater than or equal to 4-6 nm.

Description of procedure.

Preparation of the materials.

Polymer materials. The material to be examined is attached to double sided adhesive tape on a specimen stub. Conductive points such as silver or carbon suspensions are also used to attach the specimen to the sample holder. Bombardment of polymer specimens normally results in the formation of a surface charge which causes the appearance of bright spots in the image. The technique most commonly used to remove the charge is to coat the specimen with a metal. The coating of samples with a metal also reduces radiation damage and increases electron emission.

There are several kinds of coating devices. They include vacuum evaporator and sputter coaters. Vacuum evaporators are used for the thermal evaporation of metals onto specimen surfaces. They consists of a bell jar fitted on a vacuum system (rotary and diffusion pump). The time needed to perform a coating is usually about 2-3 min. after 1 hr. of pumping to attain the required vacuum for the evaporation. The metal most commonly used are gold and gold-palladium. In order to obtain a more uniform layer a specimen rotation device is required. Before evaporation the sample surface and the vacuum chamber must be clean and dry. It is advisable to fill the vacuum chamber with a dry inert gas (e.g. nitrogen) at the end of any coating process.

In sputter coating metal ions (usually gold-palladium or platinum) are dislodged by inert ions from a source (= target) and electrostatically directed onto the specimen. For the sputter units the modest vacuum in the system is established only by a rotary vacuum pump. The total time necessary for coating is about 20 min.

The heating of the sample during coating which was one of the problems of sputter coaters has been eliminated by the "cool sputter system".

In sputter coating systems care must also be taken to avoid contamination. The inert gas (argon) must be dry and the system flushed at least 10 times with the gas before evacuation and sputtering.

Both methods produce a coating layer with an average grain size of about 5-7 nm which is smaller than the resolution usually required in practice.

Sputter coating systems have the advantage of quick preparation of the samples, and they produce a uniform conductive coating layer.

Biological materials. The preparation of biological materials is more complex. The procedure generally used consists of 4 steps: 1) cleaning; 2) fixing; 3) drying and 4) coating.

ad 1)   The cleaning procedure is necessary to remove dust and fragments of cells and tissues from the samples. For dry materials, compressed air is generally used. For moist materials, the cleaning procedure must be very rapid and delicate in order to avoid alterations gently. Sprinkling the sample with an isotonic solution at a temperature compatible with the sample is normally sufficient.

ad 2)   Fixing is generally performed in two steps: the sample is placed in a solution of isotonic glutaraldehyde (2.5%) for about 12 hours at room temperature, then it is cleaned with an isotonic solution and finally it is left for about 4 hours in a solution of osmium tetroxide (1-2%) at the same temperature.

ad 3)   When fixing is completed it is possible to dehydrate the sample. This is done by successive immersion of the sample in a graded ethanol, methanol or acetone series followed by critical point drying (Hayat, 1977). The solvent-water mixtures usually used are 15, 30, 50, 70, 95 and 100% of the solvent in water; 15 min. incubation with each solution is the time required. Another way to dehydrate samples is to freeze-dry them at low pressure. This method has the disadvantage of frequently producing ice crystals that can damage the sample if the temperature is not kept below 60°C, and the freezing is not carried out rapidly.

ad 4)   Subsequent coating of the sample with metal is done as described above for polymer materials.

RESULTS

SEM analysis is particularly useful for the characterization of the polymer surface and interface, and of the morphology of the interaction between tissue and polymer. When interested in the polymer surface, the preparation of the samples is the one described for polymeric materials (see above). In the case of studies on biopolymer-tissue interface and interaction, the samples must be prepared by the procedure for biological materials. *Figs. 4-10* show SEM micrographs of polymer surfaces and polymer-tissue interface. *Figs. 4 & 5* show examples of different porous morphologies that have been obtained changing the method of fabrication. *Figs. 6 & 7* show examples of different polymer surface roughness and *Fig. 8* illustrates the morphology of proliferating endothelial cells across the inner surface of a vascular graft. *Fig. 9* illustrates the adhesion of platelets to various polymer surfaces in time for different precoating materials. By counting the number of deposited platelets at different times per unit area of the micrographs, it is possible to obtain the platelet deposition profiles shown in *Fig. 10*.

DISCUSSION

Interpretation of the results.

The morphological properties of polymer surfaces that affect biological interactions are mainly porosity and roughness. Porosity is essentially a bulk phase property, but it may also be regarded as a surface property as pore cross section is evident at the surface of a material. Pores at the surface can be classified according to size, number and shape. Porosity depends on the method of fabrication. As shown in *Fig. 4a* extruded materials have pores which are elongated in the draw direction (*Sawer & Grubb, 1987*); the pore dimensions are less than 1 µm wide and long about 2 µm. Pore volume is formed by stretching lamellae. In cast material as shown in *Fig. 4b* the structure is generally open with the polymer in the form of strings of particles. For cast materials, pore size, number and shape reflect the casting conditions (*Leenstag et al., 1988*). In mixtures which are macroscopic homogeneous like a polyurethane-poly(L-lactide), pore distribution can be adjusted by increasing the total polymer concentration. The pores in *Fig. 5a* are about 5-10 µm in size. For a sample obtained with a lower polymer concentration, the surface appears less homogeneous and contains large holes up to 75 µm (see *Fig. 5b*). Controlled porosity is important in certain applications involving growing tissue.

220

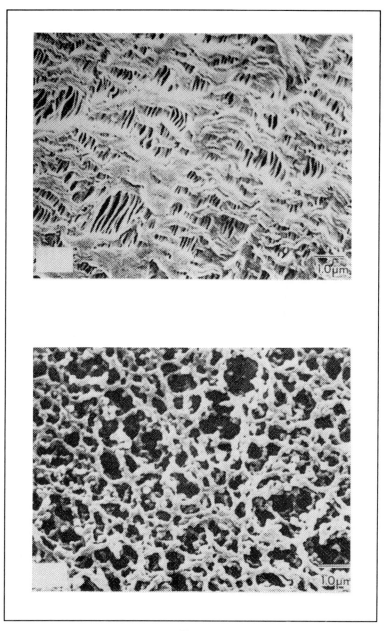

Fig. 4

*SEM micrographs of high density polyethylene (HDPE) microporous membranes obtained by extrusion (**top, a**) and by solution casting (**bottom, b**). (From L.C. Sawyer & D.T. Grubb "Polymer Microscopy" Chapman and Hall, 1987, by permission).*

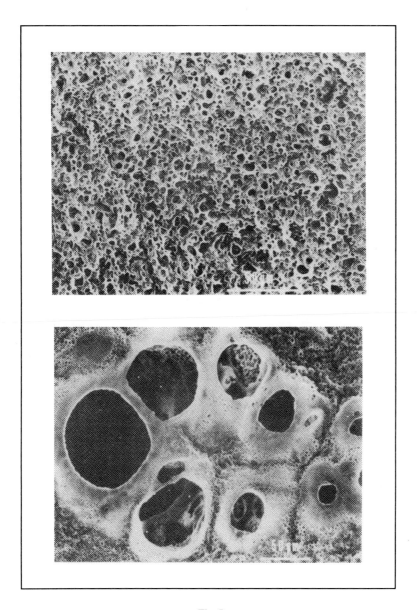

Fig. 5

*SEM micrographs of the inner surface of a polyurethane/poly(L-lactide) mixture vascular graft prepared from a 4% (**top, a**) and 8% (**bottom, b**) polymer solution in DMM/THF. (From J.W. Leenstay, M.T. Kroes, A.J. Pennings & B. van der Lei, New Polymeric Materials, 1, 111, 1988, by permission).*

Wilkes & Samuels (1973) have published the result of a systematic SEM study of series of chemically well defined polyurethanes with the aim to develop controlled pore morphology.

Fig. 6

SEM micrograph of the surface of polyethylene oxide obtained by solution casting and non isothermally crystallized.

Fig. 6 illustrates the possibility to see the roughness of a polymeric material. In this case it depends on the degree of crystallinity developed by the biopolymer (*Cimmino et al., 1988*). In crystalline polymers the surface often exhibits a spherulite-like morphology. The roughness of this material depends on the spherulite fibrils and the boundary lines between crystals. Spherulite size, number and structure depending on the crystallization conditions can readily be analysed with SEM.

The surface of amorphous materials appears smoother than that of crystalline polymers and the roughness is due to the fabrication method. It may be very important to consider the shape and size of the morphological details as an indicator of the production process.

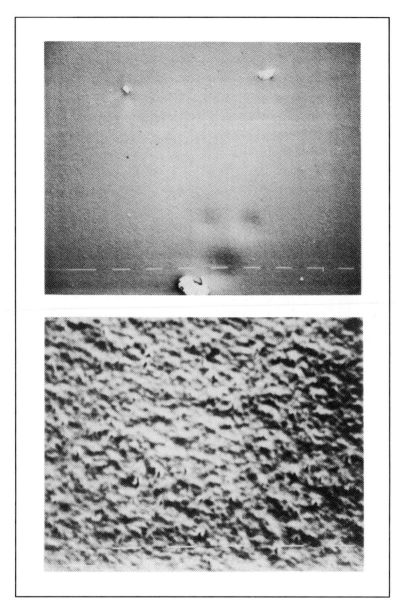

Fig. 7

*SEM micrographs of the surface of a polyurethane sample for different magnifications. Bar corresponds to 10 m (**top, a**) and 1 m (**bottom, b**).*

In *Fig. 7a* the polyurethane surface seems very smooth and homogeneous. However, at higher magnification it is seen to be full of 0.2-0.5 μm holes (*Fig. 7b*).

By SEM micrographs it is possible to monitor the kinetics of the proliferation of endothelial cells across the inner surface of a vascular graft as reported by *Leenslag et al., 1988*. Micrographs taken at appropriate time intervals (see e.g. *Fig. 8*) can illustrate the morphology and the rate of advance of the endothelial front and the time needed for the completion of the proliferation process. In the case reported the front of the endothelial cells was found to advance 6 mm/hr.

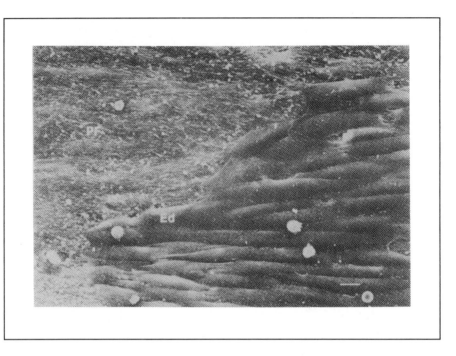

Fig. 8

SEM micrograph showing the growth zone of the endothelial cells (Ed) in a polyurethane/-poly(L-lactide) mixture vascular graft 3 weeks after implantation. (From J.W. Leenstay, M.T. Kroes, A.J. Pennings & B. van der Lei, New Polymeric Materials, 1, 111, 1988, by permission).

Application of SEM on sequential protein adsorption experiments demonstrates that at a given time platelet deposition is influenced by the nature and concentration of preadsorbed proteins and by partly known properties of the polymer surface (*Park et al., 1986*). From *Figs. 9 & 10* it is evident that on serum-coated PVC the number of particles at a given time is considerably less than for fibrinogen-coated PVC. In the case of different polymer surfaces coated with the same protein, it was found by the same authors that the maximum of platelet deposition occurred at different times of blood exposure depending on the nature of the polymer surface.

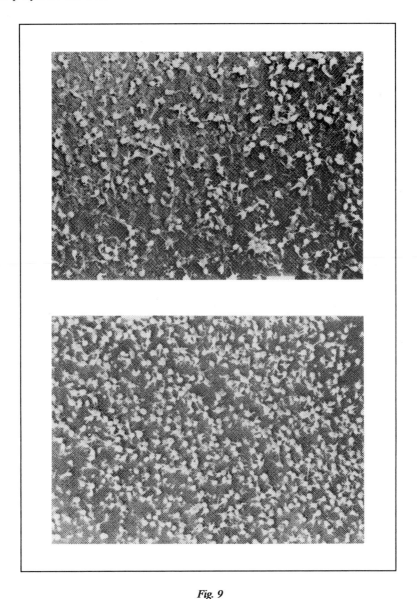

Fig. 9

*SEM micrographs of platelet deposition of PVC preadsorbed with fibrinogen (**top, a**) and with serum (**bottom, b**) after 15 minutes blood exposure.(From K. Park, D.F. Mosher & S.L. Cooper, J. Biomed. Mat. Res., 20, 589, 1986, by permission).*

Another morphological factor that can influence thrombus initiation is reported to be the distribution of the precoated protein molecules on the polymer surface (*Park et al., 1986*). This has been studied by SEM on different types of polymers using the immuno-gold bead labelling techniques.

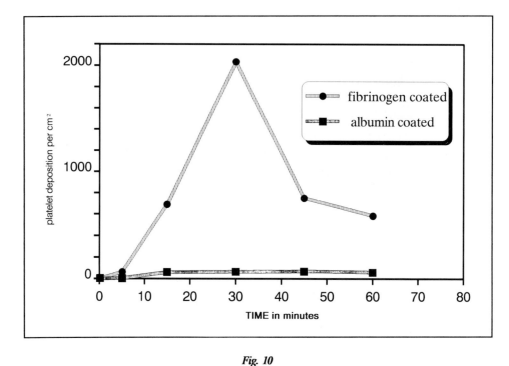

Fig. 10

Platelet deposition (average SEM) on PVC shunts preadsorbed with serum (■) and with fibrinogen (●) in relation to time.(From K. Park, D.F. Mosher & S.L. Cooper, J. Biomed. Mat. Res., 20, 589, 1986).

Limitations of the method.

Although the SEM technique is generally a simple technique and presents two main advantages:

- relative high resolution
- three dimensional images with great depth of field.

there are several potential artifacts that together with the boundaries of the resolution and the rather difficult image interpretation constitute the limitations of SEM imaging. The images should not be evaluated as macroscopic images because very different factors are important.

Artifacts are defined as any structural detail present in a micrograph, but not found in the original material. They may be produced during sample preparation or observation; paints, glues and tapes can contaminate surface imparting a false structure to the sample. Coating can introduce additional artifacts such as the formation of grain morphology. During the observation surface damage may be caused by the vacuum and the electron beam. In the case of biological materials, care should be taken to avoid artifacts caused by drying, especially with freeze drying in which formation of ice crystals will modify the morphology of the sample.

The interpretation of SEM micrographs requires much experience in looking at the images, in assessing the data in the light of potential artifacts and in specimen properties. To avoid any risk of misinterpretation and to obtain very clear pictures of the material, other microscopic (optical and transmission electron microscopy) and complementary techniques (spectroscopic and scattering techniques) should be employed in parallel to confirm the morphology of the sample.

REFERENCES

Cimmino, S., Martuscelli, E. & Silvestre, C. (1988) private communication.

Echlin, P. (1980) in Microscopia elettronica a scansione e microanalisi, Part.I, (Armigliato A. & Valdrè, U. eds.), Centro Stampa "Lo Scarabeo", Bologna.

Goldstein, J.L. & Yakowitz, H. (1975) Practical Scanning Electron Microscopy, Plenum, N.Y.

Hayat, M.A. (1977) Principles and Techniques of Electron Microscopy: Biological Application, van Nostrand Reinhold, N.Y.

Hoff, D.B., Muir, M.D., Grant, P.R. & Boswara, I.M. (1974) Quantitative Scanning Electron Microscopy, Academic Press, London.

Leenstag, J.W., Kroes, M.T., Pennings, A.J. & van der Lei, B. (1988) New Polymeric Materials, 1, 111-126.

Park, K., Mosher, D.F. & Cooper S.L. (1986) J. Biomed. Mat. Res. 20, 589-611.

Sawer, L.C. & Grubb, D.T. (1987) Polymer Microscopy ,Chapman & Hall, Publ., London.

White, J.R. & Thomas, E.L. (1984) Rubber Chem. Technology 55, 457-479.

Wilkes, R.A. & Samuels, S.L. (1973) J. Biomed. Mat. Res., 7, 541-558.

Chapter V

TEST METHODS FOR THE DETECTION OF
PROTEIN ADSORPTION ON POLYMERS

Chapter eds.: T. Beugeling[1] & S. Dawids[2]

[1]   Department of Chemical Technology, University of Twente, NL - 7500 AE Enschede

[2]   Institut of Engineering Design, Biomedical Section, Technical University of Denmark, DK - 2800 Lyngby

INTRODUCTION

Foreign material which comes into contact with blood or plasma will rapidly adsorb proteins onto the material surface, and the adsorbed protein layer determines all further events in coagulation and cellular adhesion. A theoretical treatment of protein adsorption together with detailed descriptions of experimental methods and techniques are given in survey papers (*Andrade, 1985, Brash & Horbett, 1987*). The principal phenomena following protein adsorption are blood coagulation, platelet adhesion and aggregation, and complement activation leading to leucocyte aggregation.

In the following a brief overview of the resulting events is given to enhance the understanding of the events at the blood-material interface.

Blood coagulation.

Blood coagulation is the final outcome of activation of the clotting system. Description of these complicated processes are given in text books. Thus only an outline will be given. In the presence of a foreign surface or activating factors and calcium ions the clotting factors are sequentially activated. Many of these are serine proteases present in the blood in unactivated = zymogen form. Thrombin which is the pivoting factor in coagulation converts fibrinogen into an insoluble fibrin network in which cells from the blood are entrapped. Activation by a foreign surface (= contact activation) occurs when adsorbed factor XII (Hageman factor)

S. Dawids (ed.), Test Procedures for the Blood Compatibility of Biomaterials, 229–234.
© 1993 *Kluwer Academic Publishers. Printed in the Netherlands.*

undergoes a conformational change accompanied by an activation directed against prekallekrein (PK). PK is also surface bound in a complex with high molecular weight kininogen (HMWK) which enhances the enzymatic conversion of prekallekrein into kallekrein. Kallekrein, in turn, activates adsorbed factor XII. The active factor XII (XIIa) converts factor XI (*Fig. 1*) which starts the subsequent cascade process of clotting activation. Kallekrein activates surface bound factor XII many times faster than the fluid-phase factor XII.

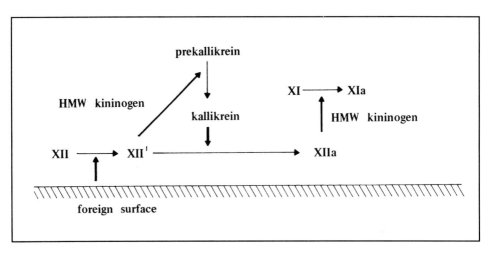

Fig. 1
Primary reactions of the intrinsic coagulation at a foreign surface.

Platelet adhesion and aggregation.

Another process which takes place at the protein coated polymer surface is the activation of platelets whereby adenosine diphosphate (ADP) and platelet factor IV (PF IV) are released. ADP activates neighbouring platelets close to the material interfase due to low fluid exchange here. Activated platelets will, in turn, induce the well known intrinsic coagulation cascade (*Fig. 2*). Platelet adhesion depends on the absorped protein. Fibrinogen on the surface seems to promote this adhesion. Other proteins like von Willebrand factor (vWF, a protein synthesized and released by endothelial cells) and fibronectin also enhance platelet adhesion. Precoating with albumin hampers the spontaneous process and delays platelet adhesion (*Poot et al., 1988*). The same effect can be elicited with preabsorbed high density lipoprotein (HDL). Thus several proteins have been identified to initiate platelet adhesion but only few which hamper the process.

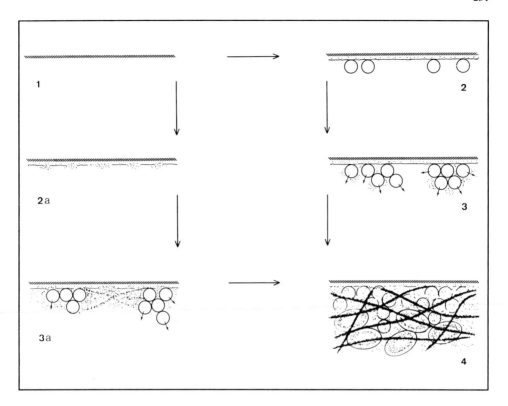

Fig. 2

Platelet adhesion/aggregation and intrinsic coagulation occurring at a foreign surface after contact with blood.

1   *Exposure of a foreign material to blood.*
2   *Adsorption of blood proteins, followed by platelet adhesion.*
2a *Surface activation of blood coagulation, initiated by the adsorption of factor XII.*
3   *Release of ADP from platelets followed by ADP induced platelet aggregation. Blood coagulation is initiated upon adhering and aggregating platelets.*
3a *Platelet aggregation induced by thrombin which is formed during blood coagulation.*
4   *Formation of an insoluble fibrin network. Fibrin strands with entrapped blood cells form a (red)thrombus.*

Complement activation and leucocyte aggregation.

The two pathways of the complement system (the classical and the alternative) are comparable with the coagulation system in structure complexity. The complement system plays an essential role in the normal immune response. Complement activation releases inter alia anaphylotoxins which elicit inflammatory reactions including aggregation of leucocytes. Foreign surfaces may activate the alternative pathway through binding of complement C3b (C3 splits into C3a and C3b). Adsorbed IgG becomes "denatured" on the material surface and may, in turn, activate the classical pathway of the complement system. It is recognised that the complement activation plays a profound clinical role in extracorporeal circulation where large foreign surfaces are exposed to blood.

Adsorption of plasma proteins.

Adsorption experiments using mixtures of albumin, IgG and fibrinogen reveal a preferential adsorption of fibrinogen. On the other hand such experiments in which protein mixtures are tested fail to predict the adsorption behaviour of hte mentioned proteins when they are adsorbed out of plasma (*Brash et al., 1981, Uniyal & Brash, 1982, Breemhaar et al., 1984*). In plasma fibrinogen is only initially adsorbed and is replaced gradually with other proteins with time. It appears that fibrinogen is gradually displaced by inter alia HMWK on mainly hydrophilic surfaces such as glass (*Vroman et al., 1980, Schmaier et al., 1983*). If adsorption of plasma proteins is studied over longer periods of time, it is found in some cases that relatively small amounts of proteins finally adsorb from plasma. In this respect it is interesting that adsorption of HDL is one possible cause for the low adsorption of proteins out of undiluted plasma onto polymers like PVC and polystyrene.

It is believed that in this competitive adsorption environment the protein molecules which are present in the highest concentrations and are of small size will first adsorb due to physical reasons. The subsequent displacement of adsorbed protein species by others are relatively inpredictable but takes place rapidly (Vroman effect). The typical sequence of displacement on glass-like surfaces with plasma seem to have the following sequence: albumin, IgG, fibrinogen, fibronectin, HMWK and factor XII (i.e. from relatively highly concentrated to less concentrated proteins). The turnover of adsorption and replacement of e.g. albumin takes place in a fraction of a second in undiluted plasma.

Protein adsorption in relation to blood-material interaction.

It has not been proven that thrombus formation and complement activation are preceded by adsorption of proteins from plasma to the polymer surface but it is suspected that platelet adhesion, surface activation of coagulations and complement system have a close causal relation to the composition of the adsorbed protein layer. It can thus be expected that artificial surfaces which demonstrate adsorption of relatively large amounts of vWF, fibrinogen, fibronectin or IgG will enhance blood-material interaction while surfaces preferentially adsorbing albumin or/and HDL or (theoretically) show no protein adsorption probably are relatively blood compatible. However, the composition of the adsorbed protein layer is time-dependent but unfortunately not much of this is known in a multicomponent mixture such as plasma. Present techniques are time-consuming in the quantitative or semiquantitative detection of a particular protein adsorbed on the surface together with many other proteins which at present makes a real time measurement almost impossible.

Detection of proteins adsorbed on foreign surface.

In this chapter the following methods for the detection of adsorbed proteins are described:

1. A uncomplicated method - Tracer labelling - in which a particular protein is labelled with radioiodine and added to the blood plasma allows one to examine the adsorption of the particular protein to foreign surfaces.

2. A method using immunoassay technique allows the specific detection of protein adsorbed to solid surfaces.

3. Identification of proteins adsorbed from blood or plasma can also be measured by eluting the adsorbed proteins from the material surface with e.g. sodium dodecylsulfate (SDS) and subsequent separation with Polyacryl-Amide Gel Electrophoresis. The method is commonly known as SDS-PAGE.

4. Reflectrometry is a relatively simple method for measuring ultrathin layers of adsorbed protein(s) (*Welin et al., 1984*) and allows a real time detection of the adsorption pattern of the total amount of proteins adsorbed from a protein solution or plasma onto thin films of a polymer, coated on a reflecting material.

5. The method of dynamic measurement on adsorption is described by using isotope tagged protein tracers, combining surface measurements on the polymer and blood concentrations as references.

234

Techniques for the study of conformational changes in adsorbed proteins have not been included in the chapter as they are not believed to be of value in this context.

REFERENCES

Andrade, J.D.(ed.) (1985) Surface and Interfacial Aspects of Biomedical Polymers vol.2, Plenum Press, New York.

Brash, J.L., Uniyal, S. & Chan, B.M.C. (1981) Interactions of protein in plasma with various foreign surfaces, Artif. Organs 5, (suppl.), 475-477.

Brash, J.L. & Horbett, T.A. (eds.) (1987), Proteins at Interfaces, ACS Symposium Series 343, Am. Chem. Soc., Washington D.C.

Breemhaar, W., Brinkman, E., Ellens, D.J., Beugeling, T. & Bantjes, A. (1984) Preferential adsorption of high density lipoprotein from blood plasma onto biomaterial surfaces, Biomaterials 5, 269-274.

Poot, A., Beugeling, T., Cazenave, J.P., Bantjes, A. & van Aken, W.G. (1988) Platelet deposition in a capillary perfusion model: Quantitative and morphological aspects, Biomaterials 9, 126-132.

Schmaier, A.H., Silver, L., Adams, A.L., Fischer, G.C., Munoz, P.C., Vroman, L. & Colman, R.W. (1983) The effect of high molecular weight kininogen on surface-adsorbed fibrinogen, Thromb. Res. 33, 51-67.

Uniyal, S. & Brash, J.L. (1982) Patterns of proteins from human plasma onto foreign surfaces, Thromb. Haemostas. 47, 285-290.

Vroman, L., Adams, A.L., Fischer, G.C. & Munoz, P.C. (1980) Interaction of high molecular weight kininogen, factor XII and fibrinogen in plasma at interfaces, Blood 55, 156-159.

Welin, S., Elwing, H., Arwin, H., Lundström, I. & Wikström, M. (1984) Reflectometry in kinetic studies of immunological and enzymatic reactions on solid surfaces, Anal. Chim. Acta, 163, 263-267.

Test for Plasma Protein Adsorption to Biomaterials using an Enzyme-Immunoassay

Authors: A. Poot & T. Beugeling

Department of Chemical Technology, University of Twente, NL - 7500AE Enschede

INTRODUCTION

Immunoglobulins (antibodies) are useful for the in vitro detection of proteins. Application of the so-called immunoassay technique shows a unique specificity and a high sensitivity to proteins adsorbed to a solid surface. By labelling immunoglobulins with enzymes a semiquantitative analysis can be obtained. Experimental details on such techniques are given in the literature (*Klebe et al., 1981, Breemhaar et al., 1982*).

Name of method:

Test of plasma protein adsorption to biomaterials using an enzyme-immunoassay.

Aim of method:

With the enzyme-immunoassay technique it is possible to study semi-quantitatively protein adsorption to a material surface from a complex protein mixture such as plasma.

Biochemical background for the method.

Development of immunoglobulins in response to exposure of a foreign macromolecular substance is well described in the literature (*Stryer, 1988*) and only an outline will be given. Antibody molecules are today produced as polyclonal antibodies, or as monoclonal antibodies derived from a single cell or a clone after using the hybrid cell technique. The most important immunoglobulins are IgG. Details of production and preparation are given in the literature (*Tijssen, 1985*).

S. Dawids (ed.), Test Procedures for the Blood Compatibility of Biomaterials, 235–245.
© 1993 *Kluwer Academic Publishers. Printed in the Netherlands.*

When a particular protein is adsorbed to a solid surface the matched antibody will combine with this protein upon exposure. To detect this complex an enzyme-labelled antibody directed towards the first antibody is added (e.g. antibodies from sheep against mice). To this last antibody an enzyme has been coupled, e.g. peroxidase.

After this second step of the assay, a substrate (hydrogen peroxide) and an indicator = leuko dye (tetramethylbenzidine) are added. The resulting reaction product (oxygen radical) will react with the leuko dye forming a coloured dye which can be measured in a spectrophotometer. The principle is given in *Fig. 1*.

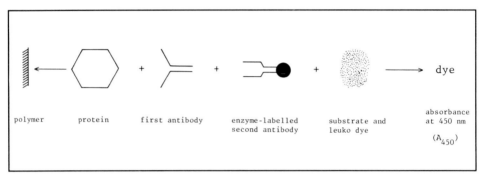

Fig 1

Principle of the applied two-step enzyme-immunoassay.

Scope of method:

The method should be viewed upon as an experimental tool until a clear knowledge of the adsorption sequence and its time frame is elucidated. At that time it may well turn into a highly predictive screening test.

DETAILED DESCRIPTION OF METHOD

Equipment.

Double distilled water
Ethanol, absolut (zur Analyse, Merck 983)
Nitric acid concentrated
Hydrochloric acid concentrated

Cleaning solution 1% v/v detergent (RBS 25, Hicol, Rotterdam, The Netherlands)
Trisodium citrate (Janssen Chimica, Beerse, Belgium)
Phosphate buffered saline (NPBI, Emmer-Compascuum, The Netherlands)
PBS with 0.005% Tween-20 (Sigma, St. Louis, Mo, USA)
Buffer solutions:

> First (antibody) buffer: 8.7 g/l NaCl, 6.1 g/l Tris (Merck, Darmstadt,
> FRG), 0.02% (v/v) Tween-20, 0.2% (w/v) gelatin (Merck), 0.5% (w/v)
> bovine serum albumin (BSA, Sigma) with pH adjusted to 7.5
> Second (conjungate) buffer: 8.7 g/l NaCl, 6.1 g/l Tris (Merck, Darmstadt,
> FRG), 0.02% (v/v) Tween-20, 0.2% (w/v) gelatin (Merck), 5% (w/v)
> bovine serum albumin (BSA, Sigma) with pH adjusted to 7.5

Leuko dye solution: 10 ml 0.11 M sodium acetate (pH 5.5), 165 μl 3,3,5,5-tetramet-
hyl benzidine (Fluka AG, Buchs, Switzerland), ad 6 mg/ml dimethylsulfoxide
(DMSO)
10 μl peroxide solution 3% (Merck)

Glass plates (2mm hard glass type 7059, Corning, New York, USA).
Sheet material of polymers.
Silastic® Medical Grade tubing 3/16 inch. ID x 5/16 inch. OD from Dow Corning
Corp., Medical Products, Midland, Michigan, USA
Antibodies against human proteins (monoclonal antibodies or rabit sera) Central
Laboratory of the Netherlands Red Cross Blood Transfusion Service (CLB),
Amsterdam, The Netherlands. Antisera are used diluted 100 fold with the first
antibody buffer.
Antibody against human HDL, i.e. apoprotein A-1 of HDL (Behringwerke AG,
Marburg, FRG) is diluted 10 fold with first antibody buffer.
Peroxidase conjugated sheep anti-rabbit IgG (United States Biochemical Corp.,
Cleveland, Ohio, USA) product number 1311. Peroxidase conjugated sheep anti-
rabbit IgG is diluted 200,000 fold with conjugate buffer before use.

24 wells test chambers (see *Fig. 2*)
Ultrasonic cleaning bath (local manufacturer)
Automatic pipettor with disposable tips (Eppendorf, Hamburg, FRG)
Cooling facilities to -30°C.
Polypropylene centrifuge tubes 50 ml (Greiner, Nurtingen, FRG)
Polypropylene reaction vessels 2.2 ml (Greiner)
Polypropylene beakers (Tamson, Zoetermeer, The Netherlands)
Polypropylene pipettor tips 200 μl (Labsystems, Helsinki, Finland; Fintip 60)

Outline of method.

The protein which has been adsorbed on the surface will - upon addition - bind specific antibodies as a first step. These may be IgG produced by animals (e.g. rabbit antibodies). Antibody labelled with enzyme (conjugate) will then bind upon the IgG. This comes from another species, e.g sheep IgG directed against rabbit IgG, and mainly directed towards the constant region of the IgG molecule. After the binding of the conjugate, the enzyme substrate is added (hydrogen peroxide) as well as the leuko dye (tetramethyl benzidine). The procedure is performed in a 24 well test chamber where the bottom of the wells consists of the polymer. Glass plates may advantageously be used as references. The test chamber is shown in *Fig. 2* together with details on the construction. The method requires sheet material.

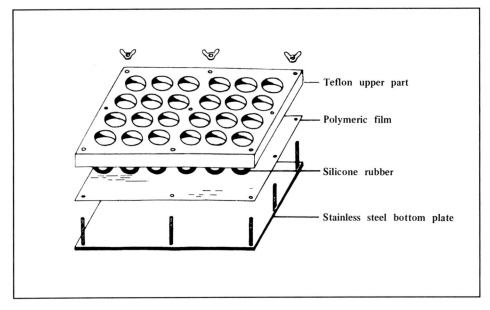

Fig. 2

Test chamber. The teflon upper part (13 x 9.5 x 1.0 cm) is provided with 24 cylindrical holes (10.0 mm ID). Each hole has a stepped recess (15.5 mm ID, depth 2.0 mm) at the bottom side in which a silicone sealing ring (10.77 ID x 2.62 mm) is placed.

It is recommended to clean the surface of the sheet carefully in the following way:

The polymeric sheet is placed for 30 min. in an ultrasonic bath with 1% (v/v) detergent solution RBS 25 followed by careful rinsing with double distilled water and ethanol (ethanol, absolut zur Analyse, Merck no 983) respectively. The sheets are then dried in dustfree air.

The glass plates are cleaned as follows: They are submerged for 15 hours in a mixture of fuming nitric acid (100 ml) and fuming hydrochloric acid (33 ml) followed by rinsing and drying as described above.

Preparation of human blood plasma. Pooled plasma should be obtained from several donors (e.g. 15 healthy persons). From each donor 100 ml of venous blood is collected with a 1.5 mm needle connected to a Silastic® tubing (length 15 cm, 3/16 inch ID) into two polypropylene centrifuge tubes (50 ml) containing 5 ml 130 mM trisodium citrate. The tube should be filled to the 50 ml mark. The tubes are immediately transferred to a centrifuge and centrifuged for 15 min. at 1570 g followed by 15 min. at 3000 g. The supernatant plasma is transferred into a polypropylene beaker (1000 ml) and mixed carefully. Portions are then transferred into polypropylene vessels of 2.2 ml and frozen at -30°C until use. Prior to the enzyme-immunoassay one or more vessels with plasma are thawed in a water bath at 37°C. Serial plasma dilutions are made with phosphate buffered saline (PBS), see later.

Description of procedure.

The 24 well test chamber is mounted with the sheet polymer. Do not touch the surface during mounting.

The 24 wells are filled as follows: 2 wells are filled with 400 μl of PBS (used as blanks) 200 μl of PBS is pipetted into each of the other wells. In two of these wells 200 μl of diluted plasma (5×10^4 dilution) is added. The end of the pipettor tip should be kept under the liquid surface of the buffer to avoid the formation of an air-liquid-solid interface. In the remaining 20 wells five other concentrations of plasma are added each in four wells: diluted 5×10^3 fold, 5×10^2 fold, 5×10^1 fold and 5 fold.

The wells are then left for adsorption to take place and should be covered with tape to avoid evaporation during this and each of the following steps.

After the desired time of exposure the contents of the wells (diluted plasma and PBS) is removed and discarded. The test surface is rinsed four times with washing buffer after this and the subsequent steps.

The first antibody solution is added (200 μl) to each well. It is left for reaction an hour. The wells are then rinsed as before. Then 200 μl of enzyme-labelled second antibody is applied. The contact time is 1 hour. After a new washing procedure the polymeric film is carefully unmounted and moved over into a clean test chamber to prevent detection of protein which has been adsorbed to the inner walls of the Teflon upper part and the Silicon rubber sealing rings.

10 ml leukodye is prepared by adding 10μl hydrogen peroxide 3% to the TMB solution in sodium acetate.

The 200 μl of substrate and leuco dye solution (freshly made) is added and the test chamber is covered. The reaction should be carried out in the dark for 30 minutes and is terminated by the addition of 100 μl of 4N sulphuric acid. Part of the yellow dye solution (250 μl) is pipetted into each of a 96 wells plate (A/S Nunc, Roskilde, Denmark) and the absorbance is measured at 450 nm with a multiscanner (Reader Micro Elisa System, Organon Teknika, Turnhout, Belgium). When the absorbance exceeds the value of 2.0 the samples should be diluted before being measured again.

Safety aspects.

Careful handling of all the chemicals, reagents and plasma is necessary. The guidelines for Good Laboratory Practice should be followed concerning, inter alia, isotopes.

RESULTS

The adsorption of human proteins: albumin (HSA), immunoglobulin G (IgG), fibrinogen (Fb), high molecular weight kininogen (HMWK) and high density lipoprotein (HDL) to polyethylene and glass is shown in *Figs 3 & 4* as a function of plasma concentration at exposure time of 1 hour. The highest plasma concentration corresponds with a ratio of plasma to PBS of 1:1.

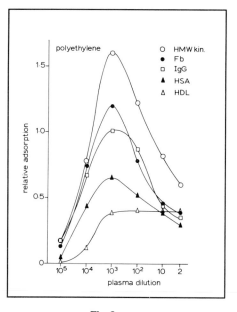

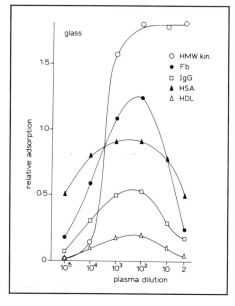

Fig. 3

Fig. 4

The adsorption of the human proteins albumin (HSA), immunoglobulin G (IgG), fibrinogen (Fb), HMW kininogen (HMW kin.) and high density liloprotein (HDL) to polyethylene as a function of plasma concentration.

The adsorption of HSA, IgG, fibrinogen (Fb), HMW kininogen (HMW kin.) and HDL to glass as a function of plasma concentration.

The measurements are not directly related to the amounts of the protein adsorbed (see *Limitations of the enzyme-immunoassay*), and are a result of the action of the enzyme fixed to the antibody which, in turn, generates the dye. Thus the amount of adsorbed protein is deduced from the absorbances of the generated dye in the enzyme-immunoassay. The standard deviation of the data is expected to be around 5%.

The adsorption pattern of Fb, HDL and HMWK to polyethylene and glass from 1:1 diluted plasma is shown in *Figs. 5 & 6* as a function of time. Even if there has been used another standard batch of plasma, the relative amounts of protein appear to be the same i.e. the adsorption pattern is the same. The amount of adsorbed proteins can be measured already after 15 sec. (*Figs. 3 & 4*). This is the practical minimum time scale with this technique and equipment.

It should be noted that the adsorption of protein increases with increasing plasma concentration with a maximum for HSA, IgG and Fb at about 1% of the normal plasma concentration using glass surface. For polyethylene the maximum is found around 0.1% of normal plasma concentration. The amounts of adsorbed proteins decrease at higher plasma concentrations. It can be seen that relatively small amounts of proteins are adsorbed from 1:1 diluted plasma. However, the important exceptions for this is HMWK adsorbed to glass and HDL adsorbed to polyethylene.

Small amounts of adsorbed HDL occurs in all dilutions when using glass surface.

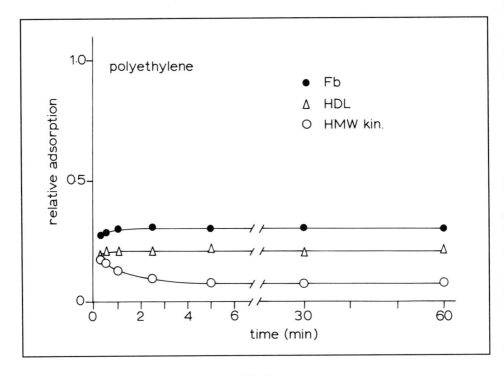

Fig. 5

The adsorption of fibrinogen (Fb), HMW kininogen (HMW kin.) and HDL from 1:1 diluted plasma to polyethylene as a function of time.

Small amounts of Fb, HDL and HMWK adsorb to polyethylene from 1:1 diluted plasma. After 15 sec. a decrease of the adsorbed amounts of fibrinogen HDL and HMWK occur. After a few minutes the adsorbed amounts are constant for the next hour.

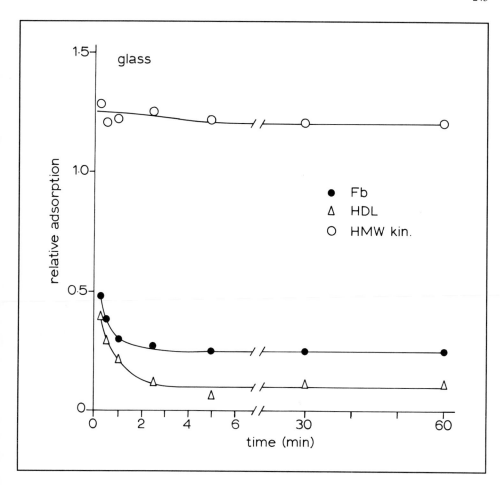

Fig. 6

The adsorption of fibrinogen (Fb), HMW kininogen (HMW kin.) and HDL from 1:1 diluted plasma to glass as a function of time.

DISCUSSION

It should be pointed out that there is only laboratory evidence that thrombus formation and complement activation are preceeded and triggered by the adsorption of specific proteins from the blood onto the surface of biomaterials. Nevertheless, both in vitro and ex vivo experiments indicate strongly that the materials which adsorb large amounts of von Willebrand factor, fibrinogen, fibronectin or IgG seem to enhance blood material interaction while surfaces which adsorb large amounts of albumin and/or HDL are relatively blood compatible. These processes take place from fractions of seconds onto at least an hour according to in vitro studies. Moreover, HMWK and/or HDL may exchange sites with fibrinogen (replace each other) on the surface. The interesting fact that some proteins adsorb stronger to the surface with increasing dilution gives information of the turnover speed of a particular protein.

The maxima of adsorbed albumin, IgG and fibrinogen on polyethylene and glass at relatively low plasma concentrations have also been found by other methods (see *Yu & Brash: Measurement of Protein Adsorption to Solid Surfaces ...*). With those methods maxima have been shown in some publications (*Horbett, 1984*). If the first antibody or conjugate has a lower activity it will not change the pattern of adsorption, but the absolute values. It does, however, underline that relative values related to reference material is important.

It has been suggested that the cause of the adsorption/desorption pattern is a characteristic feature of all mixed surfactant systems (*Horbett, 1987*). If this is correct it may explain why there is a low fibrinogen adsorption from undiluted plasma and explain the low adsorption of proteins as well.

Limitations of the enzyme-immunoassay.

The sensitivity of the method is related to measurements of the generated amount of dye from the bound enzyme. These data are thus not directly related to the amount of adsorbed protein/cm². Experimental values (*Breemhaar, 1985*) do indicate that an absorbance of 1.0 of the dye solution corresponds to approx. 70 ng adsorbed albumin/cm².

Protein molecules may undergo conformational changes during the adsorption process which may alter the antigenic determinants of the molecule due to loss of their specific structure. This phenomenon is both protein-, interface- and time-dependent.

Furthermore, the activities of different conjugates may vary mutually, and may decrease during storage for several months (at -30°C). Thus the method always will require a comparison to a reference material or be coupled to e.g. radiolabelling. This admittedly makes the test somewhat laborious.

During washing procedures, some adsorbed protein may be released although experiments have shown that this is probably very little.

It can thus be stated that it is not possible to compare data of "adsorbed amounts" of different proteins.

The standard deviation of data presented in this instruction is about 5%. However, measurements carried out at different days may vary as much as 20 - 30%.

"Real time" adsorption measurements are not possible with the enzyme-immuno-assay, partly because of the laborious procedure and partly because exposure times less than 15 sec. of a material are not easy to achieve with this technique. For such short exposure times, a flow cell has to be used.

REFERENCES

Breemhaar, W., Ellens, D.L., Beugeling, T. & Bantjes, A. (1982) A novel application of a two step enzyme-immunoassay for the investigation of blood compatibility of materials, Life Support Systems, Proc. IXth Ann. Meeting ESAO, Saunders, Eastbourne, UK.

Breemhaar, W. (1985) An enzyme-immunoassay for the detection of blood components adsorbed to polymeric surfaces, Thesis, Twente University, Enschede, The Netherlands.

Horbett, T.A. (1984) Mass action effects on competitive adsorption from hemoglobin solutions and from plasma, Thromb. Haemostas. 51, 174-181.

Horbett, T.A. (1987) Adsorption to biomaterials from protein mixtures, in Brash, J.L. & Horbett, T.A. (eds.) (1987) ACS Symposium Series 343, Proteins at interfaces, Am. Chem. Soc., Washington DC, 239-260.

Klebe, R.J., Bentley, K.L. & Schoen, R.C. (1981) Adhesive substrates for fibronectin, J. Cell. Physiol. 109, 481-488.

Stryer, L. (1988) Biochemistry, Freeman & Co., San Fransisco, USA.

Tijssen, P. (1985) in "Laboratory techniques in biochemistry and molecular biology; Practice and theory of enzyme immunoassays" (Burdon, R.H. & van Knippelberg, P.H. eds.) Elsevier, Amsterdam.

The Detection of Adsorbed Plasma Proteins from Membrane Materials Using SDS-Polyacrylamide Gel Electrophoresis of Desorbed Proteins

Author: M.C. Tanzi

Dipartimento di Bioingegneria, Politecnico di Milano, I -20133 Milano

INTRODUCTION

Name of method:

SDS-Polyacrylamide gel electrophoresis of desorbed proteins from polymer surfaces.

Aim of method:

Separation and identification of proteins desorbed from polymer surfaces.

Biophysical background of the method:

Adsorption of proteins occurs when polymers are exposed to biological fluids such as lymph, ascites in interstitial space or blood. The detection of proteins adhering to the polymer surfaces is methodologically difficult to measure quantitatively both in situ and dynamically under true circumstances, partly because adsorption occurs very rapidly and partly because conformational changes occur during the adsorption process. The amount of protein adsorption depends on many parameters of the polymer (chemical structure, manufacturing process of the polymer material, protein type and mixtures, exposure conditions).

To overcome the difficulties of measuring the adsorbed proteins in situ, a quantitative protein desorption can be carried out and the analysis can then be put into practice in many other ways. Desorption is usually done effectively by exposing the polymer to an aqueous detergent solution over some hours. A wide variety of detergents can induce desorption, and the choice of detergent should thus for practical purposes not influence the subsequent steps in protein analysis.

S. Dawids (ed.), Test Procedures for the Blood Compatibility of Biomaterials, 247–268.

This method describes the analysis of proteins in the eluates by separation in polyacrylamide gel electrophoresis (PAGE) in the presence of a detergent, sodiumdodecylsulphate (SDS). The method is usually called SDS-PAGE. The method of SDS-PAGE was initially described in 1967 as an easy method for the separation and identification of proteins in complex biological samples.

The molecular weight of popypeptide chains can be determined from the relative mobilities of their SDS complexes diffusing through the gels (*Shapiro et al., 1967, Weber & Osborn, 1969*). The sensitivity of SDS-PAGE allows detection of a few μg of proteins and is thus useful in analysing protein samples desorbed from polymeric surfaces.

The principle of the system is based on the fact that addition of SDS to protein will eliminate the specific molecular charges and induce a change of the conformity (uncoil the protein). In this way a simple relationship is achieved between the molecular weight (= size) and the electrophoretic mobility of uncharged SDS-protein complexes. The linear relation is between the logaritm of molecular weight and the electrophoretic mobility. A wide variety of proteins binds proportional amounts of amphiphile SDS with a binding ratio of about 1.4 mg of SDS per mg of protein (*Reynolds & Tanford, 1970; Dunker & Rueckert, 1969*). This binding seems to be independent of ionic strength and is primarily hydrophobic in nature. It is only the monomeric form of SDS that binds to the proteins. In turn the proteins will organise the SDS anions into a micellar complex of definite size. This causes the SDS-protein complexes to become soluble in aqueous media even if the proteins are insoluble in water in the native state. In this way almost all proteins can be analysed with this principle.

Gel electrophoresis of the proteins should be performed under reducing conditions which break the disulphide bonds and obtain polypeptide chain sub-units in a random coil conformation (*Tanford, 1968*). This can be achieved by treating the SDS-protein complexes with ß-mercapto-ethanol.

The electrophoresis is performed in a polyacrylamide gel with a predetermined concentration (= pore size) suitable for the molecular weight of the protein(s) to be separated. A fixation and colouring of the proteins are performed subsequently for their identification.

The described method is relatively simple, and the reagents and equipment for SDS-PAGE are economical and require very limited laboratory space. Not much specialised training for application is required although meticulous performance of each step in the method is necessary to obtain reproducible results.

Scope of method:

The simplicity and sensitivity of the method allows it to be used as a routine method. It is therefore suited for experimental and general screening purposes.

DETAILED DESCRIPTION OF METHOD

A number of good overviews on the method exists together with detailed description of the equipment (*Laemmli 1970, Pharmacia, 1984*). The user is advised to consult these references before implementing the method.

Equipment.

Gel electrophoresis apparatus and power supply (Bio-Rad Laboratories, Richmond, Ca., USA or Pharmacia, Uppsala, Sweden)
Gel casette kit II (Pharmacia, Uppsala, Sweden)   or
Polyacrylamide Gel Kits (Bio-Rad Laboratories, Richmond, Ca., USA)
Gel destainer GD-4 II (Pharmacia, Uppsala, Sweden)   or
Electrophoresis Stains (Bio-Rad Laboratories, Richmond, Ca., USA)
Densitometry equipment Model 620 (Bio-Rad Laboratories, Richmond, Ca., USA)
Thermostat bath facilities (4°C and - 30°C)
Ultrasonic bath Bransonic 72 (Asol snc, Milano, Italy)
Equipment for vacuum

Solutions for SDS-PAGE.
Stock solution:
    Acrylamide   > 99,9%   29.2 g
    N,N'methylene-bis-acrylamide > 99%
    to 100 ml with distilled water   0.8 g
    Filter and store at 4°C in dark (30 days max.)
Bis solution:
    N,N'methylene-bis-acrylamide > 99%   2.0 g
    to 100 ml with distilled water

Stacking buffer:

    0.83 M TRIS-HCL (100 g TRIS/l)

    buffered to   pH 6.8

    (TRIS-hydroxymethyl aminoethane · HCl)

    SDS 0.66%

Running buffer (separating gel buffer):

    1.5 M Tris-HCl (181.64 g/l)   pH 8.8

    SDS 0.40% (4g/l)

Sample buffer (working solution):

0.83 M Tris-HCl	pH 6.8	1.5 ml
SDS solution 10%		2.5 ml
Glycerol		2.0 ml
ß-mercaptoethanol		1.0 ml
Distilled water		3.0 ml
Bromophenol blue		50.0 mg

Electrophoresis buffer (final solution):

TRIS base	pH 8.3	3.0 g
Glycine		14.4 g
SDS 0.1%		1.0 l

Desorption solution:

SDS 0.5% (Sodium dodecylsulphate)	0.5 g
Water	100.0 g

Fixing solutions:

    <u>Solution 1</u>:

Trichloroacetic acid (TCA)	10.0 ml
Ethanol 33%	90.0 ml

    <u>Solution 2</u>:

Trichloroacetic acid (TCA)	5.0 ml
Ethanol 33%	95.0 ml

Coomassie Blue Staining Solution:

Coomassie Brilliant Blue	
(B-0630 Sigma Company)	0.25g
Water	50.0 g
Ethanol (96%)	40.0 g
Acetic acid 10%	10.0 g

Destaining solution:

Methanol	30.0 ml
Acetic acid	10.0 ml
Distilled water	60.0 ml

Bovine serum albumin (BSA) 0.5% solution
Bio-Rad protein assay kit (Bio-Rad Laboratories, Richmond, Ca., USA)
BCA protein assay (Pierce Europe B.V., Oud Beijerland, The Netherlands)
UV-spectrophotometer Beckman DU-6, Beckman Analytical SpA, Milano, Italy)

Outline of method.

Protein desorption.
The protein adsorbed to the polymer is desorbed by exposing the polymer to a dilute SDS solution for some time. Hereby the protein is released and SDS-protein complexes are formed in the solution.

Protein concentration determination.
Prior to the performance of SDS-PAGE, a quantitative protein analysis should be carried out using colorimetric available kits (Bio-Rad Protein Assay or BCA Protein Assay) as guideline for subsequent adjustment by dilution. Although these methods suffer from the effect of interferences, sufficiently precise estimations of the protein contents of the solution can be made. Most protein assay kits are based on the principle that a dye will react with the protein making a colorimetric estimation possible.

The Bio-Rad Protein Assay is based on the principle that light adsorption maximum for an acidic solution of Coomassie Brilliant Blue G-250 shifts from 465 nm to 595 nm when protein binding occurs (*Bradford, 1976*). The BCA Protein Assay method uses cupro ions which are generated when cupri ions reacts with proteins to form a copper/peptide chelate giving an intense purple color (at 532 nm) (*Smith et al., 1985*).

Gel electrophoresis (SDS-PAGE).
The electrophoresis in gels containing SDS is generally performed in a slab of gel (or a tube of gel) using a discontinuous system. This consists of a stacking (upper) gel and a resolving (lower) gel (*Laemmly, 1970*) (see *Fig. 3*).

Gel separation of SDS-protein complexes.

In the stacking gel, the protein-SDS complexes are concentrated into thin bands utilizing the principle of isotachophoresis (= equal travelling speed of the proteins). The initial concentration in a band enables the good separation ability of SDS-PAGE. After having travelled down to the border between the gels, the SDS-protein complexes will move down through the resolving gel with various speed depending on their molecular size. In the buffer solutions a small amount of tracking dye (Bromophenol blue) is added. The dye moves together with the front of the protein complexes. This makes it possible to follow the front protein position in the gel.

Different concentrations of polyacrylamide alters the separation properties of the gel. With a 5% gel the separation is best performed at high molecular weight proteins (350.000 Dalton). A gel with 15% will have a lower molecular sieving effect (up to 60.000 Dalton) providing it range for separation of proteins (*Dunker & Rueckert, 1969*). Concentrations between these can give other separation properties. Therefore the selection of the appropriate running gel depends very much on the protein solution to be analysed.

Fixation, Staining and analysis of protein.

At the end of the electrophoresis the cell and casting stand are carefully disassembled, and staining of the proteins is performed by immersing the gels in a solution containing Coomassie Blue, as described. Destaining procedure is carried out during a subsequent 24-48 hour period. After this, only the proteins in the gel remain stained blue. Optional fixation of proteins is sometimes performed to avoid leaching and diffusion of the proteins during subsequent procedures.

By measuring on the gels or on photographs of the gels, the migration distances are measured and plots are made of the migration distances versus the logarithm of the molecular weight. For any given running gel concentration there is an asymptotic region in which the molecular weight and migration distance are related in a linear fashion (*Shapiro & Maizel, 1969*). Plots should thus yield a straight line over a given molecular weight range which is characteristic of the gel composition. As an indication (*Dunker & Rueckert, 1969*) the following is recommended:

> 15% gel is suitable for the range 10.000 - 60.000 Dalton
> 10% gel is suitable for the range 10.000 - 100.000 Dalton
> 5% gel is suitable for the range 20.000 - 350.000 Dalton

For general application with plasma proteins, a 7-8% gel is recommended for the high molecular weight standards (50.000 - 300.000 Dalton). For the low molecular weight standards (14.000 - 100.000 Dalton) a gel with at least 12% is recommended to prevent one or more proteins from migrating to the end of the gel with the tracking dye.

Silanization.

Glass will not stick to the gel. Removal of the plates may nevertheless cause ripping of the gel. Silanization using Silane A 174 (= γ-methacryl oxypropyl trimethoxysilane) is used to bond reactive methacrylate groups to the surface glass. The acrylamide becomes chemically bound to these groups during polymerization and will firmly stick to the glass plate. In this way very thin gels can be handled safely. The methacrylate groups are unstable in the presence of moisture but if kept dry the silanized plates are stable for at least 2 months. Glass plates can be reused after scraping off the gel and thorough washing with strong sodium hydroxide solution.

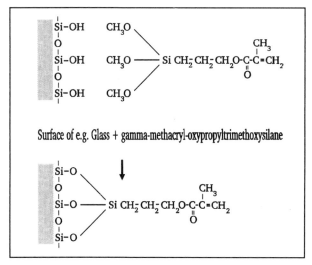

Fig. 1

Silanization: On the glass surface is grafted γ-methacryl-oxypropyl-trimethoxisilane. The gel binds to the grafting during polymerisation. This enables handling of very thin slabs of gel.

Colorimetry-Densitometry.

Densitometric analysis is a quantitative determination of the separated protein bands. The method is based on the fact that at a given wavelength one can measure the adsorption for a protein Coomassie Blue complex. The colourlines consisting of SDS-protein complexes are scanned and a chromatogram is generated.

This chromatogram is subsequently used for the determination of the precise electrophoretic mobility and the protein concentration in each band of protein. In principle, the area under each "peak" in the chromatogram indicates the total protein concentration in each protein band. For determination of the precise concentrations one applies usually a test run with standard proteins as those given in *Table I.*

Description of procedure.

1.   Preparation of protein for SDS-PAGE.

Protein desorption.
Desorption of adsorbed protein is generally accomplished by immersing the polymer surface into a 0.5% SDS-water solution for 8 hours, the temperature of the treatment varying from 30°C to 100°C depending on the strength of protein binding to the surface. The solution is gently shaken at regular intervals to stir it. The surface should not be touched (preferably use gloves or instruments in the handling of the polymer sample). Quantitative protein desorption is of importance for the subsequent SDS-PAGE application since the incomplete removal of the adsorbed protein layer might heavily influence the composition of the protein mixture. Desorption is usually complete within 3-4 hours.
After desorption it is advisable to check the polymer surface by e.g. using conventional dyes such as Coomassie Blue which binds indiscriminately to protein to get an indication of possible remnants of protein on the surface. By microscopic observation of the samples one can further identify remaining cellular components. This is also an indication of remaining protein on the material. The SDS-water solution should be filtered prior to further analysis to remove cell debris.

Preparation of reference proteins.
A reference sample protein is prepared as follows: The protein is added to a SDS-water solution and heated under reducing conditions to 100°C for a few minutes. Tracking dye 0.5 μl (Bromophenol blue 1%) is added. In some instances one must add succrose or glycerol to the sample proteins to increase the density prior to applying to the gel.

A list of the molecular weight markers commercially available is given in *Table I.*

Table I

Molecular weight markers for SDS-PAGE [*)]

Protein	Source	M_w (daltons)	Subunits
Insulin B chain		3.400	
Aprotinin	bovine lung	6.500	
Cytochrome C	horse heart	12.400	
Myoglobin (I,II,III fragments)		2.500-8.100	
Myoglobin (polypeptide backbone)		16.950	
Myoglobin (I+II fragments)		14.400	
α-Lactalbumin		14.200	
ß-Lactoglobulin			18.400
Lysozyme		14.300	
Trypsin Inhibitor, Soybean		21.500	
Chymotrypsinogen A		25.000	
Trypsinogen		24.000	
Carbonyc Anhydrase	bovine erythrocytes	29.000	
Carbonyc Anhydrase		31.000	
Glyceraldehyde-3-phosphate dehydrogenase	rabbit muscle	36.000	
RNA-polymerase, core enzyme			(α') 39.000
			(ß) 155.000
			(ß') 165.000
Pepsin		34.700	
Albumin	Hen egg	45.000	
Albumin	bovine serum	66.200	
Phosphorylase B		92.500	
Phosphorylase	rabbit muscle	97.400	
ß-Galactosidase	Escherichia coli	116.000	
Aldonase	rabbit muscle	158.000	
Myosin	rabbit muscle	205.000	
Catalase	beef liver	240.000	58.100

[*)] *different kits available from Bio-Rad, Sigma, Pierce Chem. Co.*

Total protein analysis.

Analysis of the protein contents in the desorption solution (i.e. protein desorbed in 0.5% SDS/water solution) is performed as follows: 2 ml of the BCA reagent are added to 0.1 ml of each eluate, to 0.1 of the blank (0.5% SDS/water) and to 0.1 ml aliquots containing increasing concentrations of Bovine serum albumin (from 20 to 120 μg/0.1 ml). The samples are allowed to react for 30 min. at 37°C. They are then cooled to room temperature and maintained at this temperature for 15 min. Standards are read at 562 nm in a U.V. spectrophotometer. The blank value (i.e. 0.5% SDS/water solution) is subtracted. The resulting values will generally form a linear standard graph (*see Fig. 2*) which is used for subsequent determination of the concentration of the unknown protein solution(s).

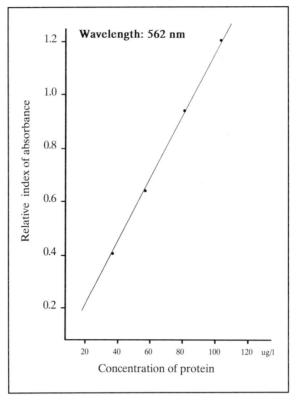

Fig. 2

Example of the standard plot obtained with the BCA protein assay (Pierce Chem. Co.) using increasing concentrations of BSA (Bovine Serum Albumin).

2. Preparation of the electrophoresis cell.

Assembly of the cell with clean glass plates, spacers and clamps should be performed according to the producer's instructions. Possibly one glass plate could be silanized (see later) to support thin slabs of gel. The cell is mounted in a casting stand according to the operating instructions of the equipment. The gels consist of a running (lower) gel and a stacking (upper) polyamide gel. These are cast by preparing with the respective reagents as given in *Table II*.

Table II

Reagents for gels preparation. The indicated amounts are enough for 1 (1.5 mm x 14 cm x 14 cm) slab.

Running gel

| | Polyacrylamide final concentration/gel | | | | |
	5%	7.5%	10%	12.5%	15%
Stock solution (ml)	4.95	7.50	9.90	12.39	15.00
Bis solution (ml)	0.60	0.90	1.20	1.50	1.80
10% (W/V) SDS/water (ml)	0.30	0.30	0.30	0.30	0.30
Running buffer (ml)	7.50	7.50	7.50	7.50	7.50
Distilled water	15.90	13.00	10.17	7.38	4.63
TEMED (μl) (tetramethylethylendiamine)	15.00	15.00	15.00	15.00	15.00
2% APS/water (fresh)(ml) (Ammonium persulphate)	0.90	0.90	0.90	0.90	0.90

Stacking gel (4.5% polyacrylamide final concentration/gel)

Stock solution	1.5 ml
10% (W/V)SDS/water	0.1 ml
Stacking buffer	1.5 ml
TEMED	10.0 μl
2% (W/V) APS/water	0.3 ml

The reagents are mixed carefully to avoid trapping of bubbles and deaerated under vacuum. An ultrasonic bath can equally do the job and should be applied for 15 min. After this amoniumpersulphate (APS) and TEMED are added immediately before casting the gels. Mixing is obtained by carefully stirring the solutions. The appropriate running gel is first poured (with a syringe) up to 3-4 cm from the top of the glass plates. Generally good operating instructions are available with the equipment. An overlaid of water is placed and the system is allowed to polymerize overnight. Next morning the water is removed and the stacking gel is cast on top of the running gel together with a "comb" to shape sampling wells. Again, one should be very careful not to trap air bubbles under the comb. The gel will polymerized in 30-45 min. The comb is slowly removed to avoid tearing the gel between the wells. A schematic representation of the electrophoresis cell is given in *Fig. 3*.

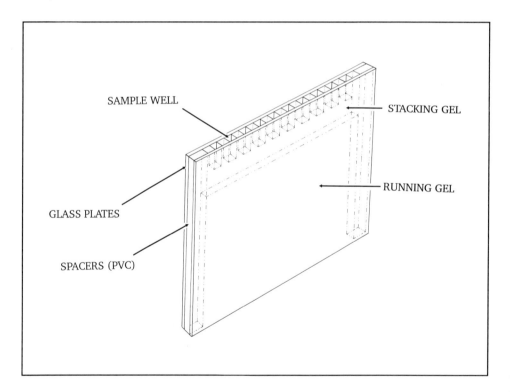

Fig. 3

Schematic representation of the electrophoresis gel.

3. **Applicationtion of samples for analysis.**

Aliquots from the protein desorption procedure should be diluted to obtain a protein concentration of approx. 200-500 μg/ml. Thus samples containing 3-5 mg/ml must be diluted at least 1:4 using the sample buffer as diluter. The best gel concentration for a specific job must be determined empirically. To each sample 5 μl Bromophemol Blue 1% is added as "tracking dye" which will travel with the protein front.

Electrophoresis.

The mounted electrophoresis cells with gels are transferred from the casting stand to the test bench. The upper chamber is fitted, and the set is mounted into the electrophoresis equipment. The proteins and reference samples are applied in the formed wells. The distribution should be even, i.e. application should initially be carried out in every second well. Free wells should be filled with pure sample buffer (without Bromophenol blue) to avoid a lateral diffusion of the sample cells. The lower end of the cell is brought into contact with the electrode buffer. For thermostate control water is circulated through a thermostate and the chamber surrounding each cell to keep the temperature uniform. Electrode buffer is carefully added on top of the samples. The lid is put on the cell, and the electrode leads are connected. If the system does not have a circulation system, it is advisable to have a small stir bar to keep the temperature uniform in the lower electrode buffer. Gels are run at 10 mA through each gel cell until the tracking dye reaches the border of the running gel. Then the current is increased to 15-20 mA. Run time is usually 4-6 hours.

First a preequilibration is performed by electrophoresis for 1 hour at a relative low voltage (70 V). Electrophoresis is then carried out at 300 V for about 10 minutes to concentrate the proteins in the border region between the gels. The electrophoresis is continued at 150 V for 30-60 min. after the tracking dye has migrated beyond the bottom of the stacking gel. The total time is generally 2 1/2 - 3 hours.

Under these conditions a protein of molecular weight of 14.000 - 15.000 Dalton will migrate close to the bottom of the stacking gel. Prolonged electrophoresis (i.e. 16 hours) will normally cause proteins with molecular weight up to 25.000 to reach the bottom of the stacking gel.

4. **Fixation of proteins.**

    After electrophoresis the protein bands in the gel can optionally be fixed. This is carried out after carefully removing one or both of the glass sides of the cell (use gloves).The proteins in the gel are fixed with TCA (Fixing solution 1) for 30 minutes. The gel is then washed in TCA (Fixing solution 2) for 2 x 45 min. or until the visible precipitate is dissolved.

5. **Staining.**

    Staining is performed with Coomassie Blue solution containing 0.2% Coomassie Brilliant Blue R 250 (or PAGE blue 83, BDH Chemicals) in methanol: acetic acid: distilled water (3:1:6) should be carried out for 3-6 hours. The staining solution should be filtered before use.

    The gel is then placed in a destaining solution of methanol:acetic acid: distilled water (3:1:6) for 30 min. before changing the bath. Destaining is repeated until the background is clear using several changes of solution. Gels which are intended to be stored are finally soaked in 4% glycerol added to the last portion of destaining solution. The gel is then dried in air and covered with a plastic sheet.

Safety aspects.

Several of the chemicals used are hazardous. Acrylamide, bis-acrylamide and ß-mercaptoethanol are quite toxic and can be adsorbed through the skin. They can also be inhaled. Gloves should be used at all times and mixing should be carried out in a hooded workbench with suction ventilation. It is advisable to use eye protectors when mixing the reagents. Furthermore, the general rules of Good Laboratory Practice should be respected. Many manufacturers provide safety guidelines for their chemicals and equipment. Electric safety regulations for the equipment must be respected.

RESULTS

The molecular weight of an unknown protein can be determined by the polypeptide bands which may be plotted in to a standard linear plot. In *Fig.4* examples of 3 standard plots are given with different running gel concentrations. For these bands the molecular weight markers listed in *Table I* are used. An improved separation and quantitation of bands may be obtained by scanning the gels with a densitometer.

As an example, *Figs. 7 & 8* report some densitometric scans obtained with a Model 620 video densitometer (Bio-Rad) at a wavelength of 220 nm.

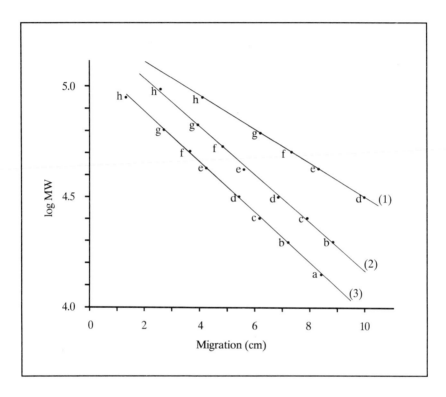

Fig. 4

Plots of migration distances obtained with different polyacrylamide concentrations in the running gel (7.5% = (1), 10% = (2), 12,5% = (3)). Markers: lysozyme (a), trypsin i (b), γ-G,L chain (c), carbonic anhydrase (d), ovalbumin (e), γ-G,H chain (f), BSA (g), phosphorylase B (h).

In *Figs. 5 - 8* some examples of SDS-PAGE of proteins eluted from haemodialysis membranes are given (Cuprophan and polyacrylonitrile-based) after clinical use. In the figures, photographs, plots and densitometric scans are compared.

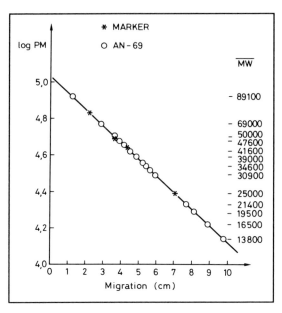

Fig. 5

*Example of standard plot obtained with SDS-PAGE (12,5% polyacrylamide running gel)
of proteins eluted from AN-69 (polyacrylonitrile-based) haemodialysis membranes.*

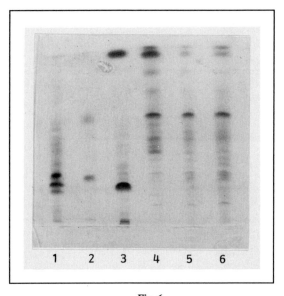

Fig. 6

*SDS-PAGE gel (12,5% running gel) showing migration of fibrinogen (α, β and γchains)
(1), γ-G, H and L chains (2), BSA + Lysozyme (3), protein eluted from polyacrylonitrile-
based haemodialysis membranes (AN-69TM = 4, AN-69S = 5 and 6).*

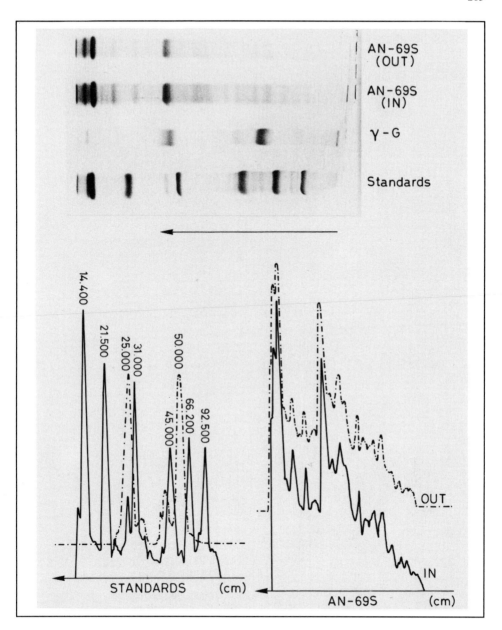

Fig. 7

Top: *SDS-PAGE gel (12,5% running gel) showing migration of standard proteins (from the left: phosphorilase B, BSA, ovalbumin, carbonyl anhydrase, soybean trypsin inhibitor, lysozyme), γ-G, H and L chains; proteins eluted from AN-69S haemodialysis membranes.*
Bottom, right: *densitrometric scans of standard proteins.*
Bottom, left: *densitrometric scans of proteins eluted from AN-69S. (IN and OUT represent the half membrane near blood inlet and outlet, respectively.*

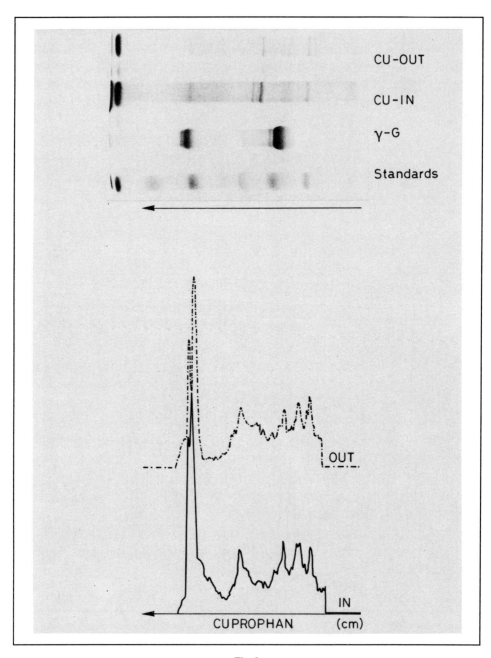

Fig. 8

Top: *SDS-PAGE gel (12,5% running gel) showing migration of standards and γ-G (see Fig. 6) and proteins eluted from CU (cuprophan) haemodialysis membranes.*
Bottom: *densitrometric scans of proteins eluted from CU. (IN and OUT repreesnt the half membrane near blood inlet and outlet, respectively).*

DISCUSSION

The presented method of SDS-PAGE provides a relative simple way of determining the molecular size of the proteins which have been released from the polymer surface. The method of desorption is very simple and can provide a quantitative estimation of the protein adsorption. One must ensure that all the adsorbed proteins are released by staining the surfaces before and after desorption to estimate colorimetrically the amount of protein. Coloration of the polymer should be measured (zero sample).

Using this method one can determine the molecular size (= weight) and quantity of a given protein. However, due to the very large number and complexity of plasma proteins the single types cannot be identified unequivocally by this technique. Similiarities in molecular weight of different proteins do not allow their absolut identification. Thus it is only useful for screening and routine testing. Nevertheless, very interesting results can be obtained and provide an indication of the sort of adsorbed protein. Positive identification can subsequently be made for albumin, γ-globulins and other proteins with characteristic molecular weight such as plasminogen and transferrin. A list of molecular weights of some proteins in human plasma is given in *Table III*.

One must realize that some proteins will be partially degraded on the surface of the material and during desorption leading to polypeptides with molecular weights different from that of the native protein. This is particularly true for fibrinogen which appears to be degraded to products with molecular weights between 43.000 and 15.000, presumably by a plasmin induced mechanism (*Brash & Thibodeau, 1986*).

One drawback of the system is the long time to carry out a full analysis which takes about 24 hours.

Related relevant testing methods.

The staining technique using Coomassie Blue R250 is one of the most commonly used methods. This approach is reliable and fairly sensitive. Its main drawback concerns the length of time required.

Table III

Molecular weight of some proteins in human plasma.

Protein	M.W.	Subunit		
Albumin	69.000			
Fibrinogen	340.000	(α)	67.000	
		(ß)	56.000	
		(δ)	47.000	
IgG (γ-Globulins)	160.000	(H) ≈ 52.000		
Transferrin	80.000			
Plasminogen	94.000			
Plasmin, Glu	91.000	(A)	66.000	
		(B)	25.000	
Plasmin, Lys	82.000	(A)	57.000	
		(B)	25.000	
Thrombin	35.000		≈ 6.000	
			≈ 29.000	
Prothrombin	75.000			
Haemoglobin	16.000			
Vitronectin	75.000			
	65.000			
α$_2$ Macroglobulin	720.000		180.000 (x 4)	
ß$_2$ Microglobulin	12.000			
α$_2$ Antiplasmin	70.000			
Hageman Factor (XII)	20.000-100.000			
Factor X	59.000		≈ 44.000	
			≈ 15.000	
Proconvertin	35.000			
	65.000			
Factor VIII	300.000-400.000			
Complement Factors:				
C3 & C5	180.000	(α)	110.000	
		(ß)	75.000	
C3a & C5a	10.000-15.000			
C3b	180.000			
C4a	6.000-8.000			

Some more rapid but less sensitive staining techniques are available. Vesterbergs stain (*Vesterberg et al., 1977*) allows detection of proteins without destaining, but the method is less sensitive. A highly sensitive method is silver stain (*Switzer et al., 1979*) and is based upon a stain originally developed for histological purposes. Is is approx. 100 times more sensitive than Coomassie Blue and equivalent in the sensitivity to autoradiography. Later refinements and modifications have been developed (*Sammons, 1981*) called the Upjohn method in which a different colour response is obtained from different proteins (blue, green, yellow and red) to allow easy discrimination of overlapping proteins. The underlying principle of the technique is not fully understood, but it appears that the variations in color are a function of different crystal sizes of silver which form complexes with the polypeptide reactive centers.

Analytical techniques can be used for specific protein identification other than protein size. Immunoblot technique is a useful test to identify a great variety of proteins (*Mulzer & Brash, 1988*). Two-dimensional peptide mapping (electrophoresis and chromatography) together with immunological studies have been developed (*Cleveland et al., 1987*) and provides very sensitive separation of different proteins of identical size. That particular method may be used to analyse a protein after separation by SDS-PAGE.

However, it involves partial enzymatic proteolysis with commonly known proteases in the presence of SDS and analysis of the cleavage products by polyacrylamide gel electrophoresis. The pattern of the obtained peptide fragments can be characterized by the protein substrate and proteolytic enzyme and seems to be highly reproducible.

REFERENCES

Bradford, M.M. (1976) Analytical Biochemistry, 72, 248-254.

Brash, J.L. & Thibodeau, J.A. (1986) J. Biomed. Mat. Res. 20, 1263-1275.

Cleveland, D.W., Fischer, S.G., Kirschner, M.W. & Laemmli, U.K. (1977) J. Biol. Chem., 1102-1106.

Dunker, A.K. & Rueckert, R.R. (1969) J. Biol. Chem., 244(18) 5074-5080.

Laemmli, U.K. (1970) Nature, 227, 680-685.

Mulzer, S.R. & Brash, J.L. (1988) III World Biomat. Congress, April 21-25, Kyoto, Abstract book p. 473.

Pharmacia (1984): Polyacrylamide Gel Electrophoresis, Laboratory Technique, Rahms, Lund, Sweden

Reynolds, J.A. & Tanford, C. (1970) Proc. Nat. Acad. of Sciences, 66(3), 1002-1007.

Sammons, D.W., Adams, L.D. & Nishizawa, E.E. (1981) Electrophoresis 2, 135-181.

Shapiro, A.L., Vinuela, E. & Maizel, J.V. (1967) Biochem. Biophys. Res. Commun. 28, 815-820.

Shapiro, A.L. & Maizel, J.V. (1969) Anal. Biochem., 29, 505-514.

Smith, P.H. et al. (1985) Anal. Biochem., 150, 76-85.

Switzer, R.C., Merril, C.R. & Shifrin, S. (1979) Anal. Biochem. 98, 231-237.

Tanford, C. (1968) Advan. Protein. Chem. 23, 121-126.

Vesterberg, O., Hansén, L. & Sjösten, A. (1977) Staining of proteins after isoelectric focusing in gels by new procedures. Biochem. Biophys. Acta 491, 160-166.

Weber, K. & Osborn, M. (1969) Biol. Chem. 244, 4406-4412.

Reflectometry as a Tool for Dynamically Measuring Protein Adsorption

Authors: W.J.M. Heuvelsland, G.C. Dubbeldam & W. Brouwer

Akzo Corporate Research, NL - Arnhem

INTRODUCTION

Name of method:

Reflectometry as a tool for dynamically measuring protein adsorption.

Aim of method:

Dynamic measurement of a protein layer being formed on a reflecting surface as a bulk interface phenomenon. It can be used for in vitro experimental and screening procedures.

Biophysical background for the method:

The reflectance of a surface can be strongly influenced by the presence of a thin transparent film (*Andrade, 1985*). With the presence of such a thin transparent film, light is reflected by two adjacent parallel surfaces. Because the path difference is very small, the interference phenomena are present. With coherent light, the amplitudes of the reflected light waves are added (rather than the intensities). Therefore the two reflected light beams can amplify or attenuate each other depending on the wavelength of the reflected light and the corresponding optical path length in the film.

This effect has been utilized to detect the adsorption and desorption of proteins (*de Feyter et al., 1978*) and to detect antibody/antigen interactions (*Nygren et al., 1985, Nygren et al., 1986*) which are applied in e.g. diagnostic tests.

269

S. Dawids (ed.), Test Procedures for the Blood Compatibility of Biomaterials, 269–285.
© 1993 *Kluwer Academic Publishers. Printed in the Netherlands.*

When only the amount of adsorbed protein is of interest and not the refractive index of the adsorbed layer, it is not necessary to use a socalled ellipsometer for the measurement. A reflectometer (*Arwin & Lundström, 1985, Mandenius et al., 1986, Welin et al., 1984*) is much simpler and contains no moving parts.

A reflectometer operates with coherent light from a He-Ne laser. The light is refracted in the protein layer and reflected off the surface of the silicon wafer. The instrument measures simultaneously the reflectance of the p- and s-polarized light. Both values are combined to one figure which is independent of the laser beam intensity and has a calibrated linear relation with the absorbed mass at the substrate (*Dubbeldam, 1988*).

The method has been used to study the adsorption kinetics of single proteins in solution such as Bovine Serum Albumin (BSA), γ-Globulin (IgG) and Fibrinogen on Silicon wafers which are oxidized (SiO_2) or coated with a polymer such as polystyrene (PS). The method can also be applied to study specific antigen/antibody interactions on polymer surfaces in another context.

Sensitivity can be increased by using label amplification.

Theory.
The reflectance of a (flat) surface is strongly influenced by the presence of a thin film of which the refractive index differs from that of the substrate. The incident light is reflected by two interfaces. The two reflected light waves have a relative phase that depends on the thickness and the refractive index of the film and on the angle of incidence. Because of interference the resulting intensity is in between a maximum and a minimum value of which the relative difference depends on the reflectance of the two interfaces. With a fixed angle of incidence the resulting intensity can be used as a measure for the thickness of a film provided that the refractive index of it is known. If that is not the case, it is still possible to calculate the mass per unit area of the film from the measured reflectance.

The refraction of an interface between two (transparent) materials can be described with Snellius law and the Fresnel equations:

$$n_0 \sin i_0 = n_1 \sin i_1 \qquad \text{(Snellius) (1)}$$

where:

n_0 and n_1 are the refractive index of the first medium (e.g. water) and the second medium (e.g. protein layer),

i_0 and i_1 are the angles from the perpendicular of the surface of incident and refracted beams.

$$r_p = \frac{tg\ (i_0 - i_1)}{tg\ (i_0 + i_1)} \qquad \text{(Fresnel)} \qquad (2)$$

$$r_s = \frac{-\ \sin\ (i_0 - i_1)}{\sin\ (i_0 + i_1)} \qquad (3)$$

where:

r_p is the reflected amplitude of parallel polarised light.

r_s is the reflected amplitude of perpendicularly polarised light.

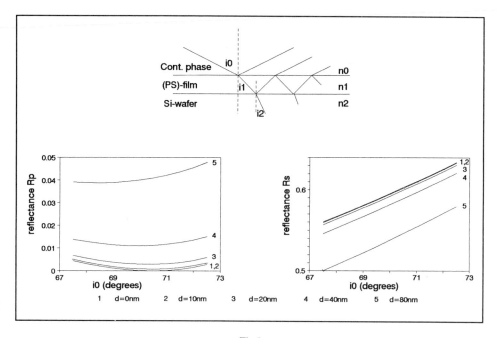

Fig.1

Reflectance versus angle of incident light (i_0) polarised in the plane of incidence at a silicon-water interface with different thickness of SiO_2 and different thickness of polystyrene (PS) coating.

The reflected amplitude of p-polarised light (r_p) and that of s-polarised light (r_s) varies as a function of the angle of incidence and of the thickness of a thin coat of film. Thus r_p and/or r_s can be used as indicators for that thickness (see *Fig. 1*). Without a film on the substrate the angle of incidence can be chosen so that $r_p = 0$. That angle is called the polarisation or Brewster angle, and it is specific for each dielectric material. For several reasons it is advantageous to choose the Brewster angle as the angle of incidence:

-    With the adjustment of the angle of incidence the minimum reflectance for p-polarised light is a useful and simple criterion.
-    The change of the reflectance with the thickness of an adsorbed film is maximal at the Brewster angle.

The reflected intensity, R, can be calculated using Drude's equation:

$$R_{s/p} = \left[\frac{r_{01}^2 + r_{12}^2 + 2\ r_{01}\ r_{12}\ \cos\ \Phi_1}{1 + r_{01}^2\ r_{12}^2 + 2\ r_{01}\ r_{12}\ \cos\ \Phi_1} \right]_{s/p} \qquad \text{(Drude)} \qquad (4)$$

and

$$\Phi_1 = \frac{(4\pi d_1\ n_1\ \cos\ i_1)}{\lambda_0} \qquad (5)$$

where:

r_{12} represents the reflected amplitude at the interface between the 1st and 2nd medium.

d_1 is the thickness of the thin film.

λ_0 is the wave length of the light in vacuum.

Because R_s and R_p change in opposite directions with the film thickness, it is possible to combine those two values to measuring value M:

$$M = \frac{(R_s - m\ R_p)}{(R_s + m\ R_p)} \qquad (6)$$

The value of M as a function of the film thickness δ is represented in *Fig. 2* for m=3 (the maximun value of R_s equals about the maximum value of $3 \cdot R_p$).

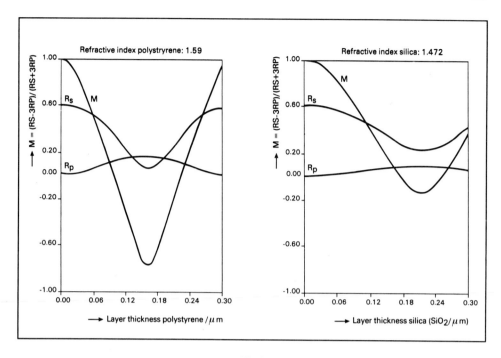

Fig. 2

Relationship between the calculated value of M and the thickness of polystyrene and SiO₂.
Brewster angle silicon/water = 70.52°, refractive index of silicon = 3.76, wavelength He-Ne
laser λ = 0.6328 nm.

The graphs of M versus δ show a relatively large linear area with a high steepness. Therefore starting with a film of about 30 nm a high sensitivity for thickness change by protein adsorption is achieved. The sensitivity is almost independent of the thickness and refractive index of the starting films within a wide range.
The value of M is a function of the film thickness δ.

Scope of method:

The method described here is very suitable to study the adsorption of single proteins and their immunological activity. Discrimination between proteins adsorbed from mixtures cannot be performed directly with the reflectometer. However, it is possible with the aid of specific antibodies.

DETAILED DESCRIPTION OF METHOD

Equipment.

Silicone wafers oriented <111>, Wacker Chemitronic, G.m.b.H., Germany
He-Ne- Laser 4 mW with polarising beam splitter, Spectra-Physics Inc., Eugene, USA
Photodetector BPY 63 P, Siemens, Germany
Reflectometer is designed locally (see *Figs. 3 & 4*)

Reagents.

Physiological phosphate buffer, pH = 7.4, NPBI, Emmer-Compascuum, The Netherlands
Phosphate buffer, pH = 8.9
   Na_2HPO_4, 2 H_2O,   11.85 g
   Aqua destilled ad 1000 g
Bovine Serum Albumin (BSA), Organon, Oss, The Netherlands
hCG and anti-hCG, Organon, Oss, The Netherlands

Software program is based on the equations given in *1-6* and is written in Turbo Pascal

Outline of method.

A diagram of the equipment and a picture are shown in *Fig. 3* and *Fig. 4*.

The light source is provided by a linear polarised He-Ne-laser. The substrate is a piece of a silicon wafer with a refractive index = 3.76 onto which a very thin polymer film has been coated. The substrate is placed in a cuvette with windows perpendicular to the incident and reflected beam. The angle of incidence should be adjusted to the Brewster angle for the interface silicon-water. The reflected beam passes a narrow band optical filter, and it is split into p- and s-polarised light with a polarising beam splitter. The intensities of each beam are measured with flat silicon photodiodes. The photocurrents i_s and i_p are equally amplified and electronically processed to the calculated value:

$$M = \frac{i_s - i_p}{i_s + i_p} \tag{7}$$

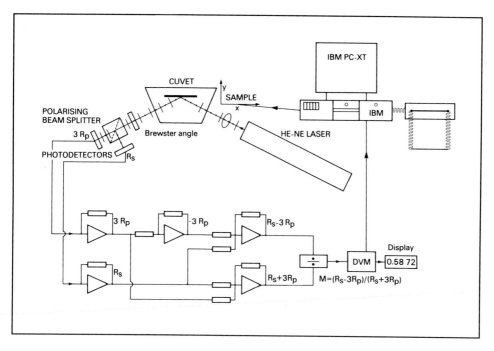

Fig. 3

Set-up for dynamic measuring the thickness of a film on a flat substrate.

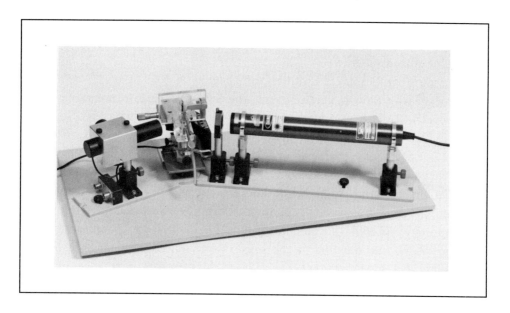

Fig. 4

Reflectometer

The factor m can be adjusted by rotating the laser about its axis so that it holds for the intensities of the p- and s-polarised parts of the incident light beam that:

$$m = \frac{I_p}{I_s} \tag{8}$$

where I_p and I_s are the intensities of the p- and s-polarised light paths.
Further it holds that i_s and i_p are proportional to $R_s \cdot I_s$ and $R_p \cdot I_p$ respectively.
Equation (7) can thus be rewritten as:

$$M = \frac{R_s - m R_p}{R_s + m R_p} \tag{9}$$

M is independent of the laser beam intensity.

Description of procedure.

Cleaning.
All materials which come into contact with protein solutions should be carefully cleaned to provide reproducibility. Cleaning with 70% HNO_3 of all glass material is recommended. The cuvettes which are made of Polyethylenetereptalate (PETP) can be cleaned with a non ionic surfactant and finally with ethanol (absolute).

Coating of silicon wafer.
Silicon wafers can be covered with SiO_2 by heating at 1000°C for 1.5 hours. The layer thickness of SiO_2 is about 50 nm. The SiO_2 surface can be modified with dimethyloctyl-Cl-silane to provide a hydrophobic property or with PROSIL-28 or with NH_4OH + H_2O to render a hydrophilic surface (*Jönsson et al., 1982*). A gradient in hydrophobic character is also possible by diffusion of dichlorodimethyl-silane from a concentrated solution over the surface of a hydrophilic wafer (*Elwing et al., 1987*). Very thin coatings of polymers can be made by dissolving the polymer in a good solvent and spincoating the solution over the silicon wafer. Styrolux from BASF (polystyrene with 18% butadiene) should be dissolved at a concentration of 1.5% in toluene and pipetted on the silicon wafer placed at a turning table. By spincoating at a speed of 2000 rpm, a layer thickness of about 50 nm can be obtained.

Adjustment of the instrument.

By means of optical light reduction filters the intensity of the light of the laser is reduced to give a suitable measuring level of the photodiodes.

The value of M (equation 9) should be recorded in distilled water for the Si-wafer with and without coating of the polymer.

After this the cuvette of the reflectometer with the coated silicon wafer is filled with the pure buffer solution. M is measured and can be correlated with the thickness or mass per area of the polymer coating before protein exposure.

The cuvette is then filled with protein solution (or if an adsorbed protein must be identified - with specific antigen solution). Adsorption (or reaction) is evaluated at the most 1 1/2 hour. The dynamic increase of mass/area can in principle be determined by on-line measuring.

The cuvette is finally flushed with buffer solution and M is measured again with buffer. From this the mass of irreversibly adsorbed protein material can be calculated.

To get accurate results it is recommended to scan over a number of positions of the surface of the wafer. At each position the value of M is measured and the adsorbed mass is calculated. In principle, scans can be made automatically logging results with a personal computer. This may be of value for overnight studies and in the calculation of the mass adsorbed according to the measured values of M. It may also provide a indication of the uniformity of the surface.

Safety aspects.

No special safety aspects are required. The handling of acids and solvents requires precautions as given in the general rules of Good Laboratory Practice.

RESULTS

Testing of the instrument.

The response of the reflectometer should be tested for reduced light intensity of the He-Ne laser. The value of the refractive index of the polystyrene and silicon dioxide layer should be measured as a function of the calculated layer thickness.

Reduced light intensity.

The responses of the reflectometer to reduced light intensity of the He-Ne laser is given in *Table I*.

Table I

Response of the reflectometer for light of reduced intensity through a polystyrene film.

Reduction factor	M	Calculated thickness /nm
1.0	0.5270	58.85
0.5	0.5305	58.55
0.2	0.5283	58.70
0.05	0.5262	58.95
0.02	0.5220	59.30
0.01	0.5180	59.55

58.87 ± 0.20nm

This table shows that a reduction of the intensity of the light of the laser beam down to 2% of the original intensity does not generate any serious error of the measured value of M. The same thickness of the polystyrene coating is calculated. Even if light is scattered much more, it can be expected that measurements are reasonable accurate. This implies that measurements can be performed even in rather turbid biological systems.

Influence of the refractive index.

The response of the reflectometer is a function of 2 main groups of parameters: the thickness and the refractive index of the layers on the wafer, but also of the refractive index of the solution in which the wafer is placed.

In *Table II* the main values are given which are in accordance with the theoretical description (*De Feyter et al., 1978, Welin et al., 1984*).

Table II

Influence of the refractive index of the solution on the value of M and δ for PS (polystyrene) and SiO$_2$ (silicon) layers.

System	Number of data	Refractive index of the liquid	Resulting variation in "M"	Constancy of δ/nm
Si + SiO$_2$	16	1.33 - 1.37	0.29 - 0.51	±0.18% = 0.23 nm
Si + PS	19	1.33 - 1.36	-0.07 - +0.15	±0.21% = 0.25 nm

By addition of sodium chloride to phosphate buffer and by using mixtures of ethanol and water as solvent, the effect of the refractive index of the medium (measured with the Abbe refractometer) on the reflectometric results can be studied. Both silicon wafers with an oxide layer or with a layer of polystyrene can be used in this test. As *Table II* shows different refractive indices result in different values of M. However, from the combination of the value of M and the measured refractive index each time the same thickness of the silicondioxide or polystyrene layer on the wafer is calculated. The refractive index of the medium does not influence the reflectometric determined thickness.

Measurements of adsorption.

To illustrate the possibilities of the method a few examples will be given.

Adsorption of Bovine Serum Albumin on silicondioxide and polystyrene.
In *Fig. 5* the influence is given of the concentration of Bovine Serum Albumin (BSA) on the adsorption on SiO$_2$ with albumin concentrations ranging between 10^{-2} and 10^{-8} g/ml. As *Fig. 5* shows, the higher the concentration of albumin the faster the maximum value is reached. The kinetics of adsorption is responsible for this behaviour. At the highest concentration of BSA (10^{-2} g/ml) a layer thickness

in solution is generated which exceeds the one in pure buffer. Probably this reflects the formation of double layers of adsorbed BSA.

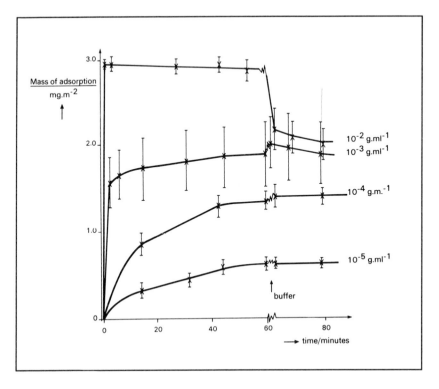

Fig. 5

Influence of the concentration of BSA (Boseral DEM) on the adsorption to SiO₂ at pH = 8.9 and 22°C.

The influence of the concentration of BSA on the mass of adsorption at silicondioxide or at polystyrene can be seen in *Fig. 6*. This figure shows that the adsorption of BSA increases almost linearly with the logarithm of the concentration. At the same concentration a larger amount of BSA is adsorbed at polystyrene compared with silicondioxide, indicating a difference in affinity. At relatively high concentrations a maximum value of adsorption is reached both at polystyrene and at silicondioxide surface. Although the maximum values of adsorption are almost equal, it is reached at a lower concentration when adsorption occurs at polystyrene surfaces.

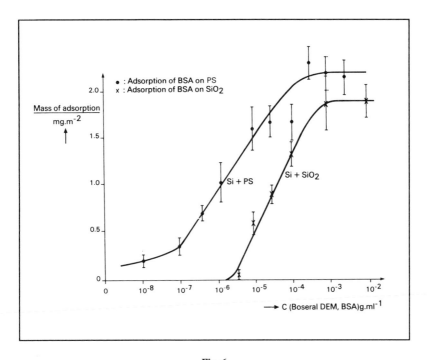

Fig. 6

Adsorption of BSA (Borseral DEM) on Silicon wafers with PS or SiO₂ after 1 hour as a function of the concentration of BSA in solution at pH = 8.9 and 22°C.

Adsorption of BSA, IgG and Fibrinogen on modified SiO₂.

It is possible to modify the surface properties of the silicon surface from hydrophobic (contact angle of water of 80°) to hydrophilic (contact angle of about 20°). Adsorption of BSA and IgG as a function of the contact angle is given in *Fig. 7*.

As this figure reflects, both proteins can adsorb on hydrophobic as well as hydrophilic substrates. At the intermediate contact angle (neither hydrophobic nor hydrophilic) less adsorption is observed. The nature of the substrate, however, influences the orientation and conformation of the protein. Bound a-hCG shows only reaction with its specific antigen (hCG) when a-hCG is adsorbed in the hydrophobic region as is illustrated in *Fig. 7*. This indicates that the active site of a-hCG (used for interaction with its antigen (hCG)) are only accessible when a-hCG is adsorbed on hydrophobic silicondioxide.

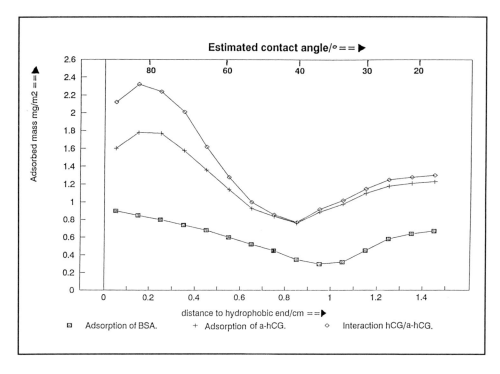

Fig. 7

Adsorption of BSA (Boseral DEM) and a-hCG on Silicon wafers with a gradient in hydrophobic character at pH=7.4. Interaction with hCG.
Concentration of BSA = 1 x 10³ g/cm³, concentration of a-hCG = 1 x 10⁴ g/cm³, concentration of hCG = 1 x 10⁴ g/cm³.

Identification of proteins with antibody antigen interaction.

Identification of proteins is possible by using the very specific antigen antibody reaction (*Nygren, 1985, 1986; Jönsson & Rönnberg, 1985*). In *Fig. 8* this is shown for a model system: reaction of anti-hCG with hCG. This figure shows the results of two tests. When first a fixed concentration of anti-hCG $(c=1x10^{-4}$ g/cm^3) is adsorbed on the silicon wafer covered with polystyrene followed by exposure with different concentrations of hCG, the lower curve is obtained. If the test is done in opposite order (first hCG at $c=1x10^{-4}$ g/cm^3 followed by anti-hCG at different concentrations) the upper curve is obtained. This difference may well be caused by the fact that anti hCG is 4 times larger than hCG. *Fig. 8* shows that in this way detection is possible up to 10^{-7} g/cm^3 which equals about 100 IU/l of hCG.

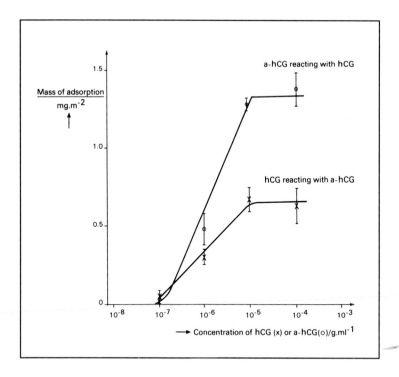

Fig. 8

Increased adsorption by the reaction of hCG with a-hCG bopund to the wafer or of a-hCG bound to the wafer at pH = 7.4 and 22°C.

DISCUSSION

Interpretation of results.

It should be realised that the presented results are not obtained under physiological conditions and therefore cannot be related directly to in vivo circumstances. However, an indication of the possibilities of adsorption and reaction can easily be obtained with reflectometry within minutes. This method may be used to develop and optimize new pharmaceutical tests and help to identify the adsorption behaviour of surfactants.

Limitations of method.

This method is an experimental approach which may also be used for protein mixtures as found in plasma. However some restrictions have to be made:

Firstly: one cannot directly identify the proteins which are adsorbed to a polymer surface.

Secondly: the polymer coating of the silicon wafer must be extremely thin (about 30 - 100 nm).

Thirdly: adsorption of mixtures of proteins are difficult to interpret and can only be identified with subsequent specific antibody reaction or by other types of protein analysis studies (e.g. isotope tagging SDS-PAGE).

Related testing methods.

The most frequently used method for performing measurements on the films at flat substrates is ellipsometry. With this method the change of the polarisation state of the reflected light is determined. The sensitivity of ellipsometry for measuring the thickness of thin adsorbed protein films is in the same order of magnitude. However, ellipsometric measurements are much more complex and thus much more costly.

Another method for measurement on thin protein films is "Surface Plasmon Resonance" (SPR). With that system the angular dependence of the reflectance of a metal (silver) film at a glass substrate is measured. The result is strongly influenced by the presence of a (protein) film at the metal surface. Also this method shows a sensitivity that is in about the same order as that of reflectometry. However, obtaining a stable silver film at the substrate is difficult, and the angle of incidence depends on the thickness of the metal film.

The main advantages of reflectometry over the two methods mentioned are:

-    The set-up is relatively simple and it comprises no moving parts.
-    No critical or complex adjustment procedures have to be carried out.
-    The output voltage of the system is linearly related with the mass per unit are of the adsorbed protein film.
-    Real-time measurements on dynamic films can be performed.

REFERENCES

Andrade, J.D. (1985) in "Surface and Interfacial Aspects of Biomedical Polymers" (Andrade, J.D. ed.), vol. 2: "Proetin Adsorption", Plenum Press, New York.

Arwin, H. & Lundström, I. (1985) Anal. Biochem. 145, 106-112.

Dubbeldam, G.C. (1988) European patent application no. 88 200 230.6

Elwing, H., Welin, S., Askendal, A., Nilsson, U. & Lundström, I. (1987) J. Colloid. Interface Sci. 119, 203-210.

De Feyter, J.A., Benjamins, J. & Veer, F.A. (1978) Biopolymers 17, 1759-1772.

Jönsson, U., Ivarsson, B., Lundström, I. & Berghem, L. (1982) J. Colloid. Interface Sci. 90, 148-163.

Jönsson, U., Rönnberg, I & Malmqvist, M. (1985) Colloids Surfaces 13, 333-339.

Jönsson, U., Lundström, I. & Rönnberg, I. (1987) J. Colloid. Interface Sci. 117, 127-138.

Mandenius, C.F., Mosbach, K., Welin, S. & Lundström, I. (1986) Anal. Biochem. 157, 283-288.

Nygren, H. & Stenberg M. (1985) J. Colloid. Interface Sci. 107, 560-566.

Nygren, H., Kaartinen, M. & Stenberg, M. (1986) J. Immunol. Methods 92, 219-225.

Welin, S., Elwing, H., Arwin, H., Lundström, I. & Wikström, M. (1984) Anal. Chim. Acta 163, 263-267.

Editors Note: This unedited contribution has been included although the authors wish to publish it in a form which does not correspond to the prescribed format.

Measurement of Protein Adsorption to Solid Surfaces in Relation to Blood Compatibility Using Radioiodine Labelling Methods

Authors: X.J. Yu & J.L. Brash

Departments of Chemical Engineering and Pathology,
McMaster University, Hamilton, Ontario, Canada

I INTRODUCTION

Radiolabelling is one of the most useful and powerful experimental methods for the study of protein adsorption at solid-solution interfaces (*Andrade, 1985; Brash & Horbett, 1987*). It consists of incorporating a radioactive nuclide into the molecular structure of the protein to be studied, and then counting the surface-bound radioactivity following contact of the material with the protein solution to which the labelled protein is added. In general the radiolabelling technique provides data of high precision on the amount of protein adsorbed under given conditions. When adsorption from plasma or blood is studied, such data may serve as a significant indication of the blood compatibility of the contacting solid surface.

Radiolabelling methods are based on the radiation phenomena that originate in certain atomic or nuclear processes (*Wang et al., 1975*). It has long been established that spontaneous transitions that occur in an unstable (radioactive) nuclide are accompanied by the release of energy in the form of radiation and/or charged particles known as alpha, beta and gamma radiations. α and ß radiations are associated with charged particles, namely the positive helium nucleus and the negative electron respectively. In contrast, γ-rays like x-rays, consists of electromagnetic radiation of high energy, and as such have no associated charge. The

S. Dawids (ed.), Test Procedures for the Blood Compatibility of Biomaterials, 287–330.
© 1993 *Kluwer Academic Publishers. Printed in the Netherlands.*

The first necessity for the application of the radiolabelling method to the study of proteins is the availability of a suitable radionuclide. In this regard, radioisotopes of iodine are probably the candidates of choice (*Bolton, 1977; Regoeczi, 1984*). As γ-emitting sources, radioiodines, in particular ^{125}I and ^{131}I, have many advantages for use in the preparation of labelled proteins. The greatest of these is the ease with which they can be counted, no special sample preparation being required. In particular with respect to adsorption studies, γ-emitting isotopes can be readily counted while attached to a solid substrate. In contrast, the ß-emitting isotopes ^{3}H and ^{14}C, which can also be used to label proteins, must generally be released (quantitatively) from the solid and counted by techniques such as liquid scintillation counting involving special sample preparation, quench corrections and the like. Another major advantage stems for the relatively simple chemical reactions underlying most protein iodination procedures.

The preparation of labelled protein to which a selected radioiodine is covalently bound constitutes the central part of the technique. Although the mechanism of iodination and the structure of iodinated proteins are not without ambiguity, it is generally believed (*Hughes, 1957*) that the incorporation of radioiodine into protein molecules results from the substitution of iodine atoms in the phenolic ring of tyrosine residues. Several additional amino acids, including cysteine, histidine, methionine, phenyalanine and trytophane, can also react with iodine. However, under normal reaction conditions the reactivity of these amino acids with iodine is relatively slow.

The use of radioiodine-labelled proteins as tracers is generally considered to be among the simplest, most quantitatively precise and reliable, and most efficient experimental methods for the study of protein adsorption. As applied to the testing of blood compatibility of materials the use of this method, and of protein adsorption methods generally, should be considered "experimental" as opposed to "clinical". A number of investigators in the past have measured adsorption of common plasma proteins from buffer solutions of the individual proteins and have tried to assign significance to the results in terms of blood compatibility. It is not likely, however, that data from this kind of experiment will bear any relation to adsorption from blood. The present authors consider that the utility of the method in relation to blood compatibility testing lies in determining adsorption of specific proteins from blood or plasma. Such data are unquestionably related to blood compatibility even if the detailed mechanisms remain obscure.

A large amount of useful protein adsorption information using radiolabelling has been generated and a number of comprehensive reviews (*Morrissey, 1977; Horbett, 1982, 1987; Brash, 1987*) are now available. In the present article the technical aspects of the method are emphasized. Two widely used procedures for the radio-iodination of proteins are described in detail, and a more general guide to other available methods is provided. Some typical experimental data, mostly from the authors' laboratory are then presented to illustrate the potential of the technique for the study of protein adsorption in the context of blood-material interactions. Finally, the merits and shortcomings of radioiodine labelling methods in comparison with some other commonly used techniques are discussed.

II METHODOLOGY

A complete experiment for measuring protein adsorption using the radioiodination method is most commonly performed as follows. The protein of interest is first radioiodinated and then added, as a tracer, to the solution to be studied. This can be the same (unlabelled) protein in buffer, or mixtures of proteins (including the protein of interest) ranging from simple (two or three proteins) to complex (e.g. plasma). The solution, thus radiolabelled, is subsequently placed in contact with the material to be tested, under appropriate experimental conditions. At the end of the adsorption step, the solid and solution phases are normally separated from each other, and the surface-bound radioactivity is counted directly using a gamma counter. The relation between the radioactivity count and the amount of protein adsorbed can readily be established by counting an aliquot of the labelled protein solution of known concentration as a callibration standard.

Clearly the radioiodination method could be used for solution depletion experiments in which the decrease in concentration of the protein solution after contact with the surface is measured. With this indirect approach to measuring adsorption, sufficient surface must be used to cause a measurable decrease in concentration, and this method is generally restricted to particulate marterials of high specific surface area. The vast majority of materials to be tested for blood compatibility are in film or tubing form and are thus of low specific surface. Solution depletion, even with the sensitivity of radiolabelling, is thus not usually suitable and the method indicated above, wherein which the surface is counted directly, must be used.

2.1 Radioiodination of proteins.

An excellent and very detailed discussion of this topic is given in a recent monograph by *Regoeczi (1984)*, and the reader intending to embark on the use of radioiodinated proteins is strongly advised to consult this reference.

2.1.1 General considerations.

In designing a protein iodination procedure, the following questions must be addressed: (i) How much radioiodine should be bound to the protein molecule? (ii) What reaction conditions should be used to achieve this degree of iodination? In all studies in which radiolabelling methods are used it must be kept in mind that one is measuring the radionuclide attached to the protein. Thus one is observing the behaviour of only the labelled molecules which may or may not reflect the behaviour of the unlabelled protein. In the first place iodination itself may influence adsorption or may denature the protein if the degree of iodination is too great. Secondly, if the labelling procedure involves severe reaction conditions such as exposure to radiation, strong oxidizing or reducing reagents or extreme pH etc., the protein structure can be significantly altered. The measurement of adsorption will then reflect the behaviour of the denatured rather than the native protein.

An ideal preparation of radioiodinated protein should contain an optimum amount of radioiodine: sufficient to provide the desired analytical sensitivity, but not enough to alter the behaviour of the protein with respect to the phenomena under investigation. The quantity of radioiodine bound to a protein is usually expressed either as specific activity (the radioactivity incorporated in unit mass of labelled material, $\mu Ci/\mu g$), or degree of substitution (number of atoms per protein molecule, I/P). In general, a low degree of substitution is preferable to minimize any effect on the behaviour of the labelled material. In this regard, a number of reports (*McFarlane, 1963; Harwig et al., 1975*) have stated that a degree of substitution of one atom of radioiodine or less per protein molecule does not significantly alter the labelled protein. A degree of substitution of about unity can generally provide sufficiently high specific activity of the labelled protein for high precision measurement of adsorption.

In addition, radioiodination of proteins should be carried out using mild reaction conditions in order to avoid denaturation. Since the early 1950s (*Hughes & Straessle, 1950; McFarlane, 1956*), a variety of methods have been developed for radioiodination of proteins, some of which are shown in *Table I*. These methods can be divided into two grops: (i) direct methods involving only a single reaction step in which iodine is directly substituted into tyrosine residues of the protein and (ii) conjugation methods consisting of iodinating and intermediary reagent, which is subsequently conjugated to a specific side chain of the protein. The selection of an appropriate method depends on the nature and amount of the protein to be labelled, and on the specific activity required.

Table I

Methods of protein iodination.

Method	Comments	References
Iodine chloride (ICl)	Simple; efficient and rapid; mild reaction conditions; low specific activity.	McFarlane, 1958, 1963 1964; Helmkamp et al. 1960.
Chloramine-T	Simple and rapid; high specific activity; relatively harsh reaction conditions.	Hunter & Greenwood, 1962; McConahey & Dixon, 1966.
Lactoperoxidase	Gentle reaction conditions; high specific activity; longer reaction time.	Marchalonis, 1969; David, 1972; David & Reisfeld, 1974.
Electrolysis	Efficient and mild; high yield and specific activity; more complex, special equipment requied.	Rosa et al., 1964
Bolton-Hunter Reagent	Labelling of tyrosine-poor or -free proteins; particularly mild; low yield and specific activity; slower more complex reaction.	Hunter & Ludwig, 1962; Bolton & Hunter, 1973.
Iodogen	Very simple and convenient; surface localized reaction; minimal exposure of protein to reagent; low losses, high yields.	Fraker & Speck, 1978.

In our research the iodine monochloride (*McFarlane, 1963*) and lactoperoxidase (*Marchalonis, 1969*) methods, both of which are direct methods, have been favoured because they are simple, rapid and non-traumatic to most of the plasma proteins

(*Regoeczi, 1984*). The experimental procedure for the iodination of fibrinogen using these two methods are described as examples. Modification may be made as necessary, to suit the particular requirements of different proteins.

2.1.2 Radioiodination using the iodine monochloride method: Application to fibrinogen.

In the ICl method radioactive ("hot") iodide ions are mixed with nonradioactive ("cold") iodine monochloride. Hot ICl is formed by rapid iodine-iodide exchange. The protein is then incubated with the ICl (mixture of hot and cold) at alkaline pH and iodine is substituted in the tyrosine residues (*Bale et al., 1966*) as shown in *Fig. 1*.

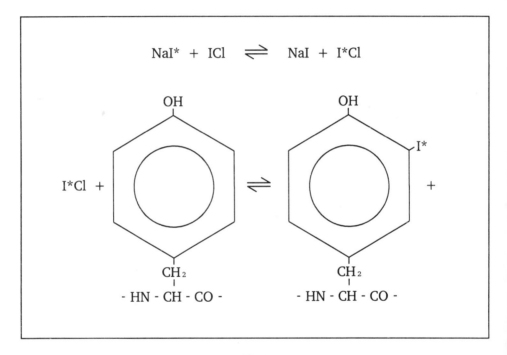

Fig. 1

Principle in iodination of protein. First "hot" iodine (I*) ions are exchanged with "cold" iodine in ICl. Tyrosine residues are then tagged with a mixture of I*Cl and ICl.

In general, the desired degree of substitution can readily be achieved by adjusting the stoichiometry, i.e. the ICl/protein ratio. The stoichiometry indicates the maximum average degree of substitution assuming 100% incorporation of iodine. All of the reactions occurring for any given protein are not known, but the incorporation of iodine is in general less than 100% presumably due to the consumption of iodine in side reactions as well as equilibrium conversions of less than 100%. Substitution can be estimated by determining the protein-associated and non protein-associated radioactivity after the labelling reaction as indicated by *Regoeczi (Regoeczi, 1984, p.35)*. It should be noted that since not all of the ICl molecules are "hot", nonradioactive as well as radioactive iodine will be attached to the protein, resulting in labelled proteins of relatively low specific radioactivity for a given degree of substitution.

Materials: Solutions of the radioiodines $Na^{125}I(\sim 20$ mCi/mL) and $Na^{131}I(\sim 20$ mCi/ml) are purchased from New England Nuclear (Boston).

Human fibrinogen is obtained from Kabi (Stockholm, Sweden). As described previously *(Chan & Brash, 1981)*, the lyophilized protein, which contains a number of salts, is dissolved in distilled water, dialysed overnight at 4°C against a suitable buffer and then stored frozen at a concentration of about 10 mg/mL. An aliquot of protein solution for labelling is quick-thawed at 37°C just before use.

Iodine monochloride stock solution, 33 mM with respect to ICl, is prepared according to the method of *McFarlane (1963)*. 150 mg NaI is dissolved in 8 mL 6M HCl, to which is added 99 mg $NaIO_3$ dissolved in 2 mL water. After mixing, the volume is made up to 40 mL with water and the solution is shaken with a few mL CCl_4. The CCl_4 is used as an indicator of free iodine, and if the CCl_4 phase is coloured faint red, it is replaced with a fresh aliquot and the shaking is repeated. The final volume of the solution, normally showing a green-yellow colour is adjusted to 45 mL. The ICl solution is extremely stable when kept in a glass container over a few mL of CCl_4 at 4°C. The "working solution" used in protein labelling is prepared from the above "stock solution" by dilution with 9 vol 2 M aqueous NaCl.

The labelling reaction is carried out in alkaline glycine buffer, pH 8.8, made by mixing 2 vol 1 M NaOH and 8 vol 1 M glycine.

Ion exchange chromatography is used to separate the labelled protein from unbound radioactive iodide. A chromatographic column (3 mL plastic syringe), is packed with anion-exchange resin (e.g. AG 1-X4 from Biorad) and equilibrated with a suitable buffer (usually the "working" buffer in which the protein is required for the adsorption measurement) before use. As well as ion-exchange chromatography *(Scott et al., 1966)*, dialysis *(Diamond & Denman, 1973)* and gel filtration *(Wilson & Greenhouse, 1976)* can be used for this purpose.

Method: 0.4 mL (4 mg, 0.012 μmol) fibrinogen solution and 0.01 mL (0.033 μmol) ICl working solution are pipetted separately into two small test tubes. The pH of the protein and ICl solutions are adjusted to be weakly alkaline using 0.1 mL and 0.05 mL glycine buffer, respectively. 5 μL (~ 0.1 m Ci) of radioiodide solution (^{125}I or ^{131}I) is then added to the ICl solution and mixed for 2-3 min. The resulting solution is then rapidly jet-mixed with the protein solution using a Pasteur pipette. The labelled material is transferred to the previously prepared ion exchange column and eluted with 5 mL working buffer. The eluate is collected in a glass test tube and kept at 4°C until used, usually later the same day. Typical counting rates for such preparations are of the order of 3 x 10^7 cpm/mg protein, with a corresponding analytical sensitivity of ~ 4 ng protein. Higher specific radioactivities (and analytical sensitivities) can be obtained by increasing the amount of radioiodide used.

The above procedure is "customized" for fibrinogen and is designed to label 4 mg protein at a molar ratio ICl-to-protein of about 3. Typical labelling efficiencies are between 60 and 70% and I/P's, therefore, between 1.8 and 2. It should be noted that this is an average value of I/P and in fact there will be a distribution of I/P in the labelled protein sample. In principle, I/P could be as high as the number of tyrosine residues and as low as zero. For any given protein to be labelled, appropriate conditions required to achieve the desired I/P must be determined by trial and error preliminary experimentation. As indicated above the equilibria for ICl-protein reactions are not generally known and reactivities for the tyrosine substitution reaction vary from protein to protein. In addition reactions of iodine with protein residues other than tyrosine are well known and can result in "consumption" of ICl without incorporation of iodine into the protein. For example sulfhydryl groups can be oxidized by iodine to form various acids while the iodine is reduced to iodide (*Trundle & Cunningham, 1969*). Thus for proteins containing free sulfhydryl groups, e.g. serum albumin, the quantity of ICl must be increased to allow for such reactions.

2.1.3 Radioiodination using the lactoperoxidase method: Application to fibrinogen.

In this method, an immobilized enzyme reagent is used consisting of lactoperoxidase and glucose oxidase immobilized on hydrophilic polyacrylamide microspheres. On the addition of glucose to a suspension of beads, containing radioiodide and protein, the immobilized glucose oxidase generates a small, steady amount of hydrogen peroxide. The lactoperoxidase catalyses the oxidation of labelled iodide to iodine by the peroxide. The radioactive iodine then labels the protein by incorporation in tyrosine residues. The overall reaction sequence is as follows:

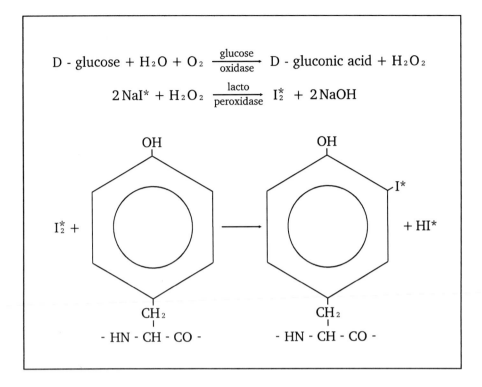

$$D\text{-glucose} + H_2O + O_2 \xrightarrow[\text{oxidase}]{\text{glucose}} D\text{-gluconic acid} + H_2O_2$$

$$2\,NaI^* + H_2O_2 \xrightarrow[\text{peroxidase}]{\text{lacto}} I_2^* + 2\,NaOH$$

Fig. 2

Immobilized enzymes (glucose oxidase and lactoperoxidase) generate small amounts of hydrogen peroxide which will catalyze conversion of iodide to iodine. The iodine labels the protein on the hydrosine residues. The method is gentle and provides high specific labelling of radioactivity.

In this gentle enzymatic method all iodine atoms are radioactive (unless a carrier is added) thus resulting in higher possible specific activities of labelled protein for a given I/P than for the ICl method. This method was introduced by Marchalonis (*Marchalonis, 1969*).

Materials: Human fibrinogen and ion exchange resin are the same as described above for the iodine monochloride method.

Radioiodides having activities of about 100 m Ci/mL are used in this procedure. These solutions have similar specific activities but about 5 times higher concentrations of iodide compared to those used in the ICl method. This helps to minimize the reaction volume during labelling.

The immobilized enzyme reagent Enzymobeads is obtained from Bio-Rad Labs. (Richmond, California). One vial of this reagent is rehydrated with 0.5 mL 0.02 M phosphate buffer, pH 7.2, at least one hour before use.

Beta-D-glucose is used as a 1% aqueous solution. 2% alpha-D-glucose solution can also be used but must be allowed to mutarotate overnight to a mixture of the α and ß forms.

Method: The following reagents are mixed in a test tube in the order given: 100 μL 0.2 M phosphate buffer, pH 7,2; 0.03 to 0.04 mL (0.3 to 0.4 mg) fibrinogen solution; 50 μL Enzymobead reagent (a suspension of microspheres); 10 μL (about 1 m Ci) radioiodide; and 50 μL 1% beta-D-glucose solution.

The mixture is incubated at room temperature for 15-25 min., and the iodination then stopped by removal of the beads from the reaction tube (centrifugation at 1000 g for 10 min.). The supernatant is applied to an anion exchange column and, following the same procedure as described above for ICl labelling, the labelled protein is eluted, collected and stored at 4°C until use. The quantities of iodide and fibrinogen used would result in a maximum I/P of about 0.6 if iodine incorporation were 100%. Incorporation under the conditions described is in general about 10%, so actual I/P values are about 0.06. Counting rates for these preparations are of the order of 3 x 10^8 cpm/mg protein.

The specific activity of protein obtained in any labelling procedure depends on a number of factors such as the stoichiometry, the labelling efficiency (iodine incorporation) and the amount of radioactivity used. If high specific activity is required to achieve adequate measurement sensitivity, then the labelling method and conditions must be optimally chosen. Of the two methods routinely used in this laboratory it is normal experience to obtain higher specific activities for the lactoperoxidase than for the ICl method under comparable conditions. The main difference is probably the fact that in the lactoperoxidase method all of the reactive iodine is radioactive whereas in the ICl method there is a mixture of hot and cold ICl.

2.1.4 Radioiodination of other plasma proteins.

The labelling procedures described above for the radioiodination of fibrinogen using the ICl and lactoperoxidase methods can be used with many other plasma proteins (see *Table II*).

Table II

Radioiodination of some plasma proteins.

Protein	Radioiodination Method	Reference
Albumin	Chloramine-T ICl Lactoperoxidase	Miles et al., 1970 McFarlane, 1958 Morrison et al., 1971
IgG	ICl Chloramine-T Lactoperoxidase	Helmkamp et al., 1960 Sonada & Schlamowitz, 1970 Marchalonis, 1969
Fibrinogen	ICl Lactoperoxidase	McFarlane, 1963 Krohn & Welch, 1974
Plasminogen	Chloramine-T Lactoperoxidase	Rabiner et al., 1969 Hatton et al., 1988
Complement C3	Lactoperoxidase	Law et al., 1979
Factor XI[a]	Chloramine-T	Bouma & Griffin, 1977
Factor XII[b]	Chloramine-T	Revak et al., 1974
Prekallekrein[c]	Chloramine-T	Mandle & Kaplan, 1977
Thrombin	Lactoperoxidase	Hatton et al., 1983
Factor VIII	ICl	Horbett & Counts, 1983

a,b,c, see also McConahey & Dixon, 1966

In the authors' laboratory, the simple and rapid ICl method is thought as the first choice, and the more complex, more expensive and longer lactoperoxidase method is used in cases where high specific activity and relatively small amounts of labelled protein (<0.1 mg) are required. For most proteins, the experimental conditions need undergo only minor modifications relative to those described for fibrinogen. Consideration should be given to the tyrosine content of proteins to be labelled. It would appear, for example, that the tyrosine content of human fibrinogen (*Cartwright & Kekwick, 1971*) is twice that of human IgG (*Heimburger et al., 1964*) and five times that of human albumin (*Peters, 1970*). On a statistical basis the existance of multiple tyrosine residues means that the labelled protein population, as already indicated, will have a distribution of iodine substitution levels, with broader distributions for proteins of higher tyrosine content. This situation is complicated by the fact that not all tyrosines are of equal reactivity.

Many plasma proteins have been successfully radioiodinated using methods other than ICl and lactoperoxidase (*Table II*). A detailed description of these methods is beyond the scope of the present article. Nevertheless, in order to facilitate the reader's further investigation of these methods, *Table II* gives a list of relevant references for the radioiodination of a few plasma proteins the adsorption of which is currently considered closely related to the blood compatibility of biomaterials.

2.1.5 Assessment of the effect of radiolabelling on protein adsorption.

As mentioned above, radiolabelling may result in alteration of the labelled protein with respect to its physicochemical and biological properties, and consequently the adsorption behaviour of the labelled and unlabelled proteins may be different. Routine, systematic checking for possible adverse effects due to radiolabelling is therefore required to validate the technique for its use in the study of protein adsorption. In this regard a simple test is used in the authors' laboratory. Adsorption is measured using a series of solutions of identical total protein concentration but of varying ratio of labelled to unlabelled protein. The results obtained for fibrinogen adsorption on a number of surfaces are shown in *Fig. 3*, and indicate that in these systems the surface concentration is independent of the ratio of labelled to unlabelled fibrinogen. Most systems examined in our laboratory over a number of years have shown no effect of labelling on adsorption. In the few cases where there is an effect, one generally finds that adsorption decreases as the fraction of labelled protein in the mixture increases, suggesting that labelling inhibits adsorption or that labelled protein occupies more surface than unlabelled. In such cases it is strictly speaking not valid to use this technique although an appropriate indication of adsorption may be gained by extrapolation of the data to zero labelled protein content. The use of a preparation with lower iodine substitution may also solve such problems.

In principle the above test is sufficient to validate the use of a given radioiodinated protein for adsorption studies, even if the protein is altered by other criteria such as in vivo clearance rate. Such in vivo studies have been used by a number of authors as the ultimate test of biologic function of iodinated proteins (*Regoeczi, 1984*).

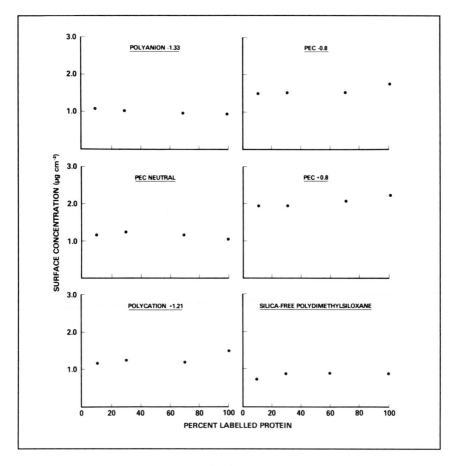

Fig. 3

Plots showing equivalence of labelled and unlabelled fibrinogen in adsorption to various surfaces. Total protein concentration 1.0 mg mL^{-1}. Polyanion - 1.33 is a sulfonated polyacrylonitrile; polycation + 1.21 is a copolymer of acrylonitrile and 1,2-dimethyl-5-vinylpyridinium iodide; PEC - 0.8, PEC neutral and PEC + 0.8 are complexes of these two polymers with ion exchange capacities of - 0.8, 0, and + 0.8 meq g^{-1} respectively. (With permission from Smith et al., 1983).

It is on the basis of such studies that the frequently cited "safe" substitution of 1 atom of iodine per molecule of protein has been derived. In vitro tests generally seem to be less sensitive to iodine-induced denaturation. Thus *Ardaillou & Larrieu (1974)* have shown that fibrinogen radioiodinated by the ICl and lactoperoxidase methods at I/P ≤ 1 behave the same as unlabelled protein with respect to clottability, SDS-PAGE (Sodium Dodecil Sulphate - PolyAcrylamide Gel Elec-trophoresis), immunoelectrophoresis and gel filtration. We have found that the clottability and glass-adsorption of fibrinogen labelled with ICl is independent of

I/P from 1 to 15. On the other hand in vivo clearance rates were found by *Mc Farlane (1963)* to increase with the degree of substitution of above 0.5, though *Harwig et al. (1975)* later found that fibrinogen clearance rates were independent of I/P up to 45. Clearly it would be unsatisfactory to use labelled proteins for adsorption studies which are known to behave abnormally in vivo, even though they satisfy a validation test. In this regard it seems prudent to avoid the use of highly iodinated proteins.

The stability of labelled proteins over time, in particular the loss of label, is also of concern. Freshly prepared radioiodinated proteins will normally contain small amounts of unbound radioactivity whether they have been treated by dialysis, ion exchange chromatography or other methods. Unbound iodine can be determined by trichloracetic acid precipitation of protein and radioactivity measurement of the supernatant. We have found that fibrinogen labelled as described above usually contains about 1 to 1.5% unbound radioactivity. It is likely that this will increase with time. *Krohn et al. (1972)* have shown that in saline at 37°C, radioiodinated fibrinogen is hydrolyzed (i.e. loses radioactivity) at rates of the order of 10 to 30% per day depending on the labelling method. Although this reaction would undoubtedly be slowed down by storing the protein frozen, it is clearly advisable in practice to use labelled proteins as soon as possible after preparing. For these reasons labelled proteins in the authors' laboratory are used on the day they are prepared. Loss of label must clearly be kept in mind if long-term adsorption experiments lasting more than a few hours are to be carried out.

2.2 Adsorption, counting and calibration.

As described previously (*Chan & Brash, 1981*), most of our experiments have utilized surfaces in the form of tubing segments of diameter in the range from 0.1 to 0.25 cm, although other solid geometries such as particles and thin films or plates are equally amenable to the methodology (*Brash & ten Hove, 1988; Horbett, 1987*). Tubing segments of about 25 cm in length are fitted at each end with three-way valves, and solutions are introduced into the segments using a syringe attached to the entrance valve. One fluid filling the tubing segment can be displaced by another without intervening air bubbles by pushing trapped air out through the "third" limb of the three-way valve. Bubbles must be avoided since they can cause Langmuir-Blodgett transfer of a protein film from the air-solution to the solid-solution interface. Tubing segments are equilibrated overnight with an appropriate buffer which should be isotonic with the protein solution to be studied. Adsorption is initiated by displacing the buffer with the protein solution. Approximately 20 segment volumes are injected over a 20s interval to achieve complete displacement. After a suitable contact time the protein solution is displaced with buffer in a similar manner. Rinsing follows a standard procedure consisting of three successive injections of 20 segment volumes of buffer allowing a 10 min incubation between each injection. Other protocols may of course be followed but the volumes of solutions used and the manner of their introduction into the tubing segments must be carefully controlled to achieve reproducible results.

Rinsing procedure necessarily involves a degree of arbitrariness, and is in some measure the "Achilles heel" of this or any other non in situ method in which the surface is evaluated directly following separation from the solution. The objective is to remove all of the solution protein from the sample but not any of the adsorbed protein. Excessive rinsing may cause a loss of adsorbed protein, especially when adsorption is rapidly reversible as is the case for many hydrophilic polymers. Clearly the experimenter must pay careful attention to the development of a rinsing protocol and must follow it rigorously if data are to be reproducible and if comparisons among experiments are to be valid.

Following rinsing the radioactivity bound to the tubing is determined. If the corresponding radioactivity of an aliquot of the labelled protein solution of known concentration is also determined under the same conditions of counting, the surface concentration of adsorbed protein can be calculated as follows:

$$\Gamma_p = \frac{A_{su} C_p}{A_{so} S}$$

where

Γ_p = surface concentration of protein ($\mu g/cm^2$)
A_{su} = radioactivity of tubing segment (counts per min, cpm)
A_{so} = radioativity of solultion (cpm/mL)
C_p = concentration of protein solution ($\mu g/mL$)
S = surface area of tubing (cm^2)

Counting can conveniently be done in most commercial gamma counters. Most of these instruments use solid scintillation detectors (typically thallium-doped sodium iodide) and have automated multiple sample handling capacity, a convenient feature for adsorption experiments where a large number of samples may have to be counted. In choosing an instrument, attention must be paid to the physical size and shape of the specimens to be counted (tubing, film etc.) and whether these are compatible with the limitations of the instrument (does the solid sample fit in the counting vial?). It is of course extremely important that the experimental specimens and calibration solutions be counted in exactly the same geometry so that the counting rates are valid for comparison in computing adsorption.

It should be noted that the use of tubing geometry as described allows adsorption to be carried out under static and well-defined flow conditions as required. It is a simple matter to place the filling syringe in a suitable syringe pump so that experiments can be done under controlled, steady flow over a wide range of flowrates.

2.3 Safety aspects.

Handling radioisotopes presents potential hazards that exist at two distinct levels, namely radiation dose rate in the lab environment, and direct physical contamination both internal and external. Radioiodine is particularly hazardous and toxic because: (a) it is volatile, (b) it can be absorbed through the skin by contact with solutions, (c) 30% of ingested iodine is concentrated in the thyroid, a relatively small organ with a mass of about 20 g, so that local radiation doses are high. The overall half-life in the body is determined by the decay half-life and the metabolic half-life. These are respectively 60 days and 120 days for ^{125}I, and 8 days and 120 days for ^{131}I, leading to overall half-lives of 40 days and 7.5 days respectively. Thus of these two commonly used isotopes, ^{125}I is the more persistent. However, due to the lower energy of ^{125}I (35 keV versus 364 keV for ^{131}I) the toxic effects of each isotope may be considered approximately equal.

In view of their hazardous nature, special precautions must be taken when using radioiodines. A detailed discussion is beyond the scope of this article and the reader is referred to other sources (*Bolton, 1977*). The following discussion provides general guidelines that should be followed for the safe handling of radioiodine. It is strongly suggested that intending users consult local experts on health physics for more detailed advice.

First, the area designated for handling and storing radioiodine, as well as the labelled materials, must be restricted, and, if possible, isolated from the rest of the laboratory. Radioiodine sources should be stored in well-closed containers and shielded by lead of about 5 cm thickness. The radiation dose rate and contamination level in the nearby environment should be monitored regularly. For example contamination of benches and other surfaces can be assessed by taking swabs of charcoal-impregnated filter paper for counting.

Secondly, because of the volatility of iodine, any procedures using unbound radioiodine must be carried out in a ventilated hood and not on the open bench. The operator should wear an additional lab coat and a pair of polyethylene gloves over a pair of surgical gloves. Transfer pipettes, spill trays and disposable absorbent bench coverings should be used to confine contamination. All vessels in contact with radioactive solutions should be clearly labelled and decontaminated after use with an appropriate decontamination solution. The waste should be isolated in sealed and labelled containers, and disposed of with the help of a specialized service.

Finally, persons frequently handling radioiodine should undergo periodic physical examination. A thyroid count is particularly useful to check that the level of body contamination is within acceptable limits.

RESULTS

A variety of investigations of protein adsorption relevant to blood compatibility can be performed using radioiodinated proteins. Some typical results of such investigations carried out in the authors' laboratory in recent years are now described.

3.1 Isotherms and kinetics of adsorption in single protein systems.

The radioiodination technique is appropriate for the accurate determination of protein adsorption isotherms (*Brash et al., 1974; Chan & Brash, 1981*). In general, an isotherm is determined by exposing a set of identical surface samples to a range of protein solutions of different concentration, until adsorption equilibrium is reached. *Fig. 4* shows the isotherm for fibrinogen adsorption on glass from 0.05 M Tris buffer, pH 7.35. The isotherm displays a well-developed plateau at a rather low bulk protein concentration (about 0.2 mg/mL), indicating a high affinity of fibrinogen for glass surface. The plateau value of 0.7 $\mu g/cm^2$ reflects the limiting capacity of the surface for protein under the experimental conditions and is in the close packed monolayer range. The same general shape has been found for the adsorption isotherms of many other protein-surface systems (*Brash & Uniyal, 1979; Schmitt et al., 1983*). Because of its high sensitivity, the radioiodination technique is suitable for determination of adsorption isotherms at low bulk concentration, say

< 0.01 mg/mL, which is not readily accessible with most other techniques (*Brash et al., 1984*).

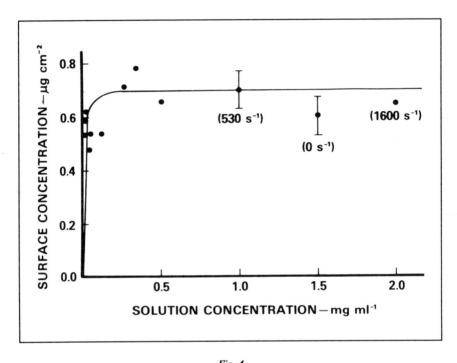

Fig. 4

Adsorption isotherm of fibrinogen on glass from 0.05 M Tris buffer, pH 7.35. Adsorption time, 3 h; temperature 23°C. Data obtained under laminar flow conditions, shear rate 1060 s⁻¹ unless otherwise noted. (With permission from Chan & Brash, 1981).

The information provided in principle by an adsorption isotherm is (a) the equilibrium constant for adsorption, and hence the Gibbs free energy of adsorption, and (b) the adsorption capacity of the surface (isotherm plateau) with, possibly, indications of the layer structure. With respect to the latter one can readily compare the plateau surface concentration with that expected for a close-packed monolayer based on known protein dimensions. For fibrinogen the expected values range from 0.2 to 1.8 $\mu g/cm^2$ depending on protein-surface orientation. In most cases of adsorption from single protein solutions monolayers or partial monolayers have been found (*Brash & Uniyal, 1979; Schmitt et al., 1983*). Cases of multilayer formation are relatively rare.

The euqilibrium constant is available from the initial, increasing part of the isotherm and examples of its determination under different assumptions can be found in the literature (*Schmitt et al., 1983; Boisson et al., 1988*). If data are available at different temperatures the heats and entropies of adsorption can be estimated by application of standard thermodynamic relations (*Boisson et al., 1988*). These interpretations imply a view of the isotherm as an equilibrium relation thus justifying the use of equilibrium thermodynamics. In turn this requires that the adsorption be reversible. However, reversibility of protein adsorption is questionable since operationally one usually cannot "descend" the isotherm on reducing the concentration (*Chan & Brash, 1981*). More will be said about this point below, but in the absence of thermodynamic rigour, isotherm slopes and their temperature dependence should probably be considered as qualitative, or at best relative indicators of binding affinity and heat of adsorption. Most values of equilibrium constants derived from isotherms correspond to free energies of a few kJ/mol (*Schmitt et al., 1983*) and suggest the predominance of physical effects such as hydrophobic interactions (*Boisson et al., 1988*).

Protein adsorption is governed by the transport of protein from the bulk to the interface followed by adsorption itself. It has generally been observed that the adsorption rate is initially controlled by transport, when surface coverage is low, then by both transport and adsorption and finally by adsorption. Much effort has been made to understand such surface-coverage-dependant adsorption kinetics, with special interest in the situation at low surface coverage (*Lok et al., 1983*). In this case, formidable experimental difficulties are encountered due to the fact that protein adsorption is in general rapid, and therefore a sensitive, in situ technique providing kinetic data with a resolution of 1 s or less, is required. In this regard, some success has been achieved using techniques such as total internal reflection fluorescence or TIRF (*Lok et al., 1983; Beissinger & Leonard, 1982*) and ellipsometry (*Cuypers et al., 1987*).

The radioiodination method used in our studies is not well adapted for investigating adsorption kinetics at short times (low surface coverage) because the procedure involves a time-consuming step for the separation of the surface from the solution before measuring, i.e. it is not an in situ technique. However, the method does give good kinetic data if lower time resolution (~30s) is sufficient. As shown in *Table III*, we have determined the kinetics of fibrinogen adsorption onto glass surface in the relatively high coverage regime. The data indicate that the adsorption is as much as 75% complete in a few minutes at a protein concentration of 1 mg/mL, and reaches an apparent equilibrium in about 1 hour. In addition, the rate appears unchanged over a range of shear rates (0 to 2100^{-1} s) under laminar flow conditions, thus confirming that under those experimental conditions the adsorption is not controlled by transport but by the surface reaction alone.

Table III

Adsorption of fibrinogen on glass at various shear rates[a,b].(With permission from Chan & Brash, 1981).

Shear rate (sec^{-1})	Contact time (min)							
	5	15	30	60	120	180	240	300
0	0.53	0.54	0.56	0.58	0.59	0.58	0.64	0.68
530	0.53	0.55	0.59	0.63	0.67	0.70	0.63	0.69
1060	0.47	0.54	0.53	0.66	0.69	0.69	0.73	0.73
1600	0.52	0.58	0.59	0.45	0.69	0.65	0.70	0.69
2100	0.53	0.52	0.59	0.57	0.63	0.74	0.75	0.72

[a] Adsorption values in $\mu g\ cm^{-2}$

[b] Fibrinogen concentration 1.0 mg mL^{-1} in 0.05 M Tris, pH 7.35

It should be re-emphasized that the technique used in our laboratory (*Section 2.2*), and practised, with variations, by other groups (*Horbett, 1984*) is a non in situ technique. The measurement of adsorption is not made "in place" but only after the surface and solution are separated. (Using radioiodinated proteins it is difficult to imagine an in situ method where adsorption is measured <u>directly</u> since the surface and solution radioactivity cannot be distinguished, and the latter is usually much greater than the former.) Separation involves rinsing the surface free of radioactive solution and this step introduces uncertainty since the loosely bound protein may be lost in varying degrees. There is thus a need to develop in situ methods, because first, rapid events occurring on the scale of 1 s cannot be followed if separations must be made and second, the separation disturbs the interaction at the solution-solid interface.

In this connection two methods using radiolabelled proteins have recently been described. The first (*Aptel et al., 1988*) uses iodinated proteins in conjunction with a "serum replacement" process, and may be described as a "kinetic solution depletion" method. It is therefore an indirect method since measurements are made on the solution, not the surface. Also the surface is in the form of particles since a high surface-to-volume ration is required as in all depletion methods.

The second method (*Baszkin et al., 1987*) involves ^{14}C-labelled proteins and is thus strictly speaking outside the scope of the present discussion, but the approach is worth mentioning. Solution radioactivity is subtracted from total radioactivity (solution plus surface) to obtain adsorption. Solution radioactivity is estimated by performing a separate experiment, identical to the adsorption, except that a different non-adsorbing ^{14}C-labelled solute of the same specific radioactivity is used. Such an approach could in principle be used with iodinated proteins. However, due to the much lower range of the soft ß-radiations from [14]C (compared to the γ-rays from ^{125}I or ^{131}I) the solution radioactivity emanates from a much thinner layer of fluid near the surface. The surface radioactivity measured using a ^{14}C label is therefore a much bigger fraction of the total than in the case of radioiodine and can be determined more accurately.

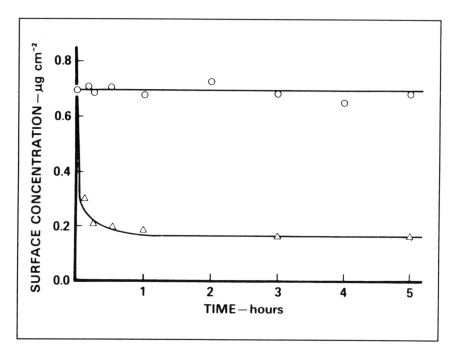

Fig. 5

Desorption of fibrinogen from glass at pH 7.35 under laminar flow conditions, shear rate 1060 s⁻¹. (O) 0.05 M Tris; (Δ) 1.0 M Tris. Adsorption was from 1.0 mg/ml fibrinogen in 0.05 M Tris, pH 7.35. (With permission from Chan & Brash, 1981).

3.2 Desorption and exchange.

Whether protein adsorption can be thought of as a thermodynamically reversible process continues to be controversial, as indicated above. We and other have discussed this problem elsewhere (*Brash, 1987; Norde et al., 1986*). Irreversibility is suggested by the fact that desorption into buffer is usually not detectable (*Lok et al., 1983; Chan & Brash, 1981; MacRitchie, 1972*) as illustrated for the glass-fibrinogen system in *Fig. 5*. However, we have observed that while desorption does not occur to any measurable extent, there may be rapid exchange of protein between solution and surface while the two remain in contact (*Brash & Samak, 1978; Chan & Brash, 1981; Brash et al., 1983*). These experiments suggest the existence of a dynamic equilibrium and thus of a reversible interaction. The use of iodine labelled proteins is particularly well adapted for this kind of study since labelled and unlabelled molecules, or molecules labelled with different isotopes can be used to follow exchange.

The surface is first allowed to equilibrate with, say, a ^{125}I-labelled protein solution so as to attain steady-state adsorption. The initial solution is then replaced with a ^{131}I-labelled protein solution of the same concentration. Exchange of the protein molecules between the solution and the surface is indicated by a simultaneous and equivalent increase of the second isotope and decrease of the first isotope. *Fig. 6* (*Chan & Brash, 1981*) shows exchange kinetics for the fibrinogen-glass system, and demonstrates that under the conditions of this experiment, about 50% of the bound fibrinogen molecules are exchangeable. From a number of such studies we have found that the rate and extent of such self-exchange vary markedly with the nature of the surface and factors such as protein concentration and flow (*Brash et al., 1983; Brash & Samak, 1978*). Other researchers have confirmed the occurrence of such exchange phenomena (*Lok et al., 1983; Chuang et al., 1978; Cheng et al., 1987*).

The occurrence of exchange in contrast to the lack of desorption into solvent is counter-intuitional since exchange perforce involves a desorption step. All the evidence, particularly concentration dependence, indicates a mechanism involving interactions between protein in solution and adsorbed protein. A likely explanation, due to Jennissen, is based on the very plausible concept of "multivalent" binding between protein and surface (*Jennissen, 1981*). Exchange is envisaged as a cooperative effect whereby a single site of an adsorbed molecule "desorbs" while a site on a solution molecule "adsorbs". Such single site exchange represents the initial step of whole molecule exchange. The concept of multivalent binding also provides an explanation for lack of desorption since the breaking of several bonds simultaneously is an unlikely occurrence. According to this explanation irreversibility is only apparent and has kinetic origins.

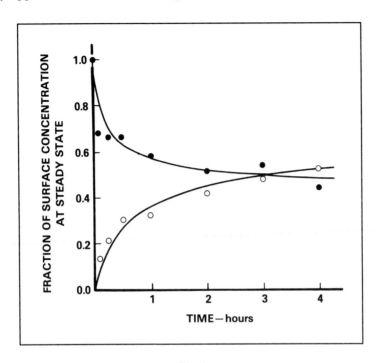

Fig. 6

Exchange kinetics of fibrinogen adsorbed on glass under laminar flow conditions (1060s⁻¹ wall shear rate). Fibrinogen concentration 2.0 mg mL⁻¹ in 0.05 M Tris, pH 7.35. (●) ¹²⁵I-labelled fibrinogen (used for initial adsorption); (O) ¹³¹-labelled fibrinogen (used for exchange). (With permission from Chan & Brash, 1981).

3.3 Competitive adsorption from mixtures of protein.

Most practical applications of protein adsorption deal with complex mixtures containing large numbers of proteins. Examples are milk, tear fluid, and saliva as well as blood which is relevant to this chapter. It is of considerable interest to know whether and how the proteins influence each other with respect to their adsorption behaviour. A major question is whether the adsorbed amount of a given protein is in proportion to its solution concentration or whether the surface "selects" some proteins in preference to others. Do the initially adsorbed proteins prevent others, arriving later at the surface, from adsorbing or can the initially adsorbed proteins be displaced? If displacement does occur, what is its rate and extent? In an attempt to answer some of these questions in the context of blood compatible materials, we have studied competitive adsorption, first from simple mixtures of fibrinogen, albumin and IgG, and, more recent, from plasma. The plasma studies, which are of special interest for blood compatibility, will be discussed in the next section.

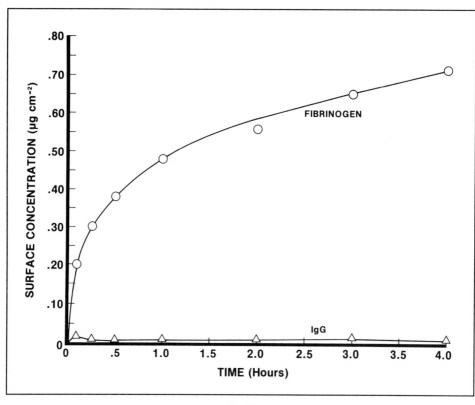

Fig. 7

Adsorption kinetics on glass from a mixture of fibrinogen, albumin and IgG with concentrations of 0.09, 1.20 and 0.36 mg mL^{-1} respectively in 0.05 M Tris, pH 7.35. Albumin was undetectable in this experiment. (With permission from Brash et al., 1984).

Radioiodination is an excellent method for detection of individual proteins adsorbing competitively from mixtures. It is capable of measuring the adsorption of two proteins simultaneously using different radioiodines. Using this methology, we have studied equilibrium competitive adsorption from the binary system albumin-fibrinogen on a variety of surfaces (*Brash & Davidson, 1976; Brash & Uniyal, 1979*). In these studies, it was found that fibrinogen is always preferentially adsorbed, in particular on glass or glass-like surfaces. Over a wide range of total and relative protein concentrations, fibrinogen is consistently enriched on the surface, but to different extents on different surfaces. Kinetic studies have also been carried out using fibrinogen-albumin-IgG mixtures having the same proportions as in plasma (*Brash et al., 1984*). *Fig. 7* shows typical data for a series of experiments in which pairs of proteins were labelled in turn. The strong preferential adsorption of fibrinogen over the whole time course from 1 min to 4h is again observed and a "deficiency" of fibrinogen in solution becomes an "excess" in the surface. Thus the conclusion from the above studies is that surfaces in general demonstrate an overwhelming preference for fibrinogen over albumin and IgG.

Several other groups have also observed preferential adsorption of fibrinogen from binary and ternary protein mixtures (*Lee et al., 1974; Horbett & Hoffman, 1975*). On the other hand, Horbett using radioiodination methods has shown that hemoglobin in either the methemoglobin or oxyhemoglobin form adsorbs preferentially to fibrinogen from binary mixtures onto polyethylene (*Horbett, 1984*).

3.4 Adsorption of proteins from plasma.

Although studies of protein adsorption from single and simple multicomponent systems have made important contributions to the understanding of blood-material interactions, only studies using plasma and blood can give information relating directly to blood compatibility. Data using simpler media are not always extrapolable to blood, a fluid infinitely more complex than a simple two- or three-component "simulant".

The use of radioiodinated proteins represent a very powerful method for measuring adsorption of individual proteins from plasma. A sample of the purified plasma protein to be studied is labelled with radioiodine and added to the plasma as a tracer of that protein. Generally, the amount of labelled protein added is less than 10% of the normal concentration of the protein in plasma so as to remain within the normal range of values. For "trace" proteins this fraction may have to be increased to be practicable, both from a handling and a measurement precision standpoint. The experiment described above (Section 2.2.) for tubing or some variant thereof to accomodate different solid samples geometries, is then performed. Adsorption can readily be studied as a function of time, plasma type (normal, normal with different anticoagulants, deficient in certain components), plasma dilution, and temperature.

Data obtained in our laboratory using this approach (*Uniyal & Brash, 1982; Brash & ten Hove, 1984; Wojciechowski et al., 1986; Brash et al., 1988*) have suggested two general conclusions. First, adsorption of the abundant proteins albumin, IgG, and particularly fibrinogen, is less from plasma than from comparable single protein solutions or from mixtures of the three, suggesting that less abundant components of plasma are significant components of the adsorbed layer. The amounts of these three proteins adsorbed, depend strongly on surface type. In general, adsorption appears to be greater on hydrophobic than on hydrophilic surfaces (although one cannot rule out the possibility that this is due to the fact that rinsing removes more adsorbed protein from hydrophilic surfaces). *Table IV* compares plasma and single protein data taken at long adsorption times for several surfaces.

Table IV

Adsorption of proteins from plasma and from 0.05 M Tris buffer (TB) [1], pH 7.4. Data in $\mu g\ cm^{-2}$. (With permission from Uniyal & Brash, 1982).

Surface	Albumin		Fibrinogen		IgG	
	TB	Plasma	TB	Plasma	TB	Plasma
PEG 1540 [2]	0.02	0	0.03	0	0.06	0.11
PEG 600 [2]	0.04	0	0.08	0	0.09	0.09
Polyethylene	0.17	0.14	1.06	transient	0.53	0.13
Siliconized glass	0.18	0.18	0.97	transient	0.55	0.08
Glass	0.20	0	0.70	0	0.35	0.13
Polystyrene	0.20	0.20	1.05	0.40	0.59	0.06
PPG 1200 [3]	0.57	0.10	1.09	0	0.61	0.12

[1] Protein concentrations, 1 mg mL^{-1}

[2] Hydrophilic segmented polyether polyurethane

[3] Hydrophobic segmented polyether polyurethane

The second conclusion from plasma experiments is that adsorption of fibrinogen is transient, i.e. it is rapidly adsorbed initially and then desorbed. Indeed in initial work we concluded that on certain surfaces, for example glass, there is no adsorption of fibrinogen from plasma. We later discovered that the adsorption transient on some surfaces occurs within one minute, and is therefore over before the first measurement is made. This transient adsorption of fibrinogen was first observed by Vroman using immunochemical methods (*Vroman & Adams, 1969*) and is referred to as the Vroman effect. Clearly such an effect could not have been predicted from experiments with simple mixtures of fibrinogen and other abundant proteins.

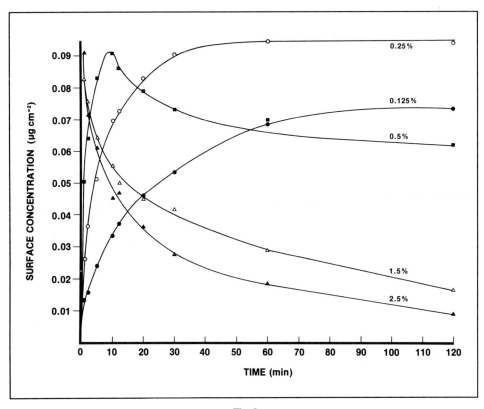

Fig. 8

Kinetics of adsorption of fibrinogen to glass from human plasma at different plasma dilutions. Plasma was diluted with isotonic Tris buffer, 7.35 and plasma "concentrations" resulting from dilution, expressed as percent normal, are shown on the curves. (With permission from Brash & ten Hove, 1984).

The Vroman effect is demonstrated for a glass surface in *Fig. 8* which shows fibrinogen adsorption kinetics for plasmas diluted to varying extents. As indicated above, when this experiment is done with undiluted plasma adsorption is zero at times longer than 1 min. As the plasma is diluted the surface residence time of adsorbed fibrinogen increases and at sufficiently high dilution becomes infinite (*Brash & ten Hove, 1984*). These data are consistent with a mechanism in which initially adsorbed protein is displaced by other components of plasma. It seems

314

likely that the displacing components will be of low concentration since low concentration components will initially be "under-represented" in the fluid boundary layer due to diffusion effects, thus favouring adsorption of more abundant components initially. At longer times the low concentration species will attain boundary layer concentrations that approach the bulk plasma values, and if their adsorption affinities are higher than those of fibrinogen, exchange should occur. The data of *Fig. 8* are consistent with such a mechanism.

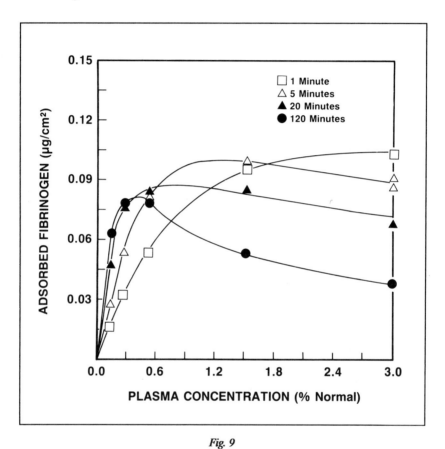

Fig. 9

Adsorption of fibrinogen from human plasma to polyethylene as a function of plasma dilution at different contact times. Plasma was diluted with isotonic Tris buffer pH 7.35. (With permission from Wojciechowski et al., 1986).

If adsorption is plotted as a function of plasma dilution at a fixed time, one should also obtain curves showing peaks as is obvious from *Fig. 8*. In a set of curves representing different times the peak moves toward the y-axis at longer times as shown in *Fig. 9* for polyethylene.

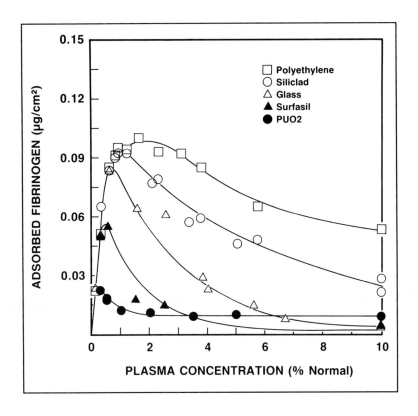

Fig. 10

Adsorption of fibrinogen from human plasma to various surfaces as a function of plasma dilution. Plasma was diluted with isotonic Tris buffer, pH 7.35 and adsorption times were 5 min. (With permission from Wojciechowski et al., 1986).

We have found that data taken at a fixed time of 5 min and plasma dilutions from 10:1 to about 500:1 provide a convenient measure of the Vroman effect on different surfaces. Typical data for five surfaces are shown in *Fig. 10*. It appears that fibrinogen adsorption is greater and more persistent the more hydrophobic the surface, presumably a reflection of fibrinogen binding affinities. The initial portion of these curves is independent of the surface, suggesting that the interaction in this region is diffusion-controlled. Application of the simple diffusion model:

$$\Gamma_F = 2 \left[\frac{Dt}{\pi} \right]^{1/2} c_F$$

to this slope indicates a diffusivity of about 10^{-7} cm^2/sec which is in reasonable agreement with literature values. (Here Γ_F is surface concentration, D is diffusivity, t is time and C_F is fibrinogen bulk concentration.)

Most surfaces so far examined show the Vroman effect in varying degrees. An exception recently found in our laboratory are polyurethanes containing an appropriate concentration of sulphonate groups. For these materials fibrinogen adsorption is high and no displacement occurs at high plasma concentration or long time. These data are as yet unpublished.

The identity of fibrinogen-displacing components is still under investigation in various laboratories. In agreement with *Vroman et al. (1980)*, we have found that HMWK, a key component of the contact phase of blood coagulation, is an important fibrinogen displacer (*Brash et al., 1988*). We also believe that factor XII is strongly implicated. It appears from these studies that the Vroman effect is related to intrinsic coagulation. From the data thus far obtained, one may speculate that surfaces on which fibrinogen is displaced rapidly by contact factors are relatively procoagulant. *Slack et al. (1987)* did not observe a particular effect of contact coagulation factors on fibrinogen adsorption and believe that displacement of fibrinogen is a composite effet involving the plasma proteins in general, and which should be observed in any complex competitive system. Breemhaar et al. have proposed that high density lipoprotein is important in fibrinogen displacement (*Breemhaar et al., 1984*).

Proteins other than fibrinogen have been found by various workers to undergo the Vroman effect. Thus *Grinnel & Phan (1983)* have observed a maximum in the adsorption of fibronectin as a function of dilution from plasma to tissue culture dishes. Also, *Vroman & Adams (1986)* have proposed that adsorption of proteins in general from plasma follows a sequence, with more abundant proteins adsorbing initially and being replaced in sequence by progressively less abundant proteins. In agreement with this we have recently observed a peak in the adsorption of IgG from plasma to glass as a function of dilution and the position of the IgG peak is at a lower plasma concentration than that for the fibrinogen peak (unpublished observations).

DISCUSSION

4.1 Interpretation of data.

The measurement of protein adsorption using radioiodinated proteins is a direct method and the interpretation of data is straightforward. With a knowledge of the radioactivity of a known quantity of protein from the same labelled preparation used for the adsorption experiment, one can unequivocally determine the quantity of protein adsorbed as described above. Of course the surface radioactivity count and solution count must be made in the same counting equipment with the same "geometry", i.e. size and shape of sample and positioning in relation to the detector. In addition, one must always bear in mind, as already stressed, the assumption inherent in the method that the labelled protein behaves as a tracer of the entire population of protein molecules of that type, both labelled and unlabelled. The investigator must be satisfied as to this assumption by performing a validation test such as described in this chapter. One does from time to time encounter systems that show preferential adsorption of either labelled or unlabelled protein. For example, in our hands iodinated hemoglobin showed preferential adsorption on several surfaces (unpublished observations). A number of critiques of the method have appeared in the literature (*Crandall et al., 1981; van der Scheer et al., 1978; Grant et al., 1977*) and these should be read by intending users of this methodology. However, any blanket condemnation of the method to the effect that labelled proteins are always altered should be treated with scepticism. Each individual system must be judged on its own merits, and in relation to adsorption, not some other aspect of protein function.

4.2 Limitations.

4.2.1 Sensitivity.

Since this method is highly sensitive it is perhaps inappropriate to consider sensitivity as a limitation. Indeed it is probably fair to say that of the methods currently available, radioiodine labelling is the most sensitive and most precise for measuring quantity of protein adsorbed. An indication of sensitivity may, however, be useful. Sensitivity depends on specific activity and with lactoperoxidase labelling and essentially carrier-free ^{125}I one can easily achieve radioactivities of about 300 cpm/ng protein at I/P ratios of about 0.1. With background counts of about 60 cpm one can therefore detect 0.5 ng with reasonable counting accuracy. If one assumes a typical measured surface of 10 cm^2 then surface concentrations of 0.05 ng/cm^2 are detectable. Monomolecular layers of protein contain of the order of 0.1

to 1 $\mu g/cm^2$ so that about 0.1% to 0.01% of a monolayer is detectable with such specific activities.

Somewhat more typically using ICl labelling, we would consider the limit of detectable surface concentration to be of the order of 1 ng/cm^2. Furthermore if all experimental error, not just counting error, is taken into account the precision of a routine adsorption measurement is likely to be in the range of ± 0.01 $\mu g/cm^2$.

4.2.2 Need for purified proteins.
It is clear that to study a particular protein whether in simple solutions or complex fluids like blood, one requires a sample of the purified protein. This requirement is particularly acute in the case of trace constituents of plasma such as the contact clotting factors. Such proteins are present in small concentrations (e.g. factor XII, ~30 $\mu g/mL$ plasma; factor XI, ~6 $\mu g/mL$) and often require multistep processes for their isolation with losses occurring at each step. In addition the process of purification may compromise the biological activity of sensitive proteins. It is undoubtedly for these reasons that the bulk of adsorption research has been done on the abundant proteins albumin, IgG and fibrinogen which are available commercially in reasonably pure form. Nonetheless it may well be that the adsorption of certain "trace" proteins (e.g. HMWK) is more important from a blood compatibility standpoint. We expect, and indeed we would hope, that research will expand in the direction of studying biologically relevant proteins such as the coagulation factors, the complement proteins, plasminogen etc.

4.2.3 Loss of label from iodinated protein.
This problem has been discussed above in section 2.1.5. The rate of loss of iodine varies with the method of labelling and with the conditions of storage (*Krohn et al., 1972*) but can be as high as 30% per day. Investigators should be aware of this phenomenon and should regularly check the loss of label in their iodinated preparations. The potential for deiodination probably sets a time limit of a few hours on experiments that can be performed at physiologic temperature.

4.2.4 Non in-situ nature of method.
As has been discussed previously, adsorption measurement using radioiodinated proteins as normally practised, is a non in-situ method wherein the protein solution or plasma must be separated from the surface, and the latter must be rinsed. Rinsing is necessarily arbitrary (even if a standard, constant procedure is used) and results in unknown losses of loosely bound protein. Also as previously indicated, the method is unsuitable for determining fast kinetics. A recent application of the

serum replacement process (*Aptel et al., 1988*) shows promise that fast kinetics may be feasible for particulate materials. However, this method is a variant of the solution depletion method and as such it is indirect.

4.2.5 Lack of information on biological/conformational status of adsorbed protein. Finally one must be aware that this method provides information only on quantity adsorbed and not on the other important aspect of protein adsorption, namely the conformational and biological status of the adsorbed protein. At best one can compare the plateau or limiting surface concentration measured to that expected for a close-packed monolayer. In the case of single proteins such a comparison can provide an indication of the area per molecule in the layer relative to the area per molecule based on the native protein's dimensions, if known (*Brash & Lyman, 1969*). This approach clearly is not valid for multiprotein systems like blood, where many other proteins are present in the layer. Methods of determining the biological activity of proteins in layers adsorbed from blood are not well developed. Specific antibody binding is perhaps the only method that provides any information on biological status but again this method has mostly been applied to single protein systems (*Anderson et al., 1987*).

Using a different experimental approach one can monitor any possible degradation of proteins adsorbed from blood using radioiodination. The labelled protein is added as a tracer to blood or plasma, the adsorption is carried out, the proteins are eluted and SDS-PAGE (SDS-polyacrylamide gel electrophoresis) is run to separate the proteins by molecular weight. The gel is stained with Coomassie Blue to reveal the entire pattern of proteins eluted. The labelled protein can be singled out by scanning the gel for radioactivity. The location of the radioactivity in relation to the known molecular weight of the protein then gives an indication of degradation if any. Using this approach we discovered that plasma-adsorbed fibrinogen remaining on glass after the Vroman "peak" is extensively lysed by plasmin (*Brash & Thibodeau, 1986*).

4.3 Other methods of measuring protein adsorption.

Several methods are available to measure adsorption although few, if any, are as well able as radioiodination to provide unequivocal quantitative information on adsorption of a given protein from blood. *Table V* provides a listing of the better known techniques, and indicates the type of information provided. Many of these methods are described in detail in an excellent recent book edited by Andrade (*Andrade, 1985*).

Table V

Experimental methods for the study of protein adsorption.

Method	Reference	Applied to Blood/Plasma?	Information		In situ?	Limitations
			Quantity Adsorbed	Other		
Ellipsometry	Cuypers et al., 1987	Yes	?	thickness, refractive index	Yes	- surface must be smooth and reflective; - not suitable for distinguishing among different proteins
Attenuated Total Internal Reflection Infrared Spectroscopy (ATIR-IR)	Chittur et al., 1987	Yes	Yes	IR Spectra; band shifts due	Yes	- distinction among proteins questionable; - possible confounding of adsorbed and bulk protein.
Visible Total Internal Reflection Fluorescence (TIRF)	Cheng et al., 1985	No (possible in principle)	Yes	fluorescence spectra, thickness	Yes	- requires attachment of fluorescent label to protein - confounding of bulk and adsorbed protein
Total Internal Reflection Intrinsic Fluorescence (TIRIF)	Hlady et al., 1985	No	Yes	fluorescence shifts, conformation	Yes	- confounding of bulk and adsorbed protein; - distinction among proteins not clear cut

Table V (continued)
Experimental methods for the study of protein adsorption.

| Method | Reference | Applied to Blood/Plasma? | Information | | | In situ? | Limitations |
			Quantity Adsorbed	Other			
Electron Spectroscopy for Chemical Analysis (ESCA)	Paynter & Ratner, 1985	No	Yes	thickness, spatial distribution	No	not applicable to mixtures
Specific Antibody binding to adsorbed proteins	Vroman et al., 1980 Giaever & Keese, 1987 Pitt et al., 1986 Anderson et al., 1987 Lindon et al., 1986 Breemhaar et al., 1984 Chuang, 1984	Yes	Yes	spatial distribution, orientation, conformation, biological activity	No	- non-specific adsorption of antibodies
Elution methods followed by SDS-PAGE	Brash & Thibodeau, 1986	Yes	No	detailed composition of adsorbed layer	No	- elution can be preferential and incomplete
Hydrodynamic capillary flow	de Baillou et al., 1984	No	No	layer thickness, conformation	Quasi in situ	- cannot distinguish among proteins
Reflectometry	Schaaf & Dejardin, 1988	No	?	thickness, refractive index	Yes	- cannot distinguish among different proteins

Although an extensive array of methods is available to study protein adsorption, no single technique is ideal in the sense of providing information on all relevnt aspects of protein-surface interaction. A number of complementary methods would have to be used to describe adsorption in a given system to the full extent presently possible. *Hlady et al. (1985)* have aptly summarized the attributes of an ideal method: quantitative; fast response; in-situ; conformationally sensitive; applicable to all surface geometries and types; applicable to competitive adsorption.

None of the methods in *Table V* has all these attributes. The spectroscopic methods based on attenuated total internal reflection (ATIR) are conformationally sensitive and provide in-situ data, but have mainly been applied to single protein or simple mixture systems. Their potential with respect to adsorption from blood remains to be exploited. Several technical difficulties must be overcome, in particular the problem of distinguishing one protein from another in blood, especially minor components. *Jakobsen & Wasacz (1987)* have undertaken a study of protein infrared spectra in aqueous solution which may be helpful for the ATIR-infrared method. Also *Chittur et al. (1987)* have described a mathematical calibration approach to this problem. However, the fact remains that the main features of protein infrared spectra do not vary much from one protein to another and one must rely on differences in weaker, secondary vibrations. Also in relation to these ATIR methods when practised in-situ, the beam "sees" part of the bulk solution protein as well as the adsorbed protein. A correction must therefore be applied which appears to be experimentally rather complex and subject to some uncertainty (*Chittur et al., 1986*). The bulk contribution to the total infrared signal may be as much as 60% in a blood contact situation (*Chittur et al., 1987*). The fluorescence methods TIRF and TIRIF have not yet been applied to adsorption from blood or plasma.

Ellipsometry is a rapid in-situ method yielding layer thickness and refractive index as primary data. These can be interpreted in terms of quantity adsorbed by the application of theoretical relations (*Cuypers et al., 1983*) but as applied to plasma or blood the most that could be expected is the total protein adsorbed or an uncalibrated signal proportional to the total protein (*Vroman & Adams, 1969*). The latter may be useful as a comparative measure of overall adsorption kinetics for a series of surfaces.

As methods that are currently well adapted to the measurement of adsorption of specified individual proteins from blood only antibody binding and radiolabelling methods should be seriously considered for the moment. Clearly it is to be hoped that the other techniques will develop in directions that make them suitable for blood or plasma studies.

Antibody binding methods are unique in giving indications of the biological status of an adsorbed protein. In order for specific binding to occur the binding site on the antigen (adsorbed protein) must be present at the surface, intact and available, implying in turn that the specific protein antigen is adsorbed, that it has not undergone significant denaturation, at least in the vicinity of the binding site, and that it is oriented on the surface so that the binding site is available. Nonspecific adsorption of antibodies to surfaces is always possible and must be prevented. This is usually done by blocking unoccupied surface sites with an inert protein such as albumin, as in protein immunoblotting (*Gershoni & Palade, 1983*).

Several methods have been used to detect antibody binding by adsorbed proteins. *Vroman & Adams (1969)* have used simple methods such as moisture adsorption (breath patterns) and staining methods. Related to these, Giaever et al. have developed the "Indium Slide" method whereby the transmission of light through an indium coated slide is altered by binding of antibody (*Giaever et al., 1984*). Other methods rely on radioiodine labelling of antibodies (*Lindon et al., 1986*) and enzyme conjugated second antibodies (*Breemhaar et al., 1984*) which can be detected by reaction with chromogenic substrates. These methods, particularly iodine labelling, can yield quantitative information of high precision. It should also be noted that ellipsometry can be used to detect the adsorption of an antibody to a previously deposited protein film (*Vroman & Adams, 1969*) and thus to detect adsorption of individual proteins from blood. Used in this way, rinsing after blood contact is required, and the in-situ nature of ellipsometry is thus lost.

REFERENCES

Anderson, A.B., Darst, S.A. & Robertson C.R. (1987) in "Proteins at Interfaces, Physicochemical and Biochemical Studies" (Brash, J.L. & Horbett, T.A. eds.) ACS Symposium Series, Vol. 343, pp. 306-323, American Chemical Society, Washington D.C.

Andrade, J.D., ed. (1985), Surface and Interfacial Aspects of Biomedical Polymers, Vol. 2, Protein Adsorption, Plenum Press, New York.

Aptel, J.D., Voegel, J.C. & Schmitt, A., (1988), Colloids Surfaces, 29, 359-371.
Ardaillou, N. & Larrieu, M.J. (1974), Thromb. Res., 5, 327-341.

Bale, W.F., Helmkamp, R.W., Davis, T.P., Izzo, M.J., Goodland, R.L., Contreras, M.A. & Spar, I.L. (1966) Proc.Soc.Exp.Biol.Med., 122, 407-414.

Baszkin, A., Deyme, M., Perez, E. & Proust, J.E. (1987) in "Proteins at Interfaces, Physicochemical and Biochemical Studies" (Brash, J.L. & Horbett, T.A. eds.) ACS Symposium Series, Vol. 343, pp. 451-467, American Chemical Society, Washington DC.

Beissinger, R.L. & Leonard, E.F. (1982), J. Colloid Interface Sci., 85, 521-533.

Boisson, C., Jozefowicz, J. & Brash, J.L. (1988), Biomaterials, 9, 47-52.

Bolton, A.E. & Hunter, W.M. (1973), Biochem. J., 133, 529-538.

Bolton, A.E. (1977), Radioiodination Techniques, Review 18, Radiochemical Center, Amersham, England.

Bouma, B.N. & Griffin, J.H. (1977) J. Biol. Chem., 252, 6432-6437.

Brash, J.L. (1987) in "Proteins at Interfaces, Physicochemical and Biochemical Studies" (Brash, J.L. & Horbett, T.A. eds.) ACS Symposium Series, Vol. 343, pp. 490-506, American Chemical Society, Washington DC.

Brash, J.L. & Davidson, V.J. (1976), Thromb. Res., 9, 249-259.

Brash, J.L. & Horbett, T.A., eds. (1987) Proteins at Interfaces, Physicochemical and Biochemical Studies, ACS Symposium Series, Vol. 343, American Chemical Society, Washington, D.C.

Brash, J.L. & Lyman, D.J. (1969), J. Biomed. Mater. Res., 3, 175-189.

Brash, J.L. & Samak, Q.M. (1978), J. Colloid Interface Sci., 65, 495-504.

Brash, J.L. & ten Hove, P. (1984), Thromb. Haemostas., 58, 326-330.

Brash, J.L. & ten Hove, P. (1988), J. Biomed. Mater. Res., in press.

Brash, J.L. & Thibodeau, J.A. (1986), J. Biomed. Mater. Res., 20, 1263-1275.

Brash, J.L. & Uniyal, S. (1979), J. Polymer Sci., C66, 377-389.

Brash, J.L., Uniyal, S. & Samak, Q. (1974), Trans. Amer. Soc. Artif. Organs, 20, 69-76.

Brash, J.L., Uniyal, S., Pusineri, C. & Schmitt, A. (1983), J. Colloid Interface Sci., 95, 28-36.

Brash, J.L., Uniyal, S., Chan, B.M.C. & Yu, A. (1984), in "Polymeric Materials and Artificial Organs" (Gebelein, C.G. ed.), ACS Symposium Series, Vol. 256, pp. 45-61, Americal Society, Washington, DC.

Brash, J.L., Scott, C.F., ten Hove, P., Wojciechowski, P.W. & Colman, R.W. (1988), Blood 71, 932-939.

Breemhaar, W., Brinkman, E., Ellens, D.J., Beugeling, T. & Bantjes, A. (1984), Biomaterials 5, 269-274.

Cartwright, T. & Kekwick, R.C.O. (1971), Biochim. Biophys. Acta 236, 550-562.

Chan, B.M.C. & Brash, J.L. (1981), J. Colloid Interface Sci. 82, 217-225.

Cheng, Y.-L., Lok, B.K. & Robertson, C.R. (1985) in "Surface and Interfacial Aspects of Biomedical Polymers", (Andrade, J.D. ed.) Vol. 2, Protein Adsorption, pp. 121-160, Plenum Press, New York and London.

Cheng, Y.-L., Darst, S.A. & Robertson, C.R. (1987), J. Colloid Interface Sci., 118, 212-223

Chittur, K.K., Fink, D.-J., Leininger, R.I. & Hutson, T.B. (1986), J. Colloid Interface Sci., 111, 419-433.

Chittur, K.K., Fink, D.J., Hutson, T.B., Gendreau, R.M., Jakobsen, R.J. & Leininger, R.I. (1987) in "Proteins at Interfaces, Physicochemical and Biochemical Studies" (Brash, J.L. & Horbett, T.A. eds.) ACS Symposium Series, Vol.343, pp. 362-377, American Chemical Society, Washington, DC.

Chuang, H.Y.K. (1984) J. Biomed. Mater. Res., 18, 547-559.

Chuang, H.Y.K., King, W.F. & Mason, R.G. (1978), J. Lab. Clin. Med., 92, 483-496.

Crandall, R.E., Janatova, J. & Andrade J.D. (1981), Prep Biochem. 11, 111-138.

Cuypers, P.A., Corsel, J., Janssen, M.P., Kop, J.M.M., Hermens, W.T. & Hemker, H.C. (1983), J. Biol. Chem., 258, 2426-2431.

Cuypers, P.A., Willems, G.M., Kop, J.M.M., Corsel, J.W., Janssen, M.P. & Hermens, W.T. (1987) in "Protein at Interfaces, Physicochemical and Biochemical Studies" (Brash, J.L & Horbett, T.A. eds.) ACS Symposium Series, Vol. 343, pp. 208-221, American Chemical Society, Washington, DC.

David, G.S. (1972), Biochem. Biophys. Res. Commun., 48, 464-471.

David, G.S. & Reisfeld, R.A. (1974), Biochemistry, 13, 1014-1021.

de Baillou, N., Dejardin, P., Schmitt, A. & Brash, J.L. (1984) J. Colloid. Interface Sci., 100, 167-174.

Diamond, P.S. & Denman, R.F. (1973), Laboratory Technique in Chemistry and Biochemistry, 2nd edn., Butterworth, London.

Fraker, P.J. & Speck, J.C., Jr. (1978), Biochem. Biophys. Res. Commun., 80, 849-857.

Gershoni, J.M. & Palade, G.E. (1983), Anal. Biochem., 131, 1-15.

Giaever, I. & Keese, C.R. (1987) in "Proteins at Interfaces, Physicochemical and Biochemical Studies" (Brash, J.L & Horbett, T.A. eds.) ACS Symposium Series, Vol. 343, pp. 582-602, American Chemical Society, Washington, D.C.

Giaever, I., Keese, C.R. & Rynes, R.I. (1984), Clin. Chem., 30, 880-883.

Grant, W.H., Smith, L.E. & Stromberg, R.R. (1977), J. Biomed Mater. Res. Symp., 8, 33-38.

Grinnell, F. & Phan, T.V. (1983), J. Cell Phys 116, 289-296.

Harwig, S.S.L., Harwig, J.F. & Coleman, R.E. (1975), Thromb. Res. 6, 375-386.

Hatton, M.W.C., Rollason, G. & Sefton, M.V. (1983) Thromb. Haemostas., 50, 873-877.

Hatton, M.W.C., Moar, S.L. & Richardson, M. (1988) Blood, 71, 1260-1267.

Heimburger, N., Heide, K., Haupt, H. & Schultze, H.E. (1964), Clin. Chim. Acta 10, 293-307.

Helmkamp, R.W., Goodland, R.L., Bale, W.F., Spar, I.L. & Matscher, L.E. (1960), Cancer Res., 20, 1495-1500.

Hlady, V., van Wagenen, R.A. & Andrade J.D. (1985) in "Surface and Interfacial Aspects of Biomedical Polymers" (Andrade, J.D. ed.) Vol. 2, Protein Adsorption, pp 81-119, Plenum Press, New York and London.

Horbett, T.A. (1982) in "Biomaterials" (Cooper, S.L. & Peppas, N.A. eds.), Adv. Chem. Ser. Vol. 199, pp. 233-244, American Chemical Society, Washington, DC.

Horbett, T.A. (1984), Thromb. Haemostas., 51, 174-181.

Horbett, T.A. (1987) in "Proteins at Interfaces, Physicochemical and Biochemical Stidues" (Brash, J.L. & Horbett, T.A. eds.), ACS Symposium Series, Vol 343, pp. 239-260, American Society, Washington, DC.

Horbett, T.A. & Counts, R.B. (1984) Thromb. Res., 36, 599-608.

Horbett, T.A. & Hoffman, A.S. (1975) in "Applied Chemistry at Protein Interfaces" (Baier, R.E. ed.) ACS Adv. Chem. Series, Vol. 145, pp. 230-254, American Chemical Society, Washington, DC.

Hughes, W.L. (1957), Ann. N.Y. Acad. Sci., 70, 3-18.

Hughes, W.L. & Straessle, R. (1950), J. Am. Chem. Soc., 72, 452-457.

Hunter, W.M. & Greenwood, F.C: (1962), Nature, 194, 495-496.

Hunter, W.M. & Ludwig, M.L. (1962) J. Am. Chem. Soc., 84, 3504-3941

Jakobsen, R.J. & Wasacz, F.M. (1987) in "Proteins at Interfaces, Physicochemical and Biochemical Studies" (Brash, J.L. & Horbett, T.A. eds.) ACS Symposium Series, Vol. 343, pp. 339-361, American Chemical Society, Washington DC.

Jennissen, H.P. (1981), Adv. Enzyme Reg., 19, 377-406.

Krohn, K.A., Sherman, L. & Welch, M. (1972), Biochim. Biophys. Acta 285, 404-413.

Krohn, K.A. & Welch, M.J. (1974) Int. J. Rad. Isotopes, 25, 315-323.

Law, S.K., Fearon, D.T. & Levine, R.P. (1979) J. Immunol., 122, 759-765.

Lee, R.G., Adamson, C.L. & Kim, S.W. (1974), Thromb. Res. 4, 485-490.

Lindon, J.N., McNamana, G., Kushner, L., Merrill, E.W. & Salzman, E.W. (1986) Blood, 68, 355-362.

Lok, B.K., Cheng, Y-L. & Robertson, C.R. (1983), J. Colloid. Interface Sci. 91, 104-116.

MacRitchie, F. (1972), J. Colloid Interface Sci., 38, 484-488.

Mandle, Jr., R. & Kaplan, A.P. (1977) J. Biol. Chem., 252, 6097-6104

Marchalonis, J.J. (1969), Biochem. J., 113, 299-305.

McConahey, P.J. & Dixon, F.J. (1966), Int. Arch. Allergy, 29, 185-189.

McFarlane, A.S. (1956), Biochem. J., 62, 135.

McFarlane, A.S. (1958), Nature, 182, 53.

McFarlane, A.S. (1963), J. Clin. Invest., 42, 346-361.

McFarlane, A.S. (1964) in "Mammalian Protein Metabolism" (Munro, H.N. & Allison, J.B. eds.) Vol.1, pp 297-341, Academic Press, New York.

Miles, D.W., Mogensen, C.W. & Gundersen, H.J. (1970) Scand.J.Clin.Lab.Invest., 26, 5-11.

Morrissey, B.W. (1977), Ann. N.Y. Acad. Sci., 288, 50-64.

Morrison, M., Bayse, G.S. & Webster, R.G. (1971) Immunochemistry, 8, 289-297.

Norde, W., MacRitchie, F., Nowicka, G. & Lyklema, J. (1986), J. Colloid Interface Sci., 112, 447-456.

Paynter, R.W. & Ratner, B.D. (1985) in "Surface and Interfacial Aspects of Biomedical Polymers" (Andrade, J.D. ed.) Vol. 2 Protein Adsorption, pp. 189-216, Plenum Press, New York and London.

Peters, T. Jr. (1970), Adv. Clin. Chem., 13, 37-111.

Pitt, W.G., Park, K. & Cooper, S.L. (1986) J. Colloid Interface Sci., 111, 343-362.

Rabiner, S.F., Goldfine, I.D., Hart, A., Summaria, L. & Robbins, K.C. (1969) J. Lab. Clin. Med., 74, 265-273.

Regoeczi, E. (1984), Iodine Labelled Plasma Proteins, Vol. 1, CRC Press, Boca Raton, Florida.

Revak, S.D., Cochrane, C.G., Johnson, A. & Hugli, T. (1974) J. Clin. Invest., 54, 619-627.

Rosa, U., Scassellati, G.A., Pennisi, F., Riccioni, N., Giagnoni, P & Giordani, R. (1964), Biochim. Biophys. Acta, 86, 519-526.

Schaaf, P. & Dejardin, P. (1988) Colloids Surface, 31, 89-103.

Schmitt, A., Varoqui, R., Uniyal, S., Brash, J.L. & Pusineri, C. (1983), J. Colloid Interface Sci., 92, 25-34.

Scott, P.J., Jones, C.T. & Simcock, J.P. (1966), Clin. Chim. Acta 13, 171-178.

Slack, S.M., Bohnert, J.L. & Horbett, T.A. (1987), Ann. N.Y. Acad. Sci., 516, 223-242.

Sonoda, S. & Schlamowitz, M. (1970) Immunochemistry, 7, 885-898.

Trundle, D. & Cunningham, L.W. (1969), Biochem., 8, 1919.

Uniyal, S. & Brash, J.L. (1982), Thromb. Haemostas., 47, 285-290.

van der Scheer, A., Feijen, J., Elhost, J.K., Krugers-Dagneaux, P.G. & Smolders, C.A. (1978), J. Colloid Interface Sci., 66, 136-145.

Vroman, L. & Adams, A.L. (1969), J. Biomed. Mater. Res., 3, 43-67.

Vroman, L., Adams, A.L., Fischer, G.C. & Munoz, P.C. (1980) Blood, 55, 156-159.

Vroman, L. & Adams, A.L. (1986), J. Colloid Interface Sci., 111, 391-402.

Wang, C.H., Willis, D.L. & Loveland, W.D. (1975), Radiotracer Methodology in the Biological, Environmental and Physical Sciences, Chapters 1-3, Prentice-Hall, Inc., Englewood Cliffs, New Jersey.

Wilson, N. & Greenhouse, V.Y. (1976), J. Chromatog. 118, 75-82.

Wojciechowski, P. ten Hove, P. & Brash, J.L. (1986) J. Colloid Interface Sci., 111, 455-465.

Chapter VI

TEST METHODS FOR THE ACTIVATION OF HAEMOSTASIS AND OTHER ACTIVATION SYSTEMS

Chapter ed.: J.-P. Cazenave

Centre Régional de Transfusion Sanguine, F - 67085 Strasbourg Cedex

INTRODUCTION

When flowing blood is exposed to artificial surfaces, a sequence of reactions is initiated: protein adsorption on the surface, platelet adhesion, secretion of biologically active products, aggregation of platelets, activation of the coagulation system and eventually activation of the fibrinolytic system to remove fibrin deposits of the mural thrombus formed.

In order to predict and evaluate the degree of thromboresistance of a biomaterial in vitro, ex vivo or in vivo, it is useful to develop methods to study analytically or globally the factors in blood (such as platelet, coagulation and fibrinolytic factors) and in the vessel wall (endothelial cells).

In this chapter we have selected useful methods to isolate and prepare platelet suspensions, to study their functions such as adhesion to surfaces, aggregation and release of α-granule contents (PF_4, ß-thromboglobulin). In addition to platelet interactions with surfaces, thrombin is a major determinant of thrombus formation. Thus thrombin generation can be followed in blood by measuring its effect on hydrolyzing specifically fibrinopeptide A from fibrinogen, by measuring its inactivation by formation of a complex with antithrombin III and by activation of protein C. Removal of fibrin thrombi by activation of the fibrinolytic system can be appreciated by measuring the components of the system. Finally, global methods, such as resonance thrombography can predict the degree of thromboresistance of a biomaterial. In order to improve the haemocompatibility of biomaterials, endothelial cell seeding appears to be a potentially useful technique. Coverage of surfaces with endothelial cells is now possible because culture techniques have progressed.

S. Dawids (ed.), Test Procedures for the Blood Compatibility of Biomaterials, 331.
© 1993 *Kluwer Academic Publishers. Printed in the Netherlands.*

Modified Resonance Thrombography

Author: W. Lemm

Dept. of Experimental Surgery, Rudolf-Virchow-Clinic, Location Charlottenburg,
D - 1000 Berlin 19

INTRODUCTION

Name of method:

Modified resonance thrombography (RTG).

Aim of method:

The modified RTG can be used in two variations as a screening test for in vitro thrombogenicity of biomaterials and in one additional variation as a screening test for the ex vivo thrombogenicity of biomaterials.

Biophysical background for the method:

The resonance thrombography (RTG) is the successor of the thrombelastography (TEG) providing an extended information on the dynamics of thrombus formation in vitro (*Hartert, 1981; Hiller, 1982*).

The modified RTG is an in vitro screening method to investigate the surface induced thrombus formation on biomaterials giving first and quick information on the degree of thrombogenicity of a synthetic surface. It is a reliable method to preselect new materials, to evaluate material modifications or surface treatment and to monitor the quality of new batches of established biomaterials (*Sprengel & Lemm, 1983*).

This RTG screening method was originally developed for clinical application to study severe coagulation disorders in human blood and to monitor an anti-coagulation therapy e.g. after the implantation of permanent vascular prosthetic

333

S. Dawids (ed.), Test Procedures for the Blood Compatibility of Biomaterials, 333–357.
© 1993 *Kluwer Academic Publishers. Printed in the Netherlands.*

devices. The RTG allows one to measure and record the different steps in the thrombus formation as shovn in *Fig. 1.*

The measurement gives information on:

- **the coagulation or reaction time (r)** which allows the diagnosis of plasmatic clotting disorders such as haemophilias and the disturbances caused by heparinisation,

- **the fibrin formation time (f)** and **the fibrin amplitude (F)** which both represent the fibrin polymerization. They are useful for the monitoring of consumption coagulopathies with an affected fibrin polymerization or of the influence of anticoagulants or of fibrinolytic drugs,

- **the platelet amplitude (P)** indicating the platelet activity which primarily depends on the number of thrombocytes able to aggregate. This parameter may indicate the functional severity of e.g. Glanzmann's thrombasthenia and of thrombocytopenias.

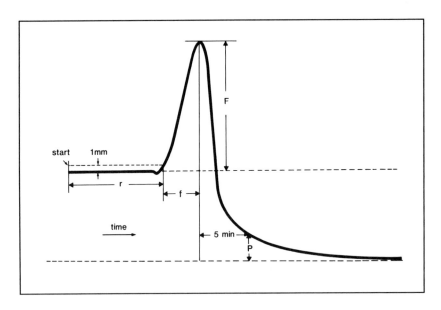

Fig. 1

Plot of a typical resonance thrombogram showing the characteristic changes during clotting.
r = clotting time (minutes), f = fibrin formation time (minutes), F = fibrin amplitude (mm),
P = platelet amplitude (mm) 5 minutes after maximum.

The RTG method simulates in vivo flow conditions: the applied flow of approximately 5 mm/sec induces shear-stresses of the blood comparable to the situation in the larger vessels in the human circulation system.

By simple modifications the RTG can be used as a screening method to test the thrombogenicity of various polymer surfaces (*Lemm & Sprengel, 1982*).

The essential part of the RTG is a pin which dips concentrically into a cylinder, both made of polished stainless steel (*Fig. 2*).

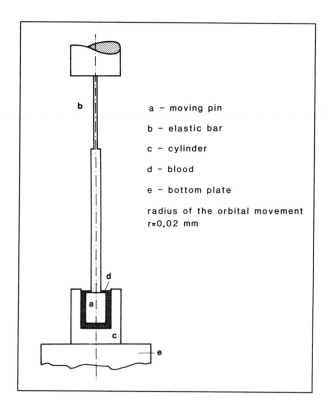

a – moving pin

b – elastic bar

c – cylinder

d – blood

e – bottom plate

radius of the orbital movement
r=0,02 mm

Fig. 2

Principle of a resonance thrombograph.

Recalcified human citrate blood is filled into the cylinder (c) immediately before each test cycle. The temperature of the cylinder and the test blood should be adjusted to 37°C. While the cylinder is fixed, the pin (a) moves along a circular pathway with a speed of 50 revolutions per second as shown in *Fig. 3* - similar to the movement of a tea spoon in a cup - driven by a system of excitation coils.

The diameter of the orbital movement is approximately 0.04 mm. The natural resonance frequency of the moving system (pin and elastic bar) is 40 Hz. This is 10 Hz below the excitation frequency.

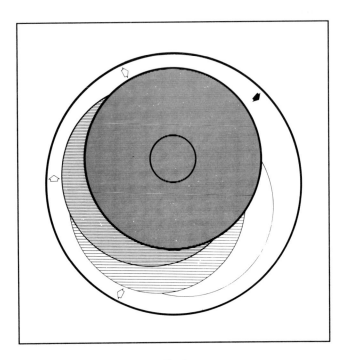

Fig. 3
Illustration of the movement of the pin in RTG.

When the blood starts to coagulate, fibrin fibrils are initially formed between the pin and the cylinder wall. The natural frequency of the system changes and approaches the excitation frequency. In the case of resonance ($f_n = f_e$) the orbital movement reaches its maximum. The thrombus gradually solidifies as the fibrin polymerization and the platelet aggregation progresses. The movement of the pin is increasingly hampered until it comes to a complete stop. The development of fibrin polymerization causes the first change in oscillation and is followed by the contribution of platelet aggregation which causes the resonance of the system to move across the excitation frequency. The typical RTG-plot in *Fig. 1* with the characteristic coagulation parameters illustrates these temporal changes. Although both sequenses of the coagulation system are mutually influenced and not strictly separated, there is a close inverse relationship between the fibrin amplitude (F) and the concentration of active fibrinogen in the test blood (*Fig. 4*).

The platelet amplitude value (P) is defined 5 min. after the curve has reached its maximum. This depends inversely on the number of active thrombocytes (*Fig. 5*).

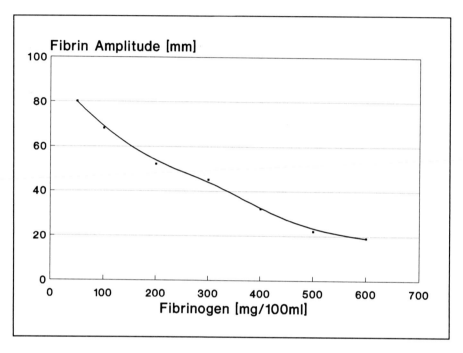

Fig. 4

The relationship between the fibrin amplitude and the fibrinogen concentration.

In order to test the thrombogenicity of various biomaterials, the RTG method is modified in three different ways:

1. **Direct in vitro** variation with polymer coatings on pins and cylinders.

2. **Indirect in vitro** variation for tubes, vascular prostheses and medical devices which will contact blood.

3. **Indirect ex vivo** variation for tubes and vascular prostheses.

Measuring the thrombogenicity with the direct in vitro variation of polymers which are soluble or are suitable for coating is possible with this method. Very careful preparation of the samples are necessary.

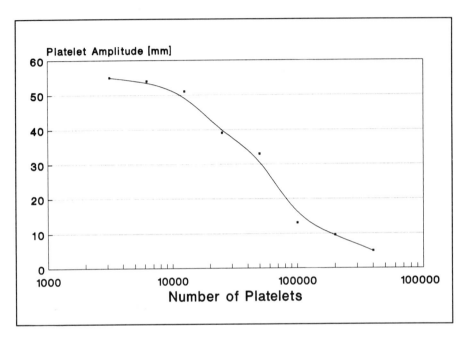

Fig. 5
The relationship between number of active platelets and the platelet amplitude P.

Measuring the thrombogenicity with the direct in vitro variation of polymers which are soluble or are suitable for coating is possible with this method. Very careful preparation of the samples are necessary.

Measuring the thrombogenicity in vitro of tubes and medical devices the first variation is suitable and recommended for screening, for preselection of new materials or material modifications and for routine quality control of batches of biomaterials. Tubes, oxygenators, dialysers, pulsatile or non-pulsatile blood pumps and similar devices can be tested with this method.

The method is also useful for adjusting and balancing e.g. the extrusion parameters during production of medical tubes for optimal surfaces.

Measuring the thrombogenicity ex vivo the advantages of in vivo and in vitro methods can be combined while at the same time avoiding the disadvantages. The principle of the bleeding test, first described by *Dudley et al. (1976)*, measures the response of native untreated blood when exposed to surfaces of tubes with small inner diameters (1.6 - 2.0 mm) (*Sprengel & Lemm, 1983*).

For the relative measurements in all 3 variations two identical RTG equipments are necessary. One for the test, the other for the refences. Normally the following values are sufficient:

Relative clotting time:	r/r_o
Relative fibrin amplitude:	F/F_o
Relative platelet amplitude:	P/P_o

Basically the clotting behaviour of healthy blood will change after passage through a medical device. Blood injuries mainly relate to the following factors:
- the chemical surface composition,
- the physical surface properties (roughness/smoothness),
- the flow conditions (shear stresses in the blood),
- the shape of the medical device proper which influence the local flow conditions (sharp edges, stagnation areas, eddy currents, vortices etc.).

The blood is affected in two ways:
1. Activation of the clotting system mainly by blood-material interaction due to:
    - adsorption of clotting factors and change of their conformation exposing activation centers within the protein molecule,
    - activation of factor XII (Hagemann factor) followed by a stepwise initiation of the coagulation cascade,
    - adhesion of platelets followed by secretion of clotting factors III and IV.

2. Inhibition of the clotting system due to:
    - adsorption of proteins and platelets leading to consumption coagulopathy,
    - damage of blood proteins and platelets rendering them inactive.

All these effects can be discerned by the RTG test.

Scope of method.

The RTG test and modifications of this can be used as general screening methods for thrombogenicity. Two identical sets of equipment are necessary to measure relative changes. The absolute values vary depending on uncontrollable parameters. The direct RTG is used in clinical routine for measurement of clotting time e.g. during extracorporeal circulation.

The method can discern between the main differences caused by changes in clotting behavior as described above.

DETAILED DESCRIPTION OF METHODS

Equipment needed.

For all 3 modifications, the following is needed:

1. Two identical sets of RTG-equipment with recorders are required (Hoerner & Sulger Co., Schwetzingen, FRG).
2. Sodium-citrate solution: 3.13 g Na-Citrate x 2 H_2O (p.a.) in 100 ml distilled water.
3. Recalcification solution: 2.50 g $CaCl_2$ x 2 H_2O (p.a.) in 1000 ml physiological NaCl-solution.
4. Ordinary plastic vials in the sizes 2 and 10 ml.

Both solutions are produced by companies and pharmacies in most countries.

Equipment especially for analysis of tubing:

1. Vertical rotation cylinder, diameter 30 cm (circumference: 94,25 cm, rotation speed: 16 - 24 rotations/min) mounted on a constant speed motor.
2. Extruded tubes for testing, length: 980 mm. Preferable inner diameter: 4.0 mm (volume = 12,32 ml, inner surface: 123,2 cm^2).

Outline of method.

A general description is given below.

From the freshly drawn citrated blood 1 ml is recalcified with 1 ml of isotonic calciumchloride solution. Immediately after recalcifiction of this mixture it is transferred to the cylinder (e.g. coated with polymer) and in the RTG equipment. The reference blood from the same batch is treated in exactly the same way. The 2 RTG equipments will automatically produce a graph. From the produced graphs (like the one given in *Fig. 1*) readings are taken for the clotting time (**r**), the fibrin formation time (**f**), the fibrin amplitude (**F**) and the platelet amplitude (**P**). The values of test and reference are then compared, and the relative changes are noted and used for the evaluation.

Description of procedure.

Prior to testing, the pins and cylinders should be carefully cleaned, washed in distilled water and air-dried (dust free).

Modification 1: Coating of pin and cylinder (*Lemm & Bücherl, 1986*).
Equipment and test procedure follow closely the general routine of RTG. For coating of the cylinders and pins, a general description cannot be given as this essentially depends on the test material itself. The following advise should be noted:

Prepare the solution of the test material under dust-free conditions using a suitable solvent. The concentration of the polymer may range between 2 and 10%, depending on the viscosity which should be 700 - 1000 mPa x sec. With this viscosity a perfect and homogeneous coating of steel pins and cylinders can be obtained. This should only be done on the parts from one RTG equipment.

The clean pins and cylinders are dipped into the polymer solution and dried immediately during slow horizontal rotation (2-6 rotation/min) in a dust-free atmosphere at temperatures between 30 - 50°C and at low humidity (risk of condensation). The temperature to dry the coatings should, however, not exceed 50% of the boiling temperture of the solvent (in C°) and the relative air humidity should not exceed 20 - 30%. Evaporation will take several hours (leave overnight).

For the preparation of the coating rapid evaporation of the solvent should be prevented as it creates surface imperfections such as "orange skin". If the solvent has a low boiling point and a low heat of evaporation, it is recommended to add the particular solvent to the atmosphere in the drying chamber to hamper the evaporation.

The coatings must not be touched at any point.

Modification 2: Preparation of tubes, vascular prostheses etc.
The relation of the blood volume to the surface area is of importance for evaluating the test results. A large surface area will affect a small volume of blood more (*Pelzer & Heimburger, 1986*).

For tubing this relation can be defined as

$$S/V_B = 4/d \quad (cm^{-1})$$

$$
\begin{aligned}
S &= \text{inner surface area of the tube } (cm^2) \\
V_B &= \text{volume of the inserted test blood } (cm^3) \\
d &= \text{inner diameter of the tube } (cm)
\end{aligned}
$$

For practical purposes S/V_B should be approximately 25 where V_B is the blood volume in the tube.

For this test 9 ml of freshly drawn human blood is mixed immediately with 1.0 ml of citrate-solution, making 10 ml mixture.

Half of this volume (5ml) is filled into the tubing which is being tested and the ends are immediately joined forming a loop. The ends of the tube are connected with a small piece of a silicone tubing (length 2 - 3 cm), swollen in ether and shrinked tightly around the tube keeping the ends safely aligned and fixed. The loop is then positioned on the vertical circular cylinder and rotated for 30 min. at a constant speed of 16 - 24 min^{-1} (*Fig. 6*).

Meanwhile the two RTG-equipments are prepared, pins and cylinders are mounted and warmed up to 37°C.

Thirty minutes later the test tubing is opened, and the blood is carefully poured into a small test vial. This blood is now considered as the test blood.

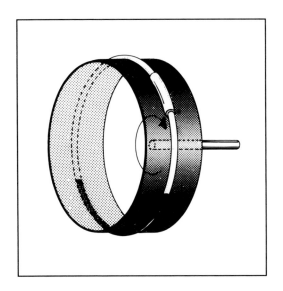

Fig. 6

Diagram of vertical rotation cylinder for test of tubing.

The 5.0 ml blood which remains in the syringe is used as the reference blood. It should be kept under the same conditions as the test blood.

The reference blood and the test blood is now measured simultaneously after recalcification in the two RTG equipments. The results are compared forming relative clotting parameters. 10.0 ml blood will be enough for at least six single RTG test cycles.

Modification 3: Preparation of animal for ex vivo test.
A goat should be preferred as blood donor in this test because its blood is known to have a very active coagulation behaviour. At the same time it is a very calm and sturdy animal to handle. The expected blood loss amounts to approximately 100 ml per test which, in practice, precludes human blood donors.

A venous cannula (Braunule® 1-G16) is inserted, filled with saline containing heparin 100 IE/ml and carefully positioned in the vena jugularis approximately two hours before the test will start. Plastic vials (polystyrene, NUNC, Roskilde, Denmark) must be prepared beforehand. The planned total volume (1 ml or 10 ml) should be clearly marked on the outside of the vials which then are filled with

10% of the intended volume of the blood mixture (sodium-citrate solution 0.1 or 1.0 ml).

The tubing to be tested is quickly and carefully connected to the venous cannula (see *Fig. 7*). The blood, which streams through the tubing being tested, is collected in a fractionated way in small plastic vials prefilled with Na-citrate solution.

The plastic vials should be calibrated to allow a precise collection of the expected volume of blood (blood/citrate solution: 9/1). During the test, the blood is fractionated into the array of prepared plastic vials, avoiding any interruption or disturbance of the blood flow.

The reference samples are stored and measured after a period identical to that of the test samples.

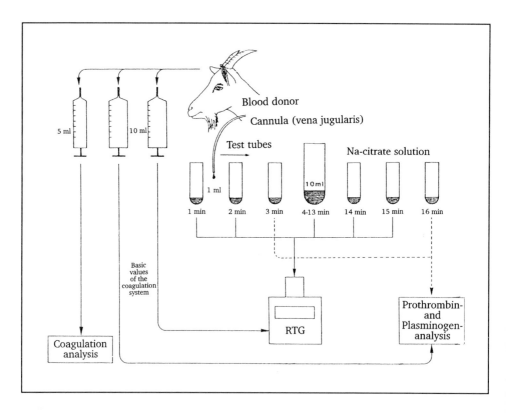

Fig. 7

Diagram of the ex vivo test procedure (modified Dudley test).

The performance of the ex vivo part of the experiment requires a quick and smooth cooperation of three persons: one who takes care of the animal, a second who quickly introduces the test tubings into the cannula, and a third who fractionates the outcoming blood in the previously prepared plastic vials. Normally the blood flow is 10 - 14 ml per min., and one test is terminated already after about 1 minute. Shifting to the next test tubing should follow immediately without bothering the animal.

Each tubing material should be tested at least four times. The blood loss of the animal should not exceed 200 ml. After each test cycle the animal has to be allowed to recover for 2-3 days to normalise the coagulation system, if the animal is intended to be used for several experiments.

Now the collected citrate-blood samples are tested after recalcification in exactly the same way as earlier described.

After the preparation of the two RTG equipments, one for the test cycles, the other for the references, 0.5 ml citrated blood (reference-blood or test-blood) is recalcified with 0.5 ml of the isotonic calcium-chloride-solution. Immediately after recalcification, 0.3 ml of this mixture is filled into the reference cylinder for the test blood, and the test cycle is started.

Only one animal is normally needed as blood donor for all the experiments. No anaesthesia or tranquilyzer is necesarry, if the animal is calm. This is normally the case, as the test is not painful or unpleasant.

The first 3 ml of blood are placed in three test tubes (1 ml in each). The next 10 ml (4-13 ml) which are bled out are collected in a single vial. The last 3 ml (14-16 ml) are collected in three separate vials, each with 1 ml of blood (see *Fig. 7*). The first two milliliters, the 4th - 13th ml (the bulk) and the 14th and 15th are subsequently recalcified and tested by the RTG method for their altered coagulation parameters after exposure to the foreign surface. The 3rd and 16th ml are analysed for changes in concentration of prothrombin and plasminogen as indicators for the ability of coagulation and fibrinolysis. A relative consumption indicates an activation of these two systems. Reference blood is taken directly from the jugular vein and mixed in the syringe. Three syringes are filled: one syringe (5ml) is used as reference measurement of coagulation and fibrinolysis. The second syringe (10 ml) is used to analyse reference levels of, inter alia, prothrombin and plasminogen. The third syringe (10 ml) is used as reference material in the RTG analysis. All the test results are thus relative to the reference blood which has not been in contact with the tube surface but only with the venous cannular.

For tubes with diameters of 1,6 - 2,0 mm the blood flow should be between 10 and 14 ml/min.

Safety aspects.

No special hazards exist. Careful cleanliness especially precleaning of the equipment is necessary to avoid erratic variations in the test measurement. Preferably work in a dust-free environment.

RESULTS

Modification 1

For a good statistical evaluation of each surface it is recommended to carry out each cycle 12 times. In general the reproducibility of the results is excellent: the standard deviation is between 5 and 10%.

Table I

Relative coagulation parameters of some biomaterials.

Material/polymer type	r/r_o	F/F_o	P/P_o
Biomer (poly-ether-urea-urethane)	1.79	0.79	0.84
Kraton 2104 (synth. rubber)	1.88	1.30	1.04
3140 RTV (PDMS)	1.85	1.12	0.89
Pellethane 2363-80AE (poly-ether-urethane)	2.05	1.48	1.12
Avcothane 51 (poly-ether-urethane-PDMS-block-copolymer)	2.83	1.64	1.23
Plathuran UM 8300 (poly-ester-urethane)	1.86	1.03	1.00
PU-Si 3 (PU-PDMS-Mixture)	1.89	1.47	0.87

Reference surface: Stainless steel

The results of tests on a selection of various polymers tested by the described method is given in *Table I*. Each result is a mean value of 12 single tests.

The following *Fig. 8* illustrates these results. High values point to surfaces with good thromboresistant properties.

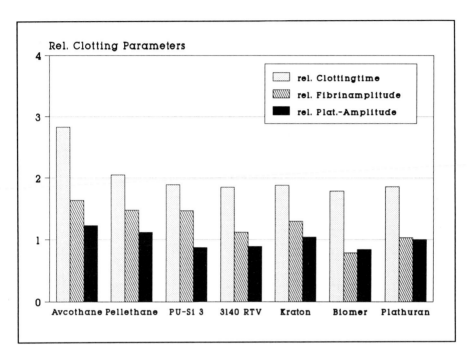

Fig. 8

Relative clotting parameters of different biomaterials.

Modification 2

Table II and *Fig. 11* compiles the results of a selection of test tubes with various chemical compositions or surface modifications, but with identical dimensions of tubing.

Table II

Relative clotting parameters.

Thrombogenecity of different tube materials in the dimensions: Length: 980 mm, Inner diameter: 4.0 mm, Volume of Tube: 12.32 ml, Inner surface area: 123.2 cm²

Material	r/r_o	F/F_o	P/P_o
Silastic	0.66	0.89	3.41
PVC soft	1.16	1.08	4.56
PVC hard	0.99	1.05	2.55
PVC-no-DOP	0.81	1.06	1.30
Synthetic caoutchouc (hard)	1.10	1.04	7.82
Polyurethane (Pellethane 2363-80AE)	0.98	1.09	1.09

Reference: Blood without foreign surface contact.

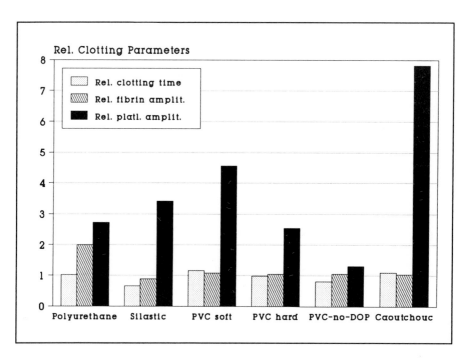

Fig. 9

In vitro thrombogenecity of different tubing materials.

Modification 3

In *Table III* the results for commonly used materials are given using different length of tubes.

The results are always relative coagulation parameters related to the reference blood. Values higher than 1.0 will indicate an inhibition of coagulation and values lower than 1.0 indicate an activation of the coagulation system. A material with no or only a slight deviation from 1.0 will prove to have the best thromboresistant properties.

The concentrations of prothrombin and plasminogen which are analysed from the 3rd and 16th ml do not normally show significant differences in comparison to the reference blood. The results do not correspond to any surface variation of the test tubes.

Table III

Test materials

Material	Inner Diameter	Length
PU (poly-ether-urethane) smooth surface	1.6 mm	100 cm
		200 cm
PU (poly-ether-urethane) rough surface	1.6 mm	100 cm
PP (poly-propylene)	1.8 mm	100 cm
		200 cm
PVC (poly-vinylchloride)	2.0 mm	100 cm
Silastic® (poly-dimethyl siloxane)	1.6 mm	100 cm

Typical relative coagulation parameters obtained by the RTG method are compiled in *Figs. 10-12*. They represent the response of the blood to the different material surfaces in the 1st ml, in the bulk and in the 15th ml.

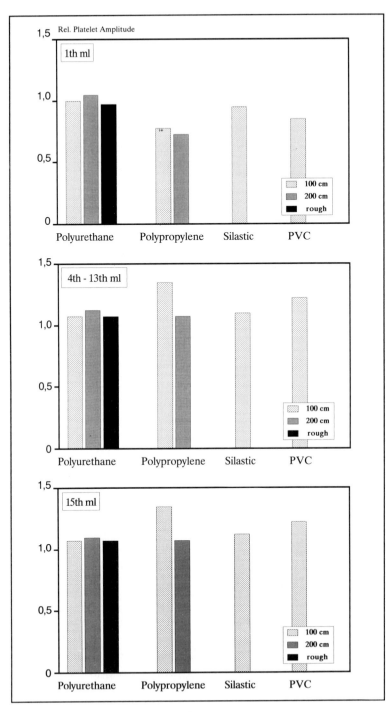

Fig. 10

The relative clotting times of the 1st, the 4th - 13th and 15th ml.

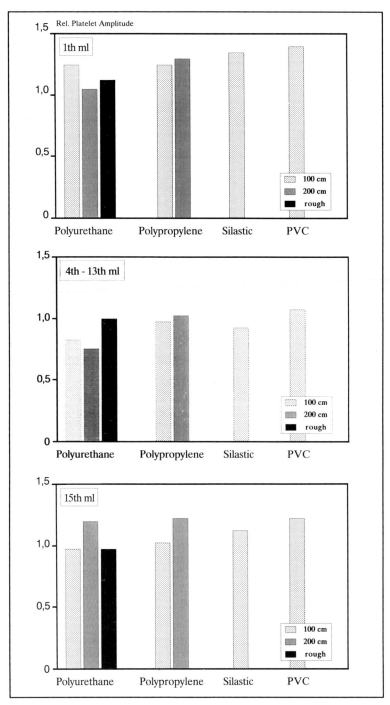

Fig. 11

The relative fibrin amplitudes of the 1st, the 4th - 13th and 15th ml.

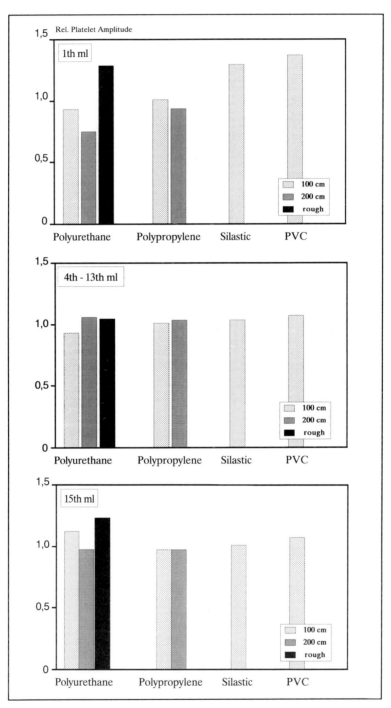

Fig. 12

The relative platelet amplitudes of the 1st, the 4th - 13th and 15th ml.

The polyurethane tubes of any modification do not influence the coagulation times in the first milliliters compared to the other materials (*Fig. 10*). The blood passes the tubes almost unaffected. In contrast, the polypropylene surface activates the coagulation system most, followed by the surfaces of PVC and Silastic®.

The relative fibrin amplitudes (*Fig. 11*) are remarkably increased for PVC, Silastic® and polypropylene in the first milliliters, being an indication for an adsorption or consumption of fibrinogen. These three materials are well known for their affinity to fibrinogen because of their hydrophobic surfaces. In the last milliliters the relative fibrin amplitudes approach the level of the reference blood. The surface is saturated with fibrinogen. No additional fibrinogen is adsorbed from the blood.

The thrombocytes are injured or adsorbed preferably by the rough polyurethane surface, followed by PVC and Silastic® which is indicated by high relative platelet amplitudes (*Fig. 12*). After about 30 sec., however, the impact of the different material surfaces on the native blood is nearly equalized.

DISCUSSION

Interpretation of results.

Modification 1.

The generally well-established Biomer®, a biomaterial for various kinds of application, activates the coagulation system in some cases more than the reference surface: stainless steel. This surprising result was confirmed by in vivo experiments: implanted blood pumps made of this Biomer® showed numerous thrombotic depositions. It could be proved by SEM analysis that these inferior results are closely related to inhomogeneities, mainly micro-bubbles (which were found in that particular Biomer® batch. Another Biomer® batch of obviously superior quality improved the relative coagulation time from 1.79 to 2.20.

To avoid erroneous results or misinterpretations the following is recommended:

1. For the preparation of the coating one should be aware that a too fast evaporation of the solvent creates surface imperfections such as "orange skin" (motted surface). In cases where a solvent with a low boiling point

and a low heat of evaporation is unavoidable, it is recommended to add solvent to the atmosphere in the drying chamber. As an example: In spite of the solubility of PVC-compositions in tetrahydrofuran these materials cannot be tested by this RTG-variation, because polymer and plasticizer (up to 40%) tend to separate when the coating dries.

2. As already mentioned, the surfaces are tested under dynamic conditions because of the blood flow. Consequently, surface defects and in-homogeneities and their influence on the thrombogenicity of the test material are recognized. For this reason a perfect quality of the coatings is of immense importance to avoid any misinterpretation of the results.

3. Other reasons for misinterpretations or even erroneous results are obtained by intrapping of air bubbles, when the blood or the pin is not properly inserted into the cylinder.

Modification 2.

The different test tubes do not differ much in the relative clotting time and in the relative fibrin amplitude. Material variations are significantly evident in their interaction with thrombocytes (P/P_o). High relative platelet amplitudes point to a desactivation or damage of thrombocytes. Concequently they have lost their aggregability. The reasons may be one or more of the following:

- The soft PVC releases parts of its plasticizer, di-(ethyl-2hexyl)-phthalat (DEHP), a chemical compound with surface-active properties. The hard PVC releases minor parts of plasticizers because of its lower content. The platelets are less influenced, the relative platelet amplitude - as can be seen in *Table II* or *Fig. 10* - being 4.56 in the first case and 2.55 in the second.

- No-DOP-PVC improves this value to 1.30, because this material is plasticized with tri-(ethyl-2hexyl)-trimellitat (TOTM), a compound with a very low migration rate.

- The polyurethane tube proves to be the one with the best throm-boresistant properties. Its influence on the blood platelets is the lowest of all tested tubes in relation to the reference blood.

- The synthetic rubber tube damages the thrombocytes most (relative platelet amplitude: 7.82). Its surface is hydrophobic and exhibits a certain roughness.

- The silicone tube activates the plasmatic part of the coagulation system considerably (relative clotting time: 0.66). The wetability of the tube surface by the blood is poor.

In general, the results are well reproducible, the standard deviation is less than 10%.

Misinterpretation of results or even contradictory observations may occur in cases where the relation of surface area and blood volume (S/V) is not strictly taken into consideration. It is recommended to compare only tubes of identical dimensions. If tubes of inner diameter smaller than 4.0 mm are subject to investigation, it should be noted that the speed of rotation of the rotary disk has to be reduced. The admitted citrate blood has to flow without impediment.

Modification 3.

As already described in the *modification 2* (in vitro test for tubes) the blood which has passed the tube is affected in two ways: 1) the coagulation system is initiated by activation of specific clotting factors and/or from shear-stressed thrombocytes which release platelet factor III and IV or 2) the coagulation system might be inhibited by consumption coagulopathy or by adhered or injured platelets having lost their aggregability. Both effects are reflected by RTG.

In this RTG variation, the fibrin amplitudes and the platelet amplitudes are the parameters of interest.

For the evaluation of the test results, the first milliliter blood proves to have a predominant significance, because of its direct interaction with the foreign surface. The subsequent blood will be exposed to a more or less proteinated surface. The first blood samples provides the most evident properties of the material.

The results prove that the blood-material interaction is obviously accomplished in its essential parts within a very short period of probably 30-60 sec. Later this first protein layer interferes with other blood components,leading to a change of the surface composition of proteins. This dynamic protein exchange is not reflected by this test procedure.

The rough polyurethane surface injures the thrombocytes seriously. However, after one minute of blood-material interaction, the fibrin and platelet amplitudes of all materials are equalized and approach the level of the reference blood.

The tube with the lowest thrombogenetic response to the blood is made from polyurethane with a smooth surface, followed by polypropylene, by Silastic®, by the rough polyurethane and finally by PVC.

The experimental efforts and costs are low compared to other ex vivo or in vivo methods.

Although the blood flow varies slightly according to variations of the veneous pressure or deviations of the tube dimensions, the relative values reflect real differences and are well reproducible.

Limitations of method.

The reference materials distributed by NIHLB (Bethesda, Md. USA) cannot be tested by the **first** RTG-modification, because the test material must be dissolved first. A reference material would be changed and loose its basic properties.

Capillary tubes will create difficulties with the **second** RTG modification and will require an active pumping of the blood (see *Capillary Perfusion Model by Mulvihill et al.*). Neither can the primary reference materials of the NHLBI be tested by this modification unless it can be made available as tubes.

Only tubes in small diameter can be tested by the **third** modification because of the limitations in allowed blood loss from the animal.

REFERENCES

Dudley, B., Williams, J.L., Able, K. & Muller, B. (1976) Trans. Am. Soc. Artif. Int. Organs (ASAIO) 22, 538-539.

Hartert, H. (1981) Biorheology 18, 693-701.

Hiller, E. (1982) Atlas of Resonance Thrombography, Verlag Hygieneplan GmbH (in English, French and German).

Lemm, W. & Sprengel, H. (1982) Proc. Eur.Soc. Artif. Organs 9, 282-286.

Lemm, W. & Bücherl, E.S. (1986) in Blood Compatible Materials and their Testing (Dawids, S. & Bantjes, A. eds.) 107-113, Martinus Nijhof Publishers, Dordrecht, Boston, Lancaster.

Pelzer, H. & Heimburger, N. (1986) J. Biomed. Mat. Res. 20, 1401-1409.

Sprengel H. & Lemm, W. (1983) Life Support Systems 1, 235-238.

Quantitative Platelet Aggregation In Vitro

Authors: J.-P. Cazenave, J.N. Mulvihill, A. Sutter-Bay, C. Gachet, F. Toti & A. Beretz

INSERM U311, Centre Régional de Transfusion Sanguine, F - 67085 Strassbourg Cédex

INTRODUCTION

Name of method:

Quantitative platelet aggregation in vitro.

Aim of method:

To standardise techniques are described for quantitative measurement of platele
aggregation *in vitro*, using citrated platelet rich plasma or washed platele
suspensions in response to various stimuli encountered *in vivo*. Studies of this typ
give valuable information concerning alteration of platelet functions induced b
contact with artificial surfaces.

Biophysical background of method:

In the interaction of platelets with biomaterials, the primary effect on platelets i
adhesion to the surface followed by activation, granule release and secondar
aggregation. The aggregates may remain attached to the surface, leading t
thrombus formation or dislodge with the blood flow to form emboli at other point
of the circulation system e.g. in the microcirculation or at joints, filters or region
of stagnation in an artificial device. Platelet aggregation in contact with foreig
surfaces may be quantitated directly by transmission electron microscop
(*Baumgartner & Muggli, 1976*) or videomicroscopy (*Adams & Feuerstein, 1981*).
However, it is simpler and more common practice to measure it indirectly by takin
blood samples from the experimental device and measuring the ability of platelet
to undergo primary aggregation in response to added aggregating agents. Result
from this type of test enable evaluation of the degree to which platelet function
are modified by interaction with the surface. The technique also allows quantitativ
investigation of the physiological and biochemical mechanisms which contro

359

S. Dawids (ed.), Test Procedures for the Blood Compatibility of Biomaterials, 359–375.

platelet aggregation and of the effects of anti-platelet drugs, under well defined conditions of pH, temperature and divalent ion concentrations.

A typical curve for platelet aggregation in response to collagen is shown in *Fig. 1*.

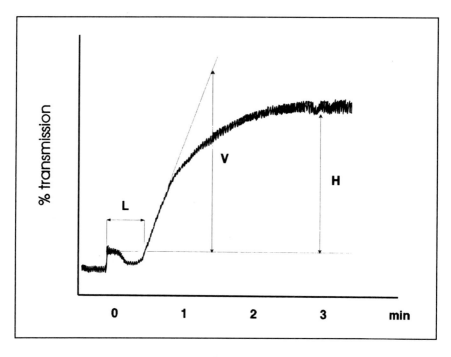

Fig. 1

Curve for platelet aggregation in citrated PRP after addition of collagen. L: lag phase, V: initial velocity of aggregation, H: heigth or amplitude of aggregation at 3 min.

Addition of the aggregating agent is accompanied by a sharp increase in light transmission caused by dilution of the PRP and is followed by a diminution in transmission reflecting the changes in platelet shape from discoid to spheroidal and the appearence of pseudopods. After a lag time, light transmission then increases steadily as the platelets aggregate to fewer but larger bodies. The oscillations of the curve which is obvious in variable amplitude with the sensitivity of the recorder, correspond to random movements of the platelets and their aggregates in the light beam. The identified phases used for calculation are as follows:

1) the lag phase (L) in seconds between addition of the aggregating agent and the beginning of aggregation,

2) the maximum initial velocity of aggregation (V) given as the slope of the curve at its origin in percent transmission per minute and

3) the maximum amplitude of heigth (H) of the aggregation curve or the amplitude at 3 min. in percent of light transmission.

Other characteristics are also noted, e.g. reversible aggregation followed by disaggregation induced by ADP or PAF-acether in washed platelet suspensions and the appearance of a second wave of aggregation in response to ADP ($\geq$ 2.5 μM) in citrated PRP.

Table I

Platelet aggregating and agglutinating agents.

Low molecular weight substances	Enzymes	Particulates and macromolecules	Agglutinating agents
ADP	Thrombin	Collagen	Ristocetin
Epinephrine	Trypsin	Viruses	Bovine factor VIII
Serotonin		Immune complexes	Zymosan
Vasopressin		Kaolin	Polylysine
PGH_2		Latex particules	Antiplatelet antibodies
Phorbol myristate acetate		Polymer microspheres	Concanavalin A
Thromboxane A_2		Endotoxin	
Arachidonic acid		Uric acid crystals	
PAF-acether		Heparin	
Ca ionophore (A23187)			
Synthetic prostaglandins			

According to Anderson & Kottke-Marchant (1987)

Platelets aggregate or agglutinate in response to numerous stimuli which may be encountered *in vivo* (*Table I, Anderson & Kottke-Marchant, 1987*). Aggregation studies *in vitro* are performed in most laboratories by a turbimetric method in a platelet aggregometer, using either citrated platelet rich plasma (PRP) or platelets resuspended in a physiological buffer medium. Rigorous standardasation of techniques is essential in order to obtain reproducible and meaningful results. The following standard procedures describe the preparation of reagents and platelet suspensions for the quantitative measurement of platelet aggregation in response to commonly employed aggregating or agglutinating agents (ADP, collagen, thrombin, arachidonic acid, ionophere A23187, PAF-acether, ristocetin) (*Cazenave et al., 1983*).

It is important to bear in mind that in vitro measurement of platelet aggregation corresponds to a "non real" situation, introducing artefacts and technical variables. Stricht control of experimental factors is therefore essential to obtain valid and reproducible results.

The principal features of platelet aggregation testing in vitro, induced by a range of commonly used aggregating or agglutinating agents, are summarised in *Table II* (*Cazenave et al., 1983*). Interpretation of aggregation studies is generally carried out by comparison of results for test and control platelets. The method was initially developed for clinical detection of platelet disorders. The aggregation curves for patient platelets are compared with those for platelets from a population of normal subjects known to be free of medication, in particular aspirin and other non-steroidal anti-inflammatory derivatives which inhibit platelet aggregation. The influence of drugs can be investigated by performing parallel tests on platelets from a single donor on the presence and absence of drug administration. Similarly, effects of biomaterials on platelet function may be evaluated by carrying out aggregation studies using a single platelet population before and after contact with the test material, or after contact with test and reference materials under the same physical and/or biological conditions. The most informative points of comparison are:

1. platelet response to ADP (see *Fig. 1*), the initial velocity (V) and maximum amplitude (H) of aggregation, the possible appearance of a second aggregation wave in PRP and the reversibility of aggregation in washed platelet suspensions,

2. after addition of collagen, the length of the lag phase (L) and the amplitude (H) of aggregation at 3 min,

3. The presence or not of aggregation in response to thrombin, arachidonic acid, ionophore A23187 or PAF-acether,

4. potentiating effects of paired stimuli and inparticular the degree of potentiation by epinephrine (10^{-1}-10^3 μM) of aggregation induced by other agents such as ADP, collagen and PAF-acether (*Lanza et al., 1988*),

5. the initial velocity (V) and amplitude (H) of agglutination at 3 min. after addition of ristocetin.

These parameters are normally determined for a range of concentrations of each aggregating agent.

The nature of the added anticoagulant has an influence on platelet aggregation. Sodium citrate (0.38%) which is the most widely employed anticoagulant, reduces sufficiently the calcium ion concentration of PRP to prevent generation of enough of thrombin to clot fibrinogen into fibrin. On the other hand, the lowered calcium levels give rise to a second irreversible wawe of aggregation in citrated PRP after addition of ADP (*Mustard et al., 1975*). This secondary aggregation which is not observed in washed platelet suspensions containing physiological ion concentrations (2 mM), results from the formation of thromboxane A_2 and is accompanied by release of the contents of dense and α-granules.

Alternative anticoagulants include EDTA, heparin and hirudin. Since only very low levels of extracellular free calcium are required for platelet aggregation (*Kinlough-Rathbone et al., 1983*), use of EDTA, a potent chelator of divalent cations, at concentrations in excess of 3 to 5 mM leads to complete inhibition of aggregation by all known agents. Collection of human blood into EDTA, however, also leads to permanent damage of labile blood clotting factors and the irreversible dissociation of the glycoprotein IIb-IIIa complex which is the platelet fibrinogen receptor. Conversely, EDTA does not inhibit the calcium independent agglutination of platelets by ristocetin in the presence of von Willebrand plasma factor. Heparin is not generally recommended for aggregation studies, as this substance has known platelet activating properties (*Salzman et al., 1980*) and has been shown to enhance aggregation in response to agents such as ADP, arachidonic acid and collagen (*Mohammad et al., 1981*). Hirudin may find more widespread future application with the recent development of readily available recombinant hirudins.

Table II

Responses in vitro of normal human platelets to aggregating and agglutinating agents.

Agent (Final concentration)	Medium	Response	Secretion
ADP (1.0-2.5 μM)	PRP	First wave, disaggregation	No
ADP ($\geq$ 2.5 μM)	PRP	First wave, second wave without disaggregation	Yes
ADP ($\geq$ 1 μM)	TA	First wave, disaggregation	No
Collagen (1.25-2.5 mg/l)	PRP or TA	Lag phase, dose dependent aggregation	Yes
Thrombin (> 0.1 U/ml)	PRP	Aggregation followed or not by plasma coagulation	Yes
Thrombin (> 0.05 U/ml)	TA	Aggregation, no disaggregation	Yes
Arachidonic acid (1.0-2.5 mM)	PRP	Dose dependent aggregation	Yes
Arachidonic acid (0.1-0.25 mM)	TA	Dose dependent aggregation	Yes
Ionophore A23187(1-5 μM)	PRP	Dose dependent aggregation	Yes
Ionophore A23187(0.5-1.0 μM)	TA	Dose dependent aggregation	Yes
PAF-acether (1-10 μM)	PRP	Aggregation, no disaggregation at high concentration	Weak
PAF-acether (1-10 μM)	TA	Aggregation, disaggregation	No
Ristocetin (1-2 g/l)	PRP or TA with EDTA 10 mM	First wave irreversible agglutination	No

PRP: Citrated platelet rich plasma   **TA**: Tyrode's-albumin buffer

Data from Cazenave et al. (1983)

Another important factor is pH. Maximum aggregation activity of platelets are observed for values in the slight alkaline range pH 7.4 to 8, with absence of aggregation and onset of platelet lysis for values below 6.4 or above 10 (*Anderson & Kottke-Marchant, 1987*). Variations in pH are minimised in the present method by conserving PRP and platelet suspensions in sealed tubes under 5% CO_2/95% air atmosphere. Aggregating agents, inhibitors and other reagents are added in minimal volumes, whenever possible in buffered solutions of pH 7-8 and osmolarity ~300mOsm. Since the platelet response to aggregating agents is gradually reduced with decreasing concentration of the suspension, platelet count is standardised at 300,000/μl, while stirring speed which controls the frequency of interplatelet collisions is maintained at 1,100 rpm, using a constant size of magnetic stirring bar. All aggregation measurements are performed at 37°C and platelets should never be exposed to low temperatures ($\leq$ 4°C) as they irreversibly change when they are rewarmed.

Mechanisms of platelet aggregation, clinical platelet disorders leading to anomalies of aggregation and pharmacological pathways of anti-platelet drugs have been extensively reviewed (*Mustard & Packham, 1975; Ludlam, 1981; Kinlough-Rathbone et al., 1983; Cazenave et al., 1983; Anderson & Kottke-Marchant, 1987*). This present chapter is more specifically concerned with platelet aggregation as a tool to follow modifications of platelet functions in the course of blood inteaction with bio-materials.

Scope of method:

The described techniques can be applied for measurement of the specific influence from foreign materials upon platelets. They can be integrated into specific experiments.

DETAILED DESCRIPTION OF METHOD

Equipment.

Two channel Payton aggregometer (Dade, France) coupled to a chart recorder.
Siliconised glass cuvettes (0.5 or 1.0 ml) (Dade, France).
A heating block maintained at 37°C.
Platelet Analyser 810 (Baker Instruments, Allentown, Pa., USA).

Sorvall RC-3B centrifuge (Sorvall Instruments, Dupont Biomedical Division, Newtown, Co., USA) with rotor H-4000.

Polycarbonate centrifuge tubes, 15 or 50 ml conical bottom (Corning Glass Works, New York, N.Y., USA).

Wide bore Pasteur pipettes (Pastettes, Biolyon, Dardilly, France).

All other material in contact with blood , PRP or platelet suspensions is in plastic or siliconised.

Reagents.

All chemicals from commercial sources (Prolabo, Paris, France; Merck, Darmstadt, FRG; Sigma Chemicals, St. Louis, Mo., USA) are of analytic grade.

Sodium citrate 3.8%

Trisodium citrate dihydrate 3.8 g in 100 ml distilled water.

Tyrode's buffer without Ca^{2+} or Mg^{2+} Plain Tyrode's buffer (stored at 4°C).

Stock solution: NaCl 16.0 g, KCl 0.4 g, NaHCO$_3$ 2.0 g, NaH$_2$PO$_4$.H$_2$O 0.116 g in a final volume of 100 ml distilled water.

Plain Tyrode's buffer (PT): stock solution 5ml, diluted to 100 ml with distilled water, pH adjusted to 7.30.

Platelet aggregating agents.

ADP (Sigma Chemicals): stock solution 5mM (0.2316 g in 100 ml distilled water) conserved in 100 μl aliquots at -30°C.

Dilutions 50, 25 and 10 μM in PT are used for aggregation tests.

Bovine collagen type I (Sigma Chemicals) is prepared as a 0.25% suspension in 0.522 M acetic acid, pH 2.8, 540 mOsm as described by *Cazenave et al. (1983)*. The stock suspension is stored at 4°C and diluted 1/100 and 1/200 in 0.9% NaCl for aggregation studies.

Bovine thrombin (Roche, Neuilly sur Seine, France): stock solution 1.000 U/ml in PT conserved in 100 μl aliquots at -30°C. Dilutions 5 and 1 U/ml in PT for aggregation tests.

Arachidonic acid (Sigma Chemicals): stock solution 50 mM (4.50 g in 275 ml Na$_2$CO$_3$ 25 mM), aliquots (100 μl) distributed under nitrogen atmosphere and conserved at -30°C. The solution is used undiluted for aggregation studies in citrated PRP, diluted to 5mM in PT for studies in washed platelet suspensions.

Ionophore A23187 (Boehringer, Mannheim, FRG): stock solution 10 mM in DMSO (10 mg in 1.9 ml DMSO) conserved in 1 ml aliquots at -30°C. Dilutions 50, 20 and 10 μM are prepared under agitation in 0.9% NaCl.

PAF-acether (Bachem, Bubendorf, Switzerland). Stock solution: 10 mM in chloroform/methanol 80/20 (5.5 mg in 1 ml) stored at 4°C. For aggregation tests, 1 μl stock solution is evaporated under nitrogen atmosphere and the residue redissolved in 1.2 or 10 ml Tyrode's-albumin buffer (10,5 or 1 μM PAF), solutions being conserved in aliquots at -30°C. Tyrode's-albumin buffer is prepared as described by *Cazenave et al. in "A centrifugation technique for the preparation of suspensions of non activated washed human platelets".*

Ristocetin sulphate (Lundbeck Co., Copenhagen, Denmark): solution 25 g/l in 0.9% NaCl stored in aliquots at -30°C.

EDTA 0.1 M: disodium salt of ethylenediamine tetra-acetic acid dihydrate 3.72 g in 100 ml distilled water, pH adjusted to 8.8 with NaOH 2M.

Human fibrinogen 4% (Grade L, Kabi, Stockholm, Sweden): lyophilised powder is dissolved in 0.9% NaCl (4g in 100 ml) and treated with diisopropylfluorophosphate (DFP, Sigma Chemicals) as described by *Cazenave et al. (1983)* to inactivate traces of contaminant plasma serine proteases. Aliquots (1 ml) are conserved at -30°C and thawed at 37°C before use.

Outline of method.

In the measurement of quantitative platelet aggregation it is important to measure well defined reactions towards aggregating agents. Aggregation of platelets are the sum of all factors, and the polymer derived stimulation of the platelets can be analysed as the increased aggregation compared to a standard. A number of aggregating agents (see list of equipment) are used in aggregation tests. Immediately before a series of aggregation tests should be performed, the aggregation agents are prepared from stock solutions at dilutions corresponding to 10 times the required final concentration after addition to the platelet suspension. The solutions are kept on ice throughout the experiment.

Fresh blood is used and centrifuged to obtain platelet rich plasma (PRP) with a predetermined concentration. Aggregating agents are then added and the platelet aggregation is followed in a Payton aggregometer for quantitative estimation of platelet aggregation using the turbimetric method. The reaction pattern of the platelets exposed to polymers and compared to the aggregating agents mentioned provides a good quantitative estimation of the activation ability of the polymer.

Description of procedure.

Blood collection and preparation of citrated platelet rich plasma (PRP).
Blood is collected by venepuncture, using a wide bore (18/10) needle mounted on a short length (10 to 20 cm) of plastic tubing, directly into a conical 15 or 50 ml centrifuge tube containing 1 volume of anticoagulant sodium citrate 3.8% for 9 blood volumes (final pH 7.5 and citrate concentration 13 mM). If the hematocrit falls outside normal limits (35 to 48%), the citrate concentration is corrected according to the relation:

$$\text{anticoagulant volume} = \text{Blood volume} \ \times \ \frac{(100\text{-hematocrit})}{4.5 \times 100}$$

After discarding the first milliliters, which might be contaminated with tissue thromboplastin, blood is allowed to flow down the tube wall to minimise air contact and is gently mixed with the anticoagulant. Immediately after the filling, the tube is stoppered of the tube. Aggregation studies should always be completed within 3 hours of blood collection.

PRP is obtained by centrifugation of anticoagulated blood in a Sorvall RC-3B centrifuge, 15 min. at 1,000 rpm (175 x g), at ambiant temperature. A plastic syringe or Pasteur pipette is used to remove the supernatant. Platelet rich plasma is then transferred while care is being taken not to aspire red blood cells or white cells localised in the "buffy coat" at the interface between plasma and packed erythrocytes. Citrated PRP is stored at room temperature in stoppered plastic tubes under a 5% CO_2/95% air atmosphere. Platelet count is adjusted to 300,000/μl with autologous platelet poor plasma (PPP), obtained by centrifugation of citrated blood or PRP for 2 min. in an Eppendorf centrifuge (9,980 x g).

Preparation of washed platelet suspensions.
Washed human platelets are isolated from blood collected into acid-citrate-dextrose anticoagulant as described in the preceeding chapter "A centrifugation technique

for the preparation of suspensions of non activted washed human platelets". When required, dense granule contents are labelled with ^{14}C-serotonin. On completion of the washing procedure, platelets are resuspended in Tyrode's-albumin buffer (Tyrode's buffer containing 2mM Ca^{2+}, 1 mM mg^{2+}, 0.1% glucose and 0.35% human serum albumin, pH 7.30, osmolarity 295 mOsm) in the presence of apyrase (2 μl/ml) and platelet count is adjusted to 300,000/μl. Addition of apyrase ensures degradation of released ADP and hence prevents loss of discoid platelet shape or the development of a refractory state. Conserved at 37°Cunder 5% CO_2/95% air atmosphere, suspensions are stable for up to 8 hours.

Measurement of platelet aggregation.
The Payton aggregometer enables quantitative estimation of platelet aggregation by a turbimetric method depending on measurement of the variations in light transmission through a cuvette containing the platelets suspended in citrated plasma or a physiological buffer medium.

Using citrated PRP, the base line (= 0% transmission) of the apparatus is adjusted with a cuvette (0.5 ml) containing 450 μl PRP and the 100% transmission with a cuvette containing 450 μl autologous PPP. The stock tube of PRP is mixed gently. From this tube, 450 μl of platelet suspension is pipetted into a cuvette, prewarmed for 3 min. in a heating block at 37°C and placed in the aggregometer. After stirring for 30 sec., 50 μl of aggregating agent is added (10 - 25 μl in the case of arachidonic acid). Light transmission is recorded for a minimum period of 3 min. and since two channels are available it is possible to observe in parallel test and control aggregation curves.

Transmission measurements with Tyrode's-albumin buffer, while prewarming is not necessary as the suspension is kept at all times at 37°C. For certain aggregating agents (ADP, PAF-acether and low concentrations of collagen), 10 μl of 4% human fibrinogen is added 30 sec. before the aggregating agent. Using platelets radiolabelled with ^{14}C-serotonin, release of dense granule contents may be quantitated by determining radioactivity both in the platelet suspension and in a supernatant obtained by centrifugation of aggregated suspension for 2 min. in an Eppendorf centrifuge (9,980 x g) as described by *Mustard et al. (1975)*.

Measurement of platelet agglutination in the presence of ristocetin.
Platelet agglutination by ristocetin is measured in citrated PRP in the presence of 10 mM EDTA. To 450 μl PRP prewarmed to 37°C in a cuvette (0.5 ml), 50 μl EDTA 0.1 M is added, the cuvette is placed in the aggregometer and after stirring

for 30 sec. ristocetin is added (20 or 25 μl at 25 g/l completed to 50 μl with PT, final concentration 1.0 or 1.25 g/l). Agglutination is followed for 3 min.

Test tubes can be coated with dissolved polymers thus providing a relative basis for compatibility testing.

In the case of washed platelet suspesions, agglutination tests are carried out in 1 ml aggregometer cuvettes using 1 ml platelet suspension, 100 μl EDTA 0.1 M and 50 μl ristocetin, the requirement for ristocetin cofactor (von Willebrand factor) being satisfied by addition of plasma pool (25 μl PPP) 30 sec. before the agglutinating agent.

Safety aspects.

Strict laboratory hygiene should be observed in the handling of blood samples. All experiments being performed with blood from healthy donors are normally screened for absence of hepatitis or human immunodeficiency virus.

RESULTS

Reference material.

The reference biomaterial of choice will depend on the type of artificial device under study. In all cases, it would be preferable to select a material of reasonable low platelet activating properties, as the test is very sensible. Suitable types of materials are found to be medical grade silicone (tubing), polyacrylonitrile or polysulfone (membranes) and polymers of enhanced gas permeability (storage bags for platelet concentrates) such as CLX (Cutter Laboratories, Berkeley, Ca., USA).

In this context the technique has been most widely employed to evaluate blood bags for platelet conservation. As an illustration, one such study in the author's laboratory concerned a series of storage packs fabricated from polyvinylchloride (PVC) containing varying proportions of silicone additives and di-2-ethylhexyl phthalate (DEHP) plastiziser. DEHP is non toxic to platelets if it leaches out of the polymer into the PRP, but during metabolism in the body it generates very toxic products (e.g. DMP).

Table III

Aggregation and agglutination tests for washed human platelets prepared from citrated PRP conserved in blood storage bags of varying polymeric composition.

Agent (Final concentration)	Parameter	Day 0	1	2	3	6
A. Polymer: PVC$_A$ (DEHP 21.38%, Silicone 10.01%)						
ADP (5 x 10^{-6}M)	V	96	72	10	0	0
	H	40	22	2	0	0
Collagen (2.5mg/l)	L	36	45	0	0	0
	H(3min)	68	72	0	0	0
Thrombin (1U/ml)	H(3min)	42	21	21	0	0
Arachidonic acid (0.1mM)	H(3min)	74	84	38	0	0
Ristocitin (1.25g/l)	H(3min)	20	30	22	0	0
B. Polymer: PVC$_B$ (DEHP 2.64%, Silicone 30.56%)						
ADP (5 x 10^{-6}M)	V	64	76	20	24	0
	H	20	20	20	6	0
Collagen (2.5mg/l)	L	33	57	90	108	0
	H(3min)	76	74	6	1	0
Thrombin (1U/ml)	H(3min)	58	58	46	32	0
Arachidonic acid (0.1mM)	H(3min)	70	64	51	38	0
Ristocetin (1.25g/l)	H(3min)	44	40	34	26	0
C. Polymer: PVC$_C$ (DEHP 31.39%, Silicone 0%)						
ADP (5 x 10^{-6}M)	V	82	60	32	14	6
	H	38	8	5	2	3
Collagen (2.5mg/l)	L	42	81	69	84	66
	H(3min)	76	30	62	4	1
Thrombin (1U/ml)	H(3min)	59	48	43	36	9
Arachidonic acid (0.1mM)	H(3min)	73	70	73	63	22
Ristocetin (1.25g/l)	H(3min)	48	16	38	31	8

V: initialvelocity (%transmission/min), **H**: maximum amplitude (% transmission)
H(3min): amplitude at 3 min (% transmission), **L**: lag phase (sec)

Results mean of three independent experiments.

Platelet concentrates ($2 \times 10^6/\mu l$) obtained by plasmapheresis were stored for 6 days at ambient temperature on a horizontal shaker. On days 0,1,2,3, and 6, samples of PRP drawn from each type of bag were used to prepare washed platelet suspensions for aggregation tests and to monitor physical properties such as pH and platelet count, mean volume and adenine nucleotide content. Storage in a first polymeric material (PVC_A in *Table III*) resulted in rapid diminution of platelet responses to all aggregating agents, zero response being attained on day 3.

A second material (PVC_B) enabled improved conservation of platelet functions over the same 3 day period, while a third polymer (PVC_C) provided the best storage conditions, aggregation inducd by ADP, thrombin and arachidonic acid and agglutination by ristocetin remaining at 8%, 15%, 30% and 17% of their respective initial values on the sixth day. These trends were confirmed by parallel results for fall in pH and platelet nucleotide (ADP, ATP content).

In a similar survey, *Quinn et al. (1986)* compared four types of plastic pack composed of PVC plasticised with DEHP (Tuta) or with trimellitate (CLX and PL 1240) or of polyolefin (PL 732) with respect to preservation of platelet functions during a 5 day storage period.

Platelet morphology and in vitro responses were generally better conserved in Tuta and CLX packs. On the fifth day, aggregation in citrated PRP induced by ADP or collagen was maintained at 50-60% of its initial amplitude in Tuta containers, as compared to 40% or less for the remaining three packs, while agglutination in the presence of ristocetin fell in the range 50-70% of initial levels for Tuta and CLX but below 30% for PL 1240 and PL 732.

DISCUSSION

In vitro aggregation tests are of value in assessing the suitability of biomaterials for fabrication of e.g. blood storage containers. Moreover, since transfusion studies have demonstrated the clinical efficieny of platelet concentrates in which pH and other in vitro parameters are maintained (*Schiffer et al., 1986*), the technique may also be employed to screen conserved concentrates of donor platelets for potential haemostatic capacity.

Platelet aggregation in vitro has also been used to follow changes in platelet functions during haemodialysis. In group of patients with chronic renal failure undergoing regular dialysis, *Levin et al. (1978)* reported loss of platelet sensitivity to ADP in citrated PRP drawn from the arterial outlet line. The maximum amplitudes of aggregation induced by 4.7 μM ADP decreased to 75%, 76% and 79% of the initial values after 30, 120 and 300 minutes of dialysis. Platelets drawn from the venous return line were aggregable only in response to a tenfold higher concentration of ADP (50 μM). Such results suggest stimulation of partial release and/or damage of platelets during passage through the dialyser fibres with retention of aggregable platelets, followed by a certain degree of recovery of the damaged cells upon return to the in vivo circulation.

Transient impairment of platelet function has been reported in cardiopulmonary bypass, as reduced response to ADP and collagen, (*Harker et al., 1980*). This was accompanied by prolongation of the bleeding time from 5.4 sec. baseline to 19 sec. after 30 min. and > 30 sec. after 2 hours of the surgical operation, despite maintenance of platelet counts above 100,000/μl. Normalisation was generally obtained some 3 to 4 hours following cessation of bypass. The process involved release and partial depletion of α-granule proteins, leading to parallel changes in plasma levels of platelet factor 4 and ß-thromboglobulin, but not of dense granule constituents. Platelet ^{14}C-serotonin content remains constant. Whenever bypass surgary was complicated by excessive abnormal bleeding, postoperative bleeding time was normalised within the usual 4 hour period. In such cases, in vitro aggregation studies may be of clinical value to confirm persistent platelet dysfunction in spite of adequate counts (> 100,000/μl), indicating the need for transfusion of donor platelets.

Limitations of the method.

Ideally, platelet aggregation should be measured in vivo to reflect the interaction of flowing blood with natural and artificial surfaces (in the absence of anti-coagulants and under physiological conditions of temperature and haemodynamics). The in vitro method inevitably introduces artefacts and experimental variables. As discussed above, in order to develop valid test procedures, it is vital to minimise the influence of purely technical factors such as anticoagulant, pH, platelet concentration, stirring speed and temperature.

The described turbimetric technique measures indirectly and is limited by its lack of sensitivity to the formation of microaggregates. Quantitative confirmation of the presence of microaggregates comprising 2 or 3 platelets may be obtained by examining a droplet of PRP or platelet suspension under the phase contrast microscope.

REFERENCES

Adams, G.A. & Feuerstein, L.A. (1981) "Kinetics of platelet adhesion and aggregation on protein coated surfaces: morphology and release from dense granules". J. Am. Soc. Artif. Intern. Organs 4, 90-99.

Anderson, J.M. & Kottke-Marchant, K. (1987) "Platelet interactions with biomaterials and artificial devices" in Blood Compatibility (Williams, D.F. ed.), vol.1, pp.103-150, CRC Press, Boca Raton.

Baumgartner, H.R. & Muggli, R. (1976) "Adhesion and aggregation: morphological demonstration and quantitation in vivo and in vitro" in Platelets in Biologi and Pathology (Gordon, J.L. ed.) pp. 23-60, Elsevier, Amsterdam.

Cazenave, J.-P., Hemmendinger, S., Beretz, A., Sutter-Bay, A. & Launay, J. (1983) "L'agrégation plaquettaire: outil d'investigation clinique et d'étude pharmacologique". Méthodologie Ann. Biol. Clin. 41, 167-179.

Harker, L.A., Malpass, T.W., Branson, H.E., Hessel, I.I.E.A. & Slichter, S.J. (1980) "Mechanisms of abnormal bleeding in patients undergoing cardiopulmonary bypass: acquired transient platelet dysfunction associated with selective α-granule release". Blood 56, 824-834.

Kinlough-Rathbone, R.L., Packham, M.A. & Mustard J.F. (1983) "Platelet aggregation" in Methods in Hematology: Measurements of Platelet Function (Harker, L.A. & Zimmerman, T.S. eds.), vol.8, pp.64-91, Churchill Livingstone, New York.

Lanza, F., Beretz, A., Stierlé, A., Hanau, D., Kubina, M. & Cazenave, J.-P. (1988) "Epinephrine potentiates human platelet activation but is not an aggregating agent". Amer.J.Physiol. 255, H1276-H1288.

Levin, R.D., Kwaan, H.C. & Ivanovich, P. (1978) "Changes in platelet function during hemodialysis". J.Lab.Clin.Med. 92, 779-786.

Ludlam, C.A. (1981) "Assessment of platelet function" in Haemostasis and Thrombosis.(Bloom, A.L. & Thomas, D.P. eds.) pp 775-795, Churchill Livingstone, New York.

Mohammad, S.F., Anderson, W.H., Smith, J.B., Chuang, H.Y.K. & Mason, R.G. (1981) "Effects of heparin on platelet aggregation and release and thromboxane A_2production". Amer.J.Pathol. 104, 132-141.

Mustard, J.F., Perry, D.W., Kinlough-Rathbone, R.L. & Packham, M.A. (1975) "Factors responsible for ADP-induced release reaction of human platelets". Amer.J. Physiol. 228, 1757-1765.

Mustard, J.F. & Packham, M.A. (1975) "Platelets, thrombosis and drugs". Drugs 9, 19-76.

Quinn, M.A., Clyne, J.H., Wolf, M.M., Cruickshank, D., Cooper, I.A., McGrath, K.M. & Morris, J. (1986) "Storage of platelet concentrates: an in vitro study of four types of plastic packs". Pathology 18, 331-335.

Salzmann, E.W., Rosenberg, R.D., Smith, M.H., Lindon, J.N. & Favreau, L. (1980) "Effect of heparin and heparin fractions on platelet aggregation". J. Clin. Invest. 65, 64-73.

Schiffer, C.A., Lee, E.J., Ness, P.M. & Reilly, J. (1986) "Clinical evaluation of platelet concentrates stored for one to five days." Blood 67, 1591-1594.

Platelet Deposition on Biomaterials
measured with a Capillary Perfusion Model

Authors:J.N. Mulvihill[1], A. Poot[2], T. Beugeling[2], A. Bantjes[2], W.B. van Aken[2], J.-P. Cazenave[1]

[1]   INSERM U 311, Centre Régional de Transfusion Sanguine, F - 67085 Strassbourg Cédex

[2]   Dept. of Chemical Technology, Univ. of Twente, NL - 7500 AE Enschede

INTRODUCTION

Name of method:

Platelet deposition on biomaterials measured with a capillary perfusion model.

Aim of method:

The described capillary perfusion model permits in vitro study of the blood interactions with artificial surfaces. The technique allows quantitative measurement of protein adsorption, platelet accumulation and activation of coagulation, fibrinolysis and complement systems under controlled, non pulsatile laminar flow. Biomaterials in the form of tubing may be tested for the effects of varying parameters on blood compatibility.

Biophysical background of method:

The capillary flow system, originally described by *Cazenave and Mulvihill (1987)* and modified by *Poot et al. (1988)* enables in vitro quantification of protein adsorption and platelet accumulation in tubes made of glass or polymeric materials under conditions of laminar flow and physiological wall shear rate ($50 - 4,000$ s^{-1}). Experiments can be performed with washed platelet suspensions or anticoagulated whole blood. Adsorption of plasma proteins can be measured using [125]Iodine labeled proteins, while platelet deposition is quantitated either with washed [111]Indium labeled platelets or, in experiments using whole blood, by surface phase radioimmunoassay with [125]Iodine labeled monoclonal antibodies against platelet membrane glycoproteins IIb-IIIa (*Mulvihill et al., 1987*). Morphology of deposited platelets may be further examined by scanning electron microscopy. Activation of

S. Dawids (ed.), Test Procedures for the Blood Compatibility of Biomaterials, 377–393.

platelets may be further examined by scanning electron microscopy. Activation of platelets and the systems of coagulation, fibrinolysis and complement during perfusion of blood or platelet suspensions may be estimated from the liberation of platelet secretion products (ß-thromboglobulin and platelet factor 4), fibrinopeptide A, fibrinogen degradation products and complement fragments (C3a and C5a).

By using [111]Indium tagged platelets it is possible to measure platelet accumulation in polymer tubes at wall shear rates in the range 150-1500 s^{-1} (*Poot et al., 1988*). Morphology of the adherent platelets can be made by scanning electron microscopy after cutting the tubes. The in vitro perfusion model in its original or modified form is well adapted to a rapid, inexpensive screening of biomaterials for blood compatibility. The effects of diverse parameters may be evaluated, including polymer composition and physical properties, protein preadsorption of the surface, shear forces, perfusion time and the presence of specific proteins, red blood cells, anticoagulants and drugs in the blood or platelet suspension

Scope of method:

The method is used for an in vitro screening of tube-shaped material under well defined conditions.

DETAILED DESCRIPTION OF METHOD

Equipment.

Two syringe pumps, model 355 (Sage Instruments, New York)
Polypropylene syringes, 50 mm (e.g. Becton & Dickinson, UK)
Three way stopcocks, glass or polyethylene, bore >1 mm (Microcaps, Drummond Scientific Co. Bromall, PA. USA)
Capillary tubes of glass, bore 0.56 mm (Microcaps, Drummond Scientific Co. Bromall, PA. USA
Glass tubes, bore 6-10 mm (standard lab. equipment)
HP-polyethylene tubes medical grade, bore 0.75 (Talas, Ommen, The Netherlands)
Thermostated hood, model ITH1 (Infors, Basel Switzerland)
Spectral ß-counter (1219 LKB-Wallac, LKB, Stockholm, Sweden)
Platelet analyser, Model 810 (Baker Instruments, Allentown, PA, USA)

Chemicals (analytic grade).
Trisodium citrate 129 mM/l (Merck, Darmstadt, FRG)
Heparin 100 U/ml (Roche, Basel, Switzerland)
Tyrode's buffer without Ca^{2+}/Mg^{2+} **or**
Hirudin 600 ATU/ml (Serva Feinbiochemica, Heidelberg, FRG) with Tyrode's buffer.
Iodogen (1,3,4,6-tetrachloro-3α,6α-diphenylglycoluril), (Pierce Chemical Co., Rockford, Ill., USA)
Scintillation liquid, Lumagel (Lumac, Schaesberg, The Netherlands) diluted 1:1 with 2.9 M/l $ZnCl_2$.
PGI_2 (Sigma Chemicals, St. Louis, MO., USA)
Apyrase (INSERM U31, Strassbourg, France)

Isotopes. (Commissariat à l'Energie Atomique, Saclay, France or from Amersham International, Amersham, UK).
^{125}Iodine, dissolved in NaOH
^{111}Indium-oxine, dissolved in NaCl

Human plasma proteins (purified).
Albumin, fibrinogen, fibronectin, immunoglobulin G, van Willebrand factor are not generally available and must be purified by standard techniques (see *Mulvihill et al., 1987, Poot et al., 1988*).

Monoclonal antibodies antihuman glycoprotein complex IIb-IIIa.
Monoclonal antibodies (INSERM U311, Centre Régional de Transfusion Sanguine, Strassbourg, France) used to quantitate platelet deposition by surface phase radioimmunoassay are directed specifically against the membrane glycoprotein complex IIb-IIIa of human platelets and do not cross-react with monocytes or other leucocytes (*Modderman et al., 1988*).

The mixtures are kept in stoppered tubes at room temperature until required for perfusion. Experiments should be completed within two hours of blood collection

Outline of method.

The perfusion apparatus (*Poot et al. 1988*) consists of: two syringe pumps each mounted with two syringes of 50 ml polypropylene in parallel. These lead via twin three way stopcocks to capillary tubes of glass or to polymer for testing. To avoid bending of the soft polymer tubes thse are placed for support inside 25 cm long glass tubes ending 2,5 cm from the entrance and exit of the capillaries. The capillaries with internal diameter of 0.56 mm are used for perfusions at shear wall rates of 800 s^{-1} (0.83 ml/min) to 4,000 s^{-1} (4.14 ml/min) while capillaries of internal diameter of 0.80 mm perform shear rates of 50 s^{-1} at perfusion of 0.15 ml/min. One of the syringes is used to perform flow experiments with perfusates or protein solutions while the other set of syringes are for rinsing the capillary tubes. The flow is regulated through three way stopcocks preferably made from polymeric material (e.g. polyethylene). The whole set up is placed in a thermostated hood to maintain temperatures of 37°C ± 0.2°C.

Testing of the ability of a polymer to activate platelets or adsorb proteins is tested in an apparatus described below and shown in *Fig. 1*. The prepared suspension of platelets which are radioactively tagged flow at a controlled velocity and shear rate through polymer capillaries. A reference glass capillary is used on the same batch of platelets. Adhesion to the wall of the polymer tube can be determined in 4 different ways:

1. Detection of radioactive platelets on the wall after removal of excess fluid.
2. Microscopic examination (SEM) of the inner surface after splitting the tube.
3. Identification of platelets using monoclonal antibodies directed against membrane glucoprotein complex IIb-IIIa of human platelets (*Modderman et al., 1988*).
4. Measurement of platelet factors released into the solution.

Detection of protein adsorption can be made by antibodies directed towards human albumin, fibrinogen, fibronectin, immunoglobulin G, van Willebrand factor.

To determine platelet deposit ([111]Indium tagged platelets), the part of each capillary is cut into six segments of 4 cm, the first cm at the entrance being discarded. Measurement of the radioactivity is made in a spectral ß-counter.

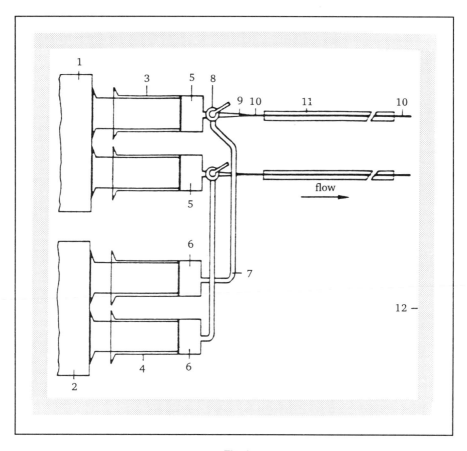

Fig. 1

Schematic diagram of the in vitro perfusion system.

1,2: Drive carriages of syringe pumps (Poot et al.); in the model of Mulvihil et al. pump 2
is replaced by a peristaltic pump. *3,4:* Polypropylene syringes (20 or 50 ml). *5:* Protein
solution, platelet suspension or anticoagulated whole blood. *6:* Rinsing buffer. *7:* Polyvinylchloride tube (ID 3mm, length 25 cm). *8:* Polycarbonate three way stopcock

Model of Poot et al.(1988), 9: Polypropylene pipette tip. *10:* Polyethylene capillary tube (ID
0.75 mm, length 30 cm). *11:* Glass tube (ID 1.5 mm, length 25 cm).

Model of Mulvihill et al.(1987), 9: Silicone connection tube (ID 1 mm, length 1 cm). *10:*
Glass capillary (ID 0.56 mm, length 4 cm or ID 0.80 mm, length 10 cm). *12:* Incubator

Description of procedure.

Prior to perfusion experiments, the capillaries are cleaned in sulphochromic acid, thoroughly rinsed in distilled water and oven dried.The tubes are stored at 22°C until required for perfusion experiments (<3 hours). Rinsing buffer from a reservoir at 37°C is led via the pump to the three way stopcocks.

Collection of blood.

Whole blood for perfusion experiments is collected with anticoagulants (9:1) in one of three mixtures: 129 mM/l trisodium citrate, 100 U/ml heparin in Ca^{2+}/Mg^{2+} free Tyrode's buffer or 600 ATU/ml hirudin in the same buffer. The stoppered tubes are kept at room temperature until required for perfusion, and experiments are always completed within two hours of blood collection.

Radiolabelling of proteins.

Proteins are radiolabelled with iodogen (*Regoeczi, 1984*) and the purity of unlabeled and labeled proteins is controlled by electrophoresis in SDS-polyacrylamide gel. Bovine collagen (type I, insoluble, Sigma Chemicals, St. Louis, MO, UAS) is prepared by the method of *Cazenave et al. (1983)*.

Preparation of human platelets (*Cazenave et al., 1983*).

Preparation of human platelets is carried out from blood collected into ACD (6:1) consisting of 66.6 mM/l citric acid, 85.0 mM/l trisodium citrate, 111.0 mM/l D(+)-glucose, pH 4.5. The entire procedure is carried out at 37°C.

Platelet rich plasma is initially obtained by centrifugation of blood at 175 g for 15 min. Platelets are allowed to recover for 30 min. and subsequently pelleted by centrifugation at 1.570 g for 13 min. The pellet is resuspended in Tyrode's albumin buffer (136.9 mM NaCl, 2.7 mM KCl, 0.4 mM NaH_2PO_4, 11.9 mM $NaHCO_3$, 1.0 mM $MgCl_2$, 2.0 mM $CaCl_2$, 5.6 mM D(+)-glucose, 5.0 mM hepes, 0.35% human serum albumin, pH 7.30, 295 mOsmol.) containing 10 U/ml heparin and 0.5 μM PGI_2 (Sigma Chemicals, St. Louis, MO, USA). After 15 min. recovery time 0.5μM PGI_2 is added. Subsequently the cells are pelleted by centrifugation at 1.100 G for 8 min. and the pellet is resuspended in Tyrode's-albumin solution containing 0.5 μM PGI_2. After 15 min. resting period a second centrifugation is performed. The pellet is resuspended in Tyrode's-albumin buffer containing 20 ml/l apyrase (INSERM U31, Centre Régional de Transfusion Sanguine, Strasbourg, France) (*Cazenave et al., 1983*). The platelet concentration is adjusted to 300,000/mm^3 (Platelet Analyser 810, Baker Instruments, Allentown, PA, USA) by adding

Tyrode's albumin buffer. The washed platelet suspension is stored at 37°C until perfusion and is stable up to three hours.

Radiolabelling of platelets.
The platelets are incubated with 10 μCi [111]Indium-oxine (0.5 mg/l) for 15 min. at 37°C (*Mulvihill et al., 1987*). Under these conditions labeling yield is of the order of 90%, and platelet aggregation with ADP, thrombin or collagen is not modified.

Preparation of red cells.
Red cells are packed by centrifugation at 1.570 g for 13 min. The cells are then washed twice at 22°C in Tyrode's buffer without Ca^{2+} or Mg^{2+} containing 5.6 mM/l D(+)-glucose (*Cazenave et al., 1983*) and resuspended in Tyrode's albumin solution containing 20 ml/l apyrase, pH 9.0. These erythrocytes are packed again by centrifugation at 1.570 g for 4 min. at 37°C and added immediately before perfusion to the suspension of washed [111]Indium labeled platelets at a volume ratio corresponding to 40% hematocrit.

Protein precoating of capillary surfaces.
In certain experiments, prior to perfusion of blood or washed platelets, capillary tubes are precoated with purified human plasma proteins or bovine collagen. Plasma proteins are diluted in Tyrode's buffer without Ca^{2+} or Mg^{2+}, pH 7.30, to solution concentrations corresponding to the plateau regions of the respective adsorption isotherms (albumin 3.5-4.0 g/l; fibrinogen, fibronectin and von Willebrand factor 0.5 g/l; immunoglobulin G 4.0 g/l while collagen is used as a 0.25% suspension in 0.522 M/l acetic acid, pH 2.8) (*Mulvihill et al., 1987, Poot et al., 1988*).

Precoating is carried out by static adsorption for one hour at 22°C (*Mulvihill et al., 1987*) or at 37°C (*Poot et al., 1988*). After rinsing with Tyrode's buffer without Ca^{2+} or Mg^{2+} for 10 min at a wall shear rate of 1,000 s^{-1}, the capillaries preadsorbed with plasma proteins are stoppered and kept filled with buffer until the perfusion experiment (<1).

Rinsed collagen coated capillaries are allowed to empty, and air-induced polymerization of the adsorbed collagen layer takes place. Care should be taken at all other times during protein precoating as in subsequent perfusion of blood or platelet suspensions, to avoid the formation of air bubbles or an unintended air-liquid interface, to prevent induced protein denaturation or platelet activation upon contact with air.

Perfusion of whole blood or washed platelets.
Capillary tubes coated or uncoated with protein are perfused at 37°C either with anticoagulated whole blood or with a suspension of washed platelets in Tyrode's-albumin buffer containing 40% washed red blood cells. Perfusion times range from 2 to 15 min. Wall shear rates may be varied from 50 to 4,000 s^{-1} in glass capillaries where 150 to 1,500 s^{-1} in polymer tubes (*Poot et al., 1988*) is more suitable. After perfusion, the capillaries are rinsed with Tyrode's buffer without Ca^{2+} or Mg^{2+} for 5 min at 1,000 s^{-1}.

Measurement of deposition.
Platelet deposition is quantitated by using [111]Indium labelled platelets in washed platelet suspensions or in experiments with whole blood by using [125]Iodine labelled anti-platelet antibodies. The surface phase radioimmunoassay with monoclonal antibodies against platelet membrane glycoproteins IIb-IIIa is performed as described by *Mulvihill et al. (1987)*. Purified [125]Iodine labeled antibody, diluted to approximately 1,000 CPM/μl in Tyrode's buffer solution without Ca^{2+} or Mg^{2+} containing 0.1% human serum albumin, is introduced into the capillary tubes by displacement of at least four times the volume of the rinsing buffer. Incubation under static conditions is continued for 30 min at 37°C. Rinsing should be done for 5 min at 1,000 s^{-1} with Tyrode's buffer without Ca^{2+} or Mg^{2+}. The capillaries are then disassembled and cut into 1.0 cm segments for radioactive counting.

[125]Iodine radioactivity is converted to the corresponding quantity of adherent platelets by means of a calibration curve, determined in double isotope experiments using capillaries perfused with heparin anticoagulated whole blood containing autologous washed [111]Indium labeled platelets. A linear correlation is obtained between deposited platelets (^{111}In) and immunoadsorbed antibody (^{125}I). Platelet accumulation may thus be quantitated from the relation:

$$\text{Adherent platelets/mm}^2 = (^{125}I_A - {}^{125}I_B) \times \frac{\text{slope}_C}{S}$$

where

$^{125}I_A$ is the measured [125]Iodine radioactivity of a capillary segment,
$^{125}I_B$ the background antibody adsorption, determined from capillaries perfused with platelet poor plasma,
slope$_C$ the slope of the linear correlation for radioimmunoassay,
S the inner surface area of a segment.

Scanning electron microscopy.

Morphological examination of the internal surface of perfused capillary tubes is performed by scanning electron microscopy (SEM). Rinsed capillaries are initially fixed with 2% glutaraldehyde in Tyrode's buffer without Ca^{2+} or Mg^{2+} and left in fixing solution for 16 hours at 4°C. The sample is dehydrated through graded ethanol solutions and dried by evaporation in a critical point apparatus. Samples are then sputter coated with approximately 10 nm gold and examined under a scanning electron microscope at 15 Kv accelerating voltage.

Safety aspects.

Normal laboratory precautions are observed in the handling of radioisotopes. Labeling of proteins and antibodies with [125]Iodine is performed behind a lead shield under a laminar flow hood, and a lead screen is likewise placed in front of the perfusion apparatus during experiments with [111]Indium labeled platelets. All radioactive waste materials are stored in lead shielded containers for three months ([111]Indium) or one year ([125]Iodine) before disposal.

RESULTS

Application of the glass capillary system of *Mulvihill et al.* and modification by *Poot et al.* to the quantitation of platelet deposition on uncoated and protein coated glass from a suspension of washed radiolabeled platelets is illustrated in *Table I*.

One can generally observe an order of surface reactivity: glass ~ albumin < glass ~ fibrinogen < glass ~ collagen. On collagen coated glass, the concentration of adherent platelets rises steadily with increasing wall shear rate from 50 to 4,000 s^{-1}, indicating that deposition is largely transport controlled, while on albumin or fibrinogen coated glass platelet accumulation peaks at 800 - 2,000 s^{-1}, suggesting that on these surfaces biochemical and physical reactions at the interface become rate limiting at shear forces in excess of ~800 s^{-1}.

Table I

Maximum platelet accumulation in glass capillaries, uncoated or precoated with albumin, fibrinogen or collagen. Washed [111]Indium labeled human platelets. Perfusion 5 min at 37°C, platelets 180,000/mm³, hematocrit 40%. Results mean of 4 independent experiments, (Mulvihill et al., 1987)

Wall shear rate (s^{-1})	Adherent platelets ($/mm^2$)			
	Protein coating			
	None	Albumin	Fibrinogen	Collagen
50	10,000	11,300	5,800	16,500
800	-	22,500	52,800	151,600
2,000	-	10,800	77,800	306,000
4,000	75,000	7,200	56,100	425,000

Typical results for perfusion of protein coated glass capillaries using heparin or citrate anticoagulated whole blood are presented in *Table II*. As compared to washed platelet suspensions, the accumulation of platelet from blood is globally reduced on all surfaces, most probably due to albumin adsorption from plasma. The observed reactivity order: albumin < fibrinogen < collagen remains unchanged. On collagen or fibrinogen coated glass, platelet deposition using heparin anticoagulation increases continuously with rising shear rate, whereas citrate anticoagulation leads to maximum surface concentrations of platelets at 800 - 2,000 s^{-1}. The inhibition of platelet functions resulting from complexing (and thus removal) of free calcium ions is reflected in the capillary system in a transition at high shear forces from transport to surface reaction controlled deposition.

Table II

Platelet accumulation in glass capillaries, precoated with albumin, fibrinogen or collagen.
Anticoagulant whole blood. Perfusion 5 min at 37°C, platelets 178,000/mm³. Platelet deposition
quantitated by radioimmunoassay with ¹²⁵Iodine labeled monoclonal antibody against platelet
membrane glycoprotein complex IIb-IIIa. Results mean of 4 independent experiments (Mulvihill
et al., 1987)

Wall shear rate (s⁻¹)	Adherent platelets (/mm²)		
	Albumin	Protein coating Fibrinogen	Collagen
A. Heparin anticoagulation (10 U/ml)			
50	90	4,500	16,000
800	150	27,200	39,200
2,000	170	25,400	43,650
4,000	100	30,900	78,950
B. Citrate anticoagulation (13mM)			
50	20	2,850	13,800
800	140	7,500	56,200
2,000	400	3,600	48,050
4,000	140	2,200	18,400

Reference materials.

The reference materials recommended for the capillary perfusion system of *Mulvihill et al.* are glass capillary tubes (Microcaps, Drummond Scientific Co., Broomall, Pa., USA) of internal diameter 0.56 mm or 0.80 mm, preadsorbed with purified human albumin or fibrinogen or bovine collagen as described under *Description of Procedure* (and by *Mulvihil et al., 1987*). Guideline results for platelet deposition from washed platelet suspensions and anticoagulated whole blood at wall shear rates from 50 to 4,000 s⁻¹ are as listed in *Tables I* and *II*. The glass capillaries, readily available, form a support of constant internal diameter (*Mulvihill et al., 1987*). Protein preadsorption is simple to perform and provides three well characterized surfaces on which platelet accumulation is reproducible and of varying magnitude: low (albumin), intermediate (fibrinogen) and high (collagen).

In the catheter perfusion model of *Poot et al.*, the recommended reference material is polyethylene tube (medical grade high pressure, Talas, Ommen, The Netherlands) of internal diameter 0.75 mm. This material is easily obtained and of high purity. Platelet accumulation for perfusion times of 5 to 15 min decreases with increasing wall shear rate from approximately $4,000/mm^2$ at $150s^{-1}$ to $2,000/mm^2$ at 500 to $1,500s^{-1}$ (*Poot et al., 1988*).

DISCUSSION

Testing of biomaterials in catheter or tube form is readily performed in the capillary system. To illustrate, results are listed in *Table III* for platelet deposition and release of ß-thromboglobulin (ßTG) in perfusion of silicone and mixed polyvinylchloride (PVC)/silicone catheters with suspensions of washed [111]Indium labeled platelets (*Cazenave and Mulvihill, 1987*). Less platelet accumulation and ßTG secretion is observed in silicone tubes consistent with the low thrombogenicity of pure silicone surfaces, compared to PVC/silicone copolymer catheters.

Table III

Platelet accumulation and release of ßTG in perfusion of silicone and PVC/silicone catheters with washed [111]Indium labeled human platelets.

Catheter material	Adherent platelets ($/mm^2$)	Released ßTG (ng/ml)
Silicone	3,200	130
PVC/15.9% silicone	4,000	170
PVC/30.5% silicone	3,900	200
PVC/26.0% silicone	3,600	180

Perfusion 5 min at 500 s^{-1}, platelets 180,000/mm^3, hematocrit 40%.
Results mean of 2 independent experiments (Cazenave and Mulvihill, 1987).

The method of *Poot et al.* has been applied in studies of platelet accumulation in polyethylene tubes from suspensions of washed [111]Indium labelled platelets. The method demonstrates the passivating effect of albumin adsorbed on artificial surfaces and in contrast the activating properties of collagen, whereas fibrinogen in this case appears to give results comparable to uncoated glass. Variations in platelet accumulation are observed as a function of wall shear rate and may be interpreted in terms of transport to the interface followed by platelet reacion at the surface (*Goldsmith and Turitto, 1986*). As compared to uncoated polyethylene, preadsorption of human vWF, fibrinogen or fibronectin is found to increase platelet deposition, whereas immunoglobulin G (IgG) gives slightly less deposition and albumin or plasma lead to surfaces almost completely devoid of adherent platelets. Results in this model thus follow the same general trend as those obtained using the technique of *Mulvihill et al.* Furthermore the data are in agreement with the platelet adhesion promoting properties of vWF, fibrinogen and fibronectin in an ex vivo canine model (*Lambrecht et al., 1986*) and with the well known surface passivating properties of albumin.

On polyethylene precoated with vWF, fibrinogen or fibronectin, the concentration of adherent platelets decreases exponentially with distance from the tube inlet according to a relation of the form:

$$\text{Platelet deposition} = \mathbf{A}^{\,\mathbf{m}},$$

where $\mathbf{A}$ is the axial distance and a constant, and $\mathbf{m}$ the power law exponent. Values of $\mathbf{m}$ after perfusion for 5 min at 500 s^{-1} are for vWF, fibrinogen and fibronectin : - 0.17 ± 0.03, - 0.19 ± 0.03 and - 0.20 ± 0.04 (±SD, n = 10) respectively. Experiments, in which the perfusion time in fibrinogen coated tubes varied from 2 to 15 min, show that results are not influenced by surface saturation. Hence, since the values of $\mathbf{m}$ are intermediate between those to be expected (*Goldsmith and Turitto, 1986*) for initial platelet adhesion transport control ($\mathbf{m}$ = - 0.33) and reaction controlled ($\mathbf{m}$ = 0), one can conclude that under these experimental conditions the platelet deposition on polyethylene tubes coated with fibrinogen and with vWF and fibronectin probably follows intermediate kinetics, depending on both cell transport and surface reactivity. Platelet accumulation in fibrinogen coated tubes seem to increase as a function of wall shear rate from 150 to 1,500 s^{-1} due to an augmentation of platelet transport with rising shear forces. In contrast to this, platelet deposition on uncoated polyethylene is almost independent of axial distance from the tube inlet and decreases with increasing wall shear rate, indicating reaction controlled adhesion (m = 0). At high shear detachment occurs of loosely surface bound platelets.

As radioisotopic techniques do not permit the distinction of platelet adhesion from surface aggregation formation, capillary tubes from a limited number of experiments should routinely be examined by scanning electron microscopy. A sparse scattering of isolated platelets without pseudopods is generally observed on albumin coated surfaces (*Mulvihill et al., 1987, Poot et al., 1988*). On polyethylene (*Poot et al., 1988*), v.WF and fibrinogen coating surfaces lead to the development of platelet pseudopods, while fibronectin induces more extensive cell spreading and IgG results in the formation of aggregates surrounded by large areas without platelet deposition. fibrinogen and collagen coated glass likewise show aggregates (*Mulvihill et al., 1987*) with platelet clustering along the collgen fibres.

The capillary perfusion system can be applied to the study of the influence of human van Willebrand factor (vWF) on platelet accumulation on artificial surfaces (*Cazenave and Mulvihill, 1987*). Using suspensions of washed [111]Indium labelled platelets through glass capillaries, uncoated or precoated with human albumin or bovine collagen, vWF is either added to the platelet suspension (5 mg/l final concentration) or preadsorbed (5 mg/l) on the capillary surface before perfusion. At low wall shear rate (50 s^{-1}), vWF platelet deposition is increased only on the albumin surface (1.2 - 1.4 times reference value in untreated glass capillaries). Conversely, high wall shear rate (2,000 s^{-1}) results in an increase in the concentration of adherent platelets in the presence of vWF on all three types of surfaces (uncoated, albumin- and collagen-coated). The values with respect to controls range from ~1.3 fold for glass and collagen to 1.87 when vWF is added to the suspension and 4.08 fold (vWF preadsorption) for albumin. Similar experiments have more recently provided evidence for the potential importance of surface generated thrombin in platelet thrombus formation on biomaterials. Preadsorption of purified human thrombin on protein coated or uncoated glass capillaries results in a three to six fold augumentaion of platelet accumulation at 2,000 s^{-1} on glass, albumin, fibrinogen and vWF, but not on fibronectin or collagen surfaces.

Limitations of the method.

The capillary perfusion technique is limited to the testing of biomaterials which are available either in catheter form or as a solution or suspension for precoating of glass or polymeric capillaries. In order to attain wall shear rates in the physiological range (50 - 4,000 s^{-1}) without perfusion of large blood volumes (< 50 ml), tubes of internal diameter 1 mm or less are required.

Borderline results may be obtained with materials of low thrombogenicity. Surfaces pretreated with albumin or plasma in particular may give very low levels of platelet accumulation, close to the detection limit of the technique (see *Table II* and *Poot et al., 1988*). Since microscopic reproducibility (SEM) in a series of results is of the order of 10% (*Mulvihill et al., 1987*), the method cannot be expected to differentiate between biomaterials of similar platelet reactivity as for example illustrated with PVC/silicone copolymers (*Table III*).

Finally, as in the case of all in vitro testing methods, only the short term blood compatibility of a biomaterial is evaluated whereas the effects of long term implantation in vivo cannot be deduced.

Related testing methods.

Comparison of results from alternative testing methods should be limited to systems of similar physiological and rheological characteristics. The two techniques for the study of blood interactions in vitro with artificial surfaces which most closely approach the capillary system of *Mulvihill et al.* are the perfusion chamber originally developed by Baumgartner and since modified by *Sakariassen et al. (1983)* and the capillary flow model of *Adams & Feuerstein (1984)*. *Fressinaud et al. (1988)* have employed the parallel plate perfusion chamber to measure platelet accumulation on enquine and human type III collagen. Cover slips spray coated with collagen ($\sim$ 30 μg/cm^2) are perfused with reconstituted human blood for 5 min at a wall shear rate of 1,600 s^{-1}, platelet deposition being evaluated morphometrically and by radioisotopic labeling (^{51}Chromium). Surface concentrations of adherent platelets lie in the range 190,000 - 223,000/mm^2. In the capillary method of *Adams and Feuerstein*, protein coated glass tubes of internal diameter 1.3 mm and length 10 cm are perfused with suspensions of washed radiolabelled platelets at low wall shear rate (80 s^{-1}) for flow times of 1 to 10 min. Platelet accumulation is strongly dependent on protein preadsorption of the glass surface. Values determined over the first 1 cm length of capillary tubing range from 39,000/mm^2 on surfaces precoated with human albumin and 48,000/mm^2 with human fibrinogen after 6 min perfusion (*Feuerstein & Skupny-Garnham, 1986*) to 240,000/mm^2 with bovine collagen (type I) after 10 min perfusion (*Adams & Feuerstein, 1984*). Results from these two alternative models are thus seen to be comparable to those obtained in the capillary system of *Mulvihill et al.* for albumin, fibrinogen and collagen coated glass (see *Table I*).

REFERENCES

Adams, G.A. & Feuerstein, I.A. (1984) "Platelet accumulation on collagen: drugs which inhibit arachidonic acid metabolism and affect intracellular cyclic AMP levels".Thromb. Haemostases 52, 45-49.

Cazenave, J.-P., Hemmendinger, S., Beretz, A., Sutter-Bay, A. & Launay, J. (1983) "L'agrégation plaquettaire: outil d'investigation clinique et d'étude pharmacologique". Méthodologie Ann. Bio. Clin. 41, 167-179.

Cazenave, J.-P. & Mulvihill, J.N. (1987) "Capillary perfusion systems for quantitative evaluation of protein adsorption and platelet adhesion to artificial surfaces" in "Proteins at Interfaces: Physicochemical and Biochemical Studies" (Brash, J.L. & Horbett, T.A. eds.), Am. Chem. Soc., Washington DC, 537-550.

Feuerstein, I.A. & Skupny-Garnham, L.E. (1986) "Adhesion and aggregation of thrombin prestimulated human platelets: Evaluation of surface-bound fibrinogen and surface-bound albumin". Thromb. Res. 43, 497-505.

Fressinaud, E., Baruch, D., Girma, J.-P., Sakariassen, K.S., Baumgartner, H.R. & Meyer, D. (1988) "von Willebrand factor-mediated platelet adhesion to collagen involves platelet membrane glycoprotein IIb-IIIa as well as glycoprotein Ib". J. Lab. Clin. Med. 112, 58-67.

Goldsmith, H.L & Turitto, V.T. (1986) "Rheological aspects of thrombosis and haemostasis: basic principles and applications". Thromb. Haemostas. 55, 415-435.

Lambrecht, L.K., Young, B.R., Stafford, R.E., Park, K., Albrecht, R.M. Mosher, D.F. & Cooper, S.L. (1986) "Influence of preadsorbed canine von Willebrand factor, fibronectin and fibrinogen on ex-vivo artificial surface-induced thrombosis". Thromb. Res. 41, 99-117.

Modderman, P.W., Huisman, H.G., van Mourik, J.A. & von dem Borne, A.E.G.K. (1988) "A monoclonal antibody to the human platelet glycoprotein IIb/IIIa complex induces platelet activation". Thromb. Haemostas. 60, 68-74.

Mulvihill, J.N., Huisman, H.G., Cazenave, J.-P., van Mourik, J.A. & van Aken, W.G. (1987) "The use of monoclonal antibodies to human platelet membrane glycoprotein IIb-IIIa to quantitate platelet deposition on artificial surfaces". Throm. Haemostas. 58, 724-731.

Poot, A., Beugeling, T., Cazenave, J.-P., Bantjes, A. & van Aken, W.G. (1988) "Platelet deposition in a capillary perfusion model: Quantitative and morphological aspects". Biomaterials 9, 126-132.

Regoeczi, E. (1984) "Iodine-Labeled Plasma Proteins". CRC Press, Fla., 49-56.

Sakariassen, K.S., Aarts, P.A., de Groot, P.G., Houdijk, W.P. & Sixma, J.J. (1983) "A perfusion chamber developed to investigate platelet interaction in flowing blood with human vessel wall cells, their extracellular matrix and purified components". J. Lab.

Enzyme Immunoassay Determination of Specific Proteins in Plasma

Author: H. Pelzer

Dept. of Bloodcoagulation and Fibrinolysis, Behringwerke AG, D - 3550 Marburg

INTRODUCTION

Name of method:

Enzyme immunoassay determination of specific proteins in plasma.

Aim of method:

Quantitative immunochemical determination of proteins: Human ß-Thromboglobulin (ß-TG), Human platelet Factor 4 (PF4) and Thrombin/Antithrombin III complex (TAT).

Biochemical background for the method:

A number of proteins can be identified in the blood as indicators for activation of different phases of the coagulation system. Identification is made by the Enzyme-Linked Immuno-Sorbent Assay test (ELISA) which provides very high sensitivity in the analysis. Three different proteins are of importance.

ß-Thromboglobulin (ß-TG) is a protein released by platelets upon activation (*Lane et al., 1984*). This occurs at an early stage of the activation (*Files et al., 1981*). It appears that the platelets are highly sensitive to foreign surfaces and release ß-TG as one of the first factors.

Human platelet factor 4 (PF 4) is likewise released from the blood platelets upon activation together with ß-TG. The physiological importance of PF 4 is not completely clear, but it can be regarded as a very sensitive platelet activation marker.

S. Dawids (ed.), Test Procedures for the Blood Compatibility of Biomaterials, 395–408.
© 1993 *Kluwer Academic Publishers. Printed in the Netherlands.*

<u>Antithrombin III (AT III)</u> is a potent inhibitor of the coagulation cascade through inactivation of the serine proteases one of which is thrombin released through activation of prothrombin (*Blanke et al., 1987*). Thrombin has a central function in the coagulation process. Its activity is inhibited by activated AT III which binds to the proteases and forms an inactive protease inhibitor complex. When blood is in contact with an artificial surface, plasma proteins bind immediately followed by adhesion of platelets and other blood cells forming a protein cell layer at the interface. Upon generation Thrombin will combine with circulating AT III forming a Thrombin/Antithrombin III complex (TAT). Thus increased concentration of TAT (and Thrombin) reflects an activation of the last stages of the coagulation system. In pathophysiological systems it may indicate a state of hypercoagulation or a (pre)thrombotic state with risk of thromboembolic phenomenon.

Determination of liberated ß-TG and PF 4 will thus reflect the activation of platelets. Likewise determination of TAT in plasma reflects the functional state of the coagulation system. All three indicators represent a diagnostic tool for detection of a parameter in the coagulation activation process.

Scope of the method.

The method is suitable for general screening investigations and as a clinical routine procedure for monitoring the surface properties of polymers.

DETAILED DESCRIPTION OF METHOD

Equipment.

Most manufacturers of equipment for ELISA (mainly the monoclonal antibodies) deliver all necessary reagents including a clear description of assay performance etc.

General equipment is:

> Washing device such as a 2 ml dispenser and a liquid aspiration pump (water-jet pump)
> Washing equipment for microwell plates or strips
> Water bath ($+20°$-$+25°C$ or $+37°C$)

Photometer suitable for 1ml sample volumes with a 1 cm path length, wavelength 492 nm (Model EL 309, BIO-TEK Instruments, Germany)
Plate rader set at 492 nm

For ß-TG:

Equipment: Precision pipettes (multichannel pipettes), 50 μl, 100 μl and 200 μl

Reagents: Anti-ß-TG microwells
Anti-ß-TG/POD conjugate
Conjugate buffer (ß-TG)
Sample buffer (ß-TG) (used for dilution of plasma)
Reference ß-TG
Washing solution
Chromogen POD
Stopping solution POD
Adhesive foils
ß-TG sampling medium (used for blood collection)

For PF 4:

Equipment: Precision pipettes, 100 μl, 200 μl and 1000 μl

Reagents: Anti-PF 4 tubes
Anti-PF4/POD conjugate
Conjugate buffer (PF 4)
PF 4 standard plasma (S1 to S4)
PF 4 control plasma
Sample buffer (PF 4) (used for dilution of plasma)
Washing solution POD
Buffer/substrate POD
Chromogen POD
Stopping solution POD
Adhesive foils
PF 4 sampling medium (used for blood collection)

For TAT:

Equipment: Precision pipettes, 100 μl, 200 μl and 1000 μl

Reagents: Anti-Thrombin tubes
Anti-AT III/POD conjugate
Conjugate buffer (TAT)
TAT Standard Plasma (S1 to S4)
TAT Control Plasma

Sample buffer (TAT) (used for dilution of plasma)
Washing solution POD
Buffer/substrate POD
Chromogen POD
Stopping solution POD
Adhesive foils
TAT sampling medium (used for blood collection)

Testing conditions.

The assay should be performed at a fixed temperature.
For ß-TG: 18°-25°C.
For PF 4: 20°-25°C or 37°C.
For TAT 20°-25°C or 37°C.
See further in *Description of procedure*.

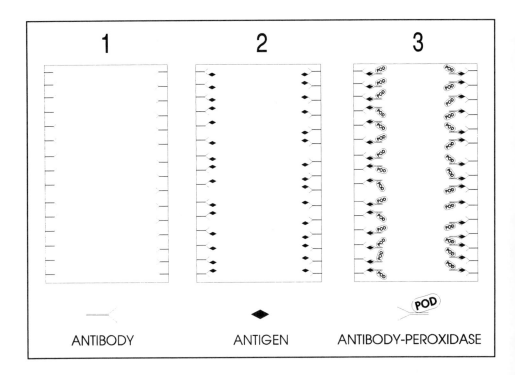

Fig. 1

Test principle of ELISA.

1. *Preadsorbed antibody on the microwell surface*
2. *Binding of TAT antigen*
3. *Binding of peroxidase conjugated antibodies.*

Outline of the method.

The ELISA method is designed as a sandwich enzyme immunoassay: During the first incubation the molecule in question (ß-TG, PF 4, TAT) in the plasma sample is bound to specific antibodies fixed to a plastic surface (microwell). After washing, peroxidase conjugated antibodies (towards ß-TG, PF 4 or TAT) are bound in a second reaction to those antigenic determinants which are still free. The excess enzyme conjugated antibodies are washed off and the bound enzyme activity is subsequently determined. The enzymatic reaction of hydrogen peroxide and o-phenylenediamine (chromogen) is stopped after a preset time by the addition of dilute sulfuric acid. The color intensity is proportional to the concentration of the molecule (ß-TG, PF 4 or TAT) and is determined photometrically. The principle in ELISA is shown in *Fig. 1*.

Description of the procedure.

To avoid activation of the clotting system and the platelets, the collection of blood is critical. A careful clean venipuncture is absolutely essential. For the analysis of ß-TG and PF 4, the specimen must be refrigerated at $+2°$ to $+8°C$ for at least 30 min. before centrifugation.

Collecting the specimen for ß-TG and PF 4.
- Fill a disposable syringe with an initial volume of the sampling medium (ß-TG sampling medium or PF 4 sampling medium) followed by the blood: 1 Vol. of sampling medium to 9 Vol whole blood (preferably 0.5 ml to 4.5 ml).
- Administer a pressure of about 5.3 kg/Pasc. (40 mm Hg) on the patient's upper arm for as short time as possible to perform an exact venipuncture using a butterfly needle of 1.1 mm diameter. Do not move the needle once it is in the vein.
- Draw the blood sample into a disposable syringe and transfer within 10 seconds the blood to a polystyrene tube. Store this tube immediately in an icebath for 30 min. The blood sample can be stored at $+2°$ to $+8°C$ for up to 2 hours (icebath or refrigirator).

Preparation of the specimen for ß-TG and PF 4

Procedure A 2-step centrifugation at 2000 x G and 8000 x G	Procedure B 1-step refrigirated centrifugation at 2000 x G
1. Centrifuge refrigerated sample at 2000 x G at a temperature between +2° and +25°C for 15 min.	1. Centrifuge reefrigerated sample at 2000 x G at a temperature between +2°and +8°C for 60 min.
2. Remove middle layer of plasma supernatant and centrifuge at 8000 x G at a temperature between +2° and +25°C for 15 min.	2. Remove middle layer of plasma supernatant and transfer to a polystyrene tube.
3. Remove middle layer of plasma supernatant and transfer to a polystyrene tube.	3. Use as test material in the enzyme immunoassay.
4. Use as test material in the enzyme immunoassay.	
Stability of samples: up to 24 hours at +2° to +8°C and 4 weeks at -20°C.	

Any deviation from this procedure can result in falsely elevated PF 4 values.

Collecting the specimen for TAT.

Using a clean sterile syringe, draw up 1 part of 0.11 mol/l sodium citrate solution, then 9 parts of venous blood from a clean venipuncture. Mix thoroughly, avoiding the formation of foam and transfer the mixture to a clean plastic tube free of thrombin and detergents. Centrifuge within 2 hours at approximately 3000 rpm (1500 x g) for 10 min. and remove the plasma supernatant. - Samples are stable for 8 hours at either +2° to +8°C or +15° to +25°C and for 2 month at -20°C. Inadequate mixing of venous blood with the sodium citrate solution may lead to falsely elevated TAT values.

Assay procedures for:

ß-TG.

Notes:

- All samples should be tested at 1:10 dilution (1 vol plasma + 9 vol dilution buffer).
- Each sample (standard, unknowns) must be assayed in duplicate.
- Reference ß-TG is reconstitued in dilution buffer and dilutions between 10 and 200 IU/ml are prepared.

Test Scheme: Pipette as follows

1st pipetting	Reference ß-TG dilution or patient sample dilution	200 µl
1st incubation	1 hour at +18° to +25°C. Cover microwell strips with adhesive foil. Wash 5 times with washing solution.	
2nd pipetting	Conjugate solution. Cover with adhesive foil and incubate:	200 µl
2nd incubation	1 hour at +18° to +25°C. Wash 5 times with washing solution.	
3rd pipetting	Chromogen-buffer/substrate solution	200 µl
3rd incubation	Incubate exactly 3 min. then add:	
4th pipetting	Stopping solution Measure within 1 hour with a reader (wavelength 492 nm).	50 µl

Additional notes are described under *Assay procedure of PF 4.*

PF 4.

Notes:

- Allow all reagents and samples to warm to +20° to +25°C before beginning the assay. Failure to do so may cause erroneous results.

- Each sample (standards, controls, all unknowns) must be assayed in duplicate.
- Do not incubate samples in tubes without sample buffer. The time required for addition of all samples to the tubes should not exceed 5 min.
- The incubation period timing should begin when the first tube is filled and end with the first aspiration. The exact incubation period may vary within stated limits, but should be the same for all tubes in a series.
- Start with the preparatory work: reconstitute the standards and control plasma. Dilute the Antibody/POD conjugate (Anti-ß-TG/POD conjugate, Anti-PF 4/POD conjugate and Anti-AT III/POD conjugate) and the washing solution POD concentration according to the directions in the kit.

Test Scheme: Pipette as follows

1st pipetting	Sample buffer	100 µl
2nd pipetting	Patient sample (or standard or control plasma). Cover with self-adhesive foil and incubate:	100 µl
1st incubation	2 hours ± 10 min. at +20° to +25°C or 1 hour ± 5 min. at +37°C (water bath). Wash twice with washing solution POD (2 ml for each washing and tube).	
3rd pipetting	Conjugate solution Cover with new self-adhesive foil and incubate:	200 µl
2nd incubation	2 hours ± 10 min. at +20° to +25°C or 1 hour ± 5 min. at +37°C (water bath) Wash 3 times with washing solution POD (2 ml for each washing and tube).	

4th pipetting	Chromogen-buffer/substrate solution	200 µl
	Cover with new self-adhesive foil.	
	Note: Prepare chromogen-buffer/substrate	
	solution just before incubation time is over.	

3rd incubation	Incubate protected from light (in a
	closed cupboard) 30 min. ± 2 min. at +20° to
	25°C.

5th pipetting	Stopping solution	1000 µl
	Add the solution to each tube in the same	
	timing sequence followed by adding the	
	chromogen-buffer/substrate.	

Measure within 1 hour in a photometer
(wavlength 492 nm) against a water blank.

TAT.

Notes:

- Allow all reagents and samples to warm to +20° to +25°C before beginning the assay. Failure to do so may cause erroneous results.
- Each sample (the standards, the control, all unknowns) must be assayed in duplicate.
- Do not incubate samples in tubes without sample buffer (TAT). The time required for addition of all samples to the tubes should not exceed 5 min.
- The incubation period timing should begin when the first tube is filled and end with the first aspiration. The exact incubation period may vary within stated limits, but should be the same for all tubes in the series.
- Start with the preparatory work: reconstitute the standards and the control plasma. Dilute the Anti-At III/POD conjugate and the washing solution POD concentration according to the directions in the kit.

Test Scheme: Pipette as follows

| 1st pipetting | Sample buffer | 100 µl |

2nd pipetting	Patient sample (or standard or control	100 µl
	plasma). Shake and cover with adhesive	
	foil and incubate:	

1st incubation	30 min. ± 2 min. at +37°C (water bath). Wash twice with washing solution POD (2ml for each washing and tube).	
3rd pipetting	Conjugate solution. Cover with new adhesive foil and incubate:	200 μl
2nd incubation	30 min. ± 2 min. at +37°C (water bath). Wash 3 times with washing solution POD (2 ml for each washing and tube).	
4th pipetting	Chromogen-buffer/substrate solution. Cover with new adhesive foil. Note: Prepare chromogen-buffer/substrate solution just before incubation time is over.	200 μl
3rd incubation	Incubate protected from light (in a closed cupboard) 30 min. ± 2 min. at +20° to +25°C.	
5th pipetting	Stopping solution Add the solution to each tube in the same timing sequence followed by adding the chromogen-buffer/substrate.	1000 μl
	Measure within 1 hour in a photometer (wavelength 492 nm) against a water blank.	

1st and 2nd incubation may alternatively be performed at +20° to +25°C for 1 hour ± 5 min.

Calibration and Calculation.
- Standard plasma solutions (ß-TG, PF 4 or TAT) S1 to S4 are assayed exactly as described above.
- Calculate the mean absorbance value from the duplicate absorbance measurements.
- Plot the mean absorbance value for each of the 4 standards (S1 to S4) against the known concentration on log/log graph paper.
- Draw a reference graph by joining the points with a smooth line.

- The control plasma (ß-TG, PF 4 or TAT) must be assayed together with each batch of unknown specimens. The result measured should fall within the confidence range.
- Calculate the mean absorbance from the duplicate readings for each unknown sample.
- Using the mean absorbance, interpolate the reference curve to determine the concentration (in µg/l) in the unknown plasma specimen.
- If the unknown absorbance exceeds the upper limit of the reference curve, the sample must be diluted with sample buffer and reassayed.

Safety aspects.

Care should be taken that the standard plasma may be in infectious (they are normally only examined for hepatitis B/HBsAg and antibody to HIV). Thus it is potentially biohazardous material.

The stopping solution and the chromogen are caustic. Handle with care. Do not allow the chromogen, chromogen-buffer/substrate or stopping solution to contact the skin.

Do not allow the chromogen-buffer/substrate solution to contact metals or oxidizing substances.

RESULTS

Estimation of the results should be performed as described under *Assay procedure*. When plasma samples from apparently healthy adults are assayed with the ELISA technique, the median values are as given below:

ß-TG-ELISA 24 IU/ml. Reference range 16 - 32 IU/ml
PF 4-ELISA 2.5 µg/l. Reference range 1.4 - 6.1 µg/l
TAT-ELISA 1.5 µg/l. Reference range 1.0 - 4.1 µg/l

DISCUSSION

Interpretation of results and limitations of method.

The results of the test should be interpreted respecting the individual basic values of each person and in the light of other laboratory and/or clinical findings (*Pelzer & Heimburger, 1986*).

ß-TG.

Normal ß-TG concentrations in the range of 15 - 40 IU/ml reflect that no abnormal platelet activation takes place. Concentrations of > 40 IU/ml indicates an activated state. False positive results (> 40 IU/ml) might be induced in vitro by platelet activation during specimen collection and/or centrifugation procedure. Therefore the initial steps (careful venipuncture, rapid blood sampling and processing) is absolutely essential. The half-life of elimination from plasma is difficult to estimate as there is an affinity to endothelial cells and a potential release by heparin (*Daves et al., 1978*). In vitro activation of platelet with subsequent release of ß-TG does take place in a small scale. It should be noted that induced release of ß-TG takes place in case of foreign implants and devices. This high sensitivity of platelets towards foreign surfaces can lead to activation during blood sampling. During ex vivo and in vivo tests the platelet activation should be avoided by other means (*Green et al., 1980; Pumphrey et al., 1981; Conard et al., 1984*).

PF 4.

Normal PF 4 concentrations are in the range of 1.0 - 6.0 μg/l and reflect no abnormal platelet activation whereas a concentration of > 6.0 μg/l indicates an activated state. False positive results (> 6.0 μg/l) might be induced by in vitro platelet activation during specimen collection and/or centrifugation procedure. Therefore the same aspects on procedures as mentioned for ß-TG are valid. Especially in experiments where in vitro or ex vivo PF 4 release can be expected it is important to compare the results with appropriate control experiments (relative value changes). As in the other methods the ELISA can give a quantitative determination of the protein. Several investigations indicate that different biomaterials exhibit distinct properties regarding to platelet activation. In vitro experiments with anticoagulated whole blood demonstrate that polyether-urethane and polyvinyl-chloride induce the release of PF4 at different levels (*Pelzer et al.,1986*). Similar effects can be obtained with the same materials using non-anticoagulated native blood (*Pelzer, 1986*). It thus seems as if a pattern of the compatibility of organic polymers agree in ex vivo and in vitro. PF 4 seems to be a valuable marker in the characterization of dialysis membranes in respect to platelet activation (*Pelzer, 1984; Pelzer et al., 1982*).

TAT.

Normal TAT concentrations in the range of 1.0 - 4.0 $\mu g/l$ indicate no activation of the coagulation system (*Pelzer et al., 1988*). On the other hand, increased concentrations > 4.0 $\mu g/l$ indicates activated state, i.e. a continuously formation of Thrombin-Antithrombin III complex. TAT generation may be induced through intravasal (i.e. in vivo) disorders of the coagulation pathway (*Hoeck et al., 1988*). In experimental procedures TAT formation is generally detectable. In any case elevated TAT values due to blood collection and/or sample preparation must be avoided by the abovementioned precautions. In experiments where in vitro or ex vivo TAT formation can be expected, the obtained results from an investigation should be compared to appropriate control experiments to obtain relative values.

REFERENCES

Blanke, H. et al. (1987) "Die Bedeutung des Thrombin-Antithrombin III Komplexes in der Diagnostik der Lungeembolie und der tiefen Venenthrombose - Vergleich mit Fibrinopeptid A, Plättchenfaktor 4 und β-Thromboglobulin", Klin. Wochenschrift, 65, 757-763.

Conard, J. et al. (1984) "Plasma β-thromboglobulin in patients with valvular heart disease with or without value replacement: Relationship with thromboembolic accidents", Eur. Heart Journal (Suppl.), 5, 13-18.

Dawes, J. et al. (1978) "The release, distribution and clearance of human β-thromboglobulin and platelet factor 4", Thromb. Res., 12, 851-861.

Files, J.C. er al. (1981) "Studies of human platelet alfa-granule release in vivo", Blood, 58, 607-618.

Green, D. et al. (1980) "Elevated β-thromboglobulin in patients with chronic renal failure: effect of hemodialysis", J. Lab. Clin. Med., 95, 679-685.

Hoek, J.A. et al. (1987) "Laboratory and clinical evaluation of an assay of thrombin-antithrombin III complexes in plasma", Clin. Chem., 34, 2058-2062.

Lane et al. (1984) "Detection of enhanced in vitro platelet alfa-granule release in different patient groups-comparison of β-thromboglobulin, platelet factor 4 and thrombospodin assays", Thromb. Haemost., 52, 183-187.

Pelzer, H., Fuhge, P., Michalik, R., Lange, H. & Heimburger, N. (1982) "Interactions between dialysis membranes and the coagulation system during hemodialysis", Proc. IX Ann. Meeting, ESAO, Brussels.

Pelzer, H. (1984) "Biochemical methods to evaluate blood compatibility of biomaterials" in "Polyurethanes in Biomed. Engineering", (H. Planck, E. Egbers, J. Syre eds.) Progr. Biomed. Eng., 1, Elsevier, Amsterdam.

Pelzer, H. (1986) "Assessment of in vitro/ex vivo blood compatibility of biomaterials" in "Blood Compatible Materials and their Testing" (S. Dawids & A. Bantjes, eds.) Martinus Nijhof Publishers, Dordrecht.

Pelzer, H. & Heimburger, N. (1986) "Evaluation of the in vitro and ex vivo blood compatibility of primary reference materials", J. Biomed. Mat. Res., 20, 1401-1409.

Pelzer, H., Schwarz, A. & Heimburger N. (1988) "Determination of human thrombin-antithrombin III complex in plasma with an enzyme-linked immunosorbent assay", Thromb. and Haemost., 59 (1), 101-106.

Pumphrey, C.W. et al. (1981) "Elevation of plasma βthromboglobulin in patients with prosthetic cardiac values", Thromb. Res., 22, 147-155.

Human Adult Endothelial Cell Culture Conditions for Optimal Colonization of Biomaterials Used in Vascular Surgery

Authors: C. Klein-Soyer, J.-P. Mazzucotelli, A. Beretz, G. Archipoff, C. Brisson, J.-P. Cazenave

INSERM U311, Centre Régional de Transfusion Sanguine, F - 67085 Strassbourg Cedex

INTRODUCTION

Name of method:

Human adult endothelial cell culture conditions for optimal colonization of biomaterials used in vascular surgery.

Aim of method:

The appropriate culture conditions for human endothelial cells (EC) have been determined in order to obtain optimal cell adhesion, spreading and growth on biomaterials currently used in vascular surgery.

Biophysical background for the method:

Human EC have for a long time been known as difficult cells to grow in vitro. The principal hindrance to the establishment of a culture was to isolate homogeneous populations of EC and to define culture conditions which allowed easy propagation of the cells while maintaining their essential characteristic properties: contact inhibition, polygonal morphology, non thrombogenecity, synthesis of von Willebrand factor and presence of the specific endothelial organelles, the Weibel-Palade bodies. In 1973, (*Jaffe et al., 1973; Gimbrone et al., 1973*) it was shown that it is possible to grow repeatedly and to characterize EC obtained from human umbilical cord veins. However, mostly primary cultures were used. The introduction of precoating of surfaces for cell culture with adhesive proteins which mimic the extracellular matrix prior to cell seeding (*Gospodarowicz et al., 1979*) and the introduction of growth factors or high concentrations of human serum now allows

409

S. Dawids (ed.), Test Procedures for the Blood Compatibility of Biomaterials, 409–423.
© 1993 *Kluwer Academic Publishers. Printed in the Netherlands.*

serial propagation of EC (*Thornton et al., 1983*). Although the replacement of pathological portions of vessels of large caliber by biomaterial conduits is fairly successful, the replacement of small caliber vessels (diameter ≤4mm) remains disappointing. The biomaterials most commonly used in vascular surgery are PETP (polyethylene terephtalate or Dacron®) and e-PTFE (expanded polytetrafluoro-ethylene or Gore-Tex®), but their patency has proven limited in time. The use of small caliber grafts with these biomaterials is not feasible because of their rapid thrombotic occlusion. The vascular endothelium lining the cardiovascular system represents the best possible haemocompatible surface. Unfortunately, no true endothelial lining develops in humans on these grafts as long as 5-11 years after surgery, except at the anastomotic sites (*Berger et al., 1972; Camilleri et al., 1985*). Thus the development of an endothelial lining on the inner surface of the graft to implantation should improve their patency.

This implies defining optimal conditions of cell harvest and culture from human vessel fragments and determining conditions of EC adhesion and growth on biomaterials used in vascular surgery. The establishment of a complete functional endothelial lining on vascular grafts depends mainly on the surface characteristics of the polymer such as wettability, charge, roughness, porosity and permeability (*van Wachem et al., 1985*). Biomaterials are designed to be biocompatible or haemocompatible, that is to say to minimize interactions with blood cells and plasma proteins such as coagulation, fibrinolysis or complement factors. In counterpart, EC would not easily adhere and spread on such materials without any treatment providing adhesion sites. Several treatments of biomaterials like glow discharge (*Amstein et al., 1975*), radio-frequence plasma etching (*Chinn et al., 1988*) and precoating with adhesive proteins have been proposed to favor cell adhesion (*van Wachem et al., 1987*).

Scope of method:

This method is still experimental, but is aiming at clinical application.

DETAILED DESCRIPTION OF METHOD

Equipment.

Laminar flow bench and aseptic facilities are necessary.

Tissue culture grade polystyrene dishes 35 mm (2 dishes in a vessel of 4 cm length),
(Corning, New York, N.Y. USA)

Cryotubes 2 ml (NUNC, Roskilde, Denmark)

Pasteur pipettes, sterile, disposable

For siliconizing: Rhodorsil emulsion E1P (Prolabo, Paris, France)

Silicone polymer Rhodorsil RTV 141 (Rhône-Poulenc, Paris, France)

Reagents:

Hank's balanced salt solution without Ca^{2+} or Mg^{2+}

Culture medium M199/RPMI 1640 v/v, added (*Klein-Soyer et al., 1986*):

-   10 mMol Hepes
-   2 mMol L-glutamine
-   penicillin (100 U/ml)
-   streptomycin (100 µg/ml)
-   fungizone (2.5 µg/ml), (GIBCO, Paisley, UK)
-   30% pooled human serum, (Centre Régionale de Transfusion Sanguine
    (CRTS) Strassbourg, France)

Purified human fibronectin (FN) prepared according to *Ruoslahti et al., 1978*

Factor VIII intermediate purity concentrate (CRTS, Strassbourg, France)

Cell detachment solution:

-   isotonic sodium chloride
-   trypsin 0.05%
-   EDTA 0.02%

Dulbecco's rinsing solution:

-   phosphate buffered saline (pH 7.4) containing Mg^{2+} and Ca^{2+}

Freezing solution:

-   90% culture medium
-   10% DMSO (dimethyl sulfoxide)

Crude FN (cFN), a side-product of plasma cryoprecipitate containing 50% fibronectin (FN), fibrinogen, albumin and immunoglobulins. It is at present not
commercial. The description is given in the literature (*van Wachem et al.,
1987*)

Factor VIII concentrate (Serum Institute, Copenhagen, Denmark)

Transglutine® (TGL), a biological glue (CRTS, Strassbourg, France)

Outline of method.

Human endothelial cells can be harvested from venous fragments left over from
patients undergoing coronary bypass surgery. After gentle washing the cells are

carefully scraped from the surface and transferred to a culture medium where they are dispersed. The cells are then seeded in tissue culture dishes where the culture medium is regularly renewed. The vital cells will fix onto the surface and resist removal. They are then detached by treament with trypsin and subcultured. These cultures will reach confluence in 5-7 days after splitting. All the established primary cultures will be cryopreserved after the second passage. When they are to be used they are thawed in a water bath and suspended in culture medium. Before they are seeded the viability of the thawed cells is ensured. The grafts of polymer material are precoated with different types of adhesive protein mixtures e.g.fibronectin (FN), fibrinogen and Factor VIII.

Description of procedure.

Human endothelial cells (EC) are obtained from fragments (2-10 cm in length) of apparently healthy saphenous veins from patients undergoing coronary bypass surgery. The vessels are usually kept at room temperature if they are to be processed at once (or 4°C if processing is delayed for more than 12 hours). The vessels are submerged in Hank's balanced salt solution without Ca^{2+} or Ma^{2+}. To this solution is added penicillin (100 U/ml), streptomycin (100 μg/ml), fungizone (2.5 μg/ml) and human serum albumin ad 1%.

Sample processing.

The vascular segment is processed under sterile conditions after surgery, preferably within 4-6 hours after extraction. The vessel is mounted with sterile pins on a Petri dish containing 1 cm thick silicone polymer. The wall is opened with fine scissors, and the lumen is carefully rinsed with Dulbecco's solution to remove blood using a Pasteur pipette. The luminal surface is gently scraped once is a scalpel blade no. 10. The clumps of cells which are retained on the blade are transferred to the culture medium (*Klein-Soyer et al., 1986*) by shaking the blade in the culture medium. The clumps are then dispersed by repeated aspiration with a siliconized Pasteur pipette.

Cell culture.

The suspension of endothelial cells is seeded in 35 mm diameter tissue culture dishes, precoated for adhesion with purified human fibronectin (FN). The inner surface is precoated for adhesion with fibronectin or Factor VIII. Adhesion is prepared by exposing the surface to the solution for a few minutes. The medium is changed within 2 hours in order to prevent potential smooth muscle cell adhesion and to remove non-adherent material. Subsequent changes take place twice weekly.

The number of adherent cells varies from 3×10^3 to 5×10^3 cells/cm^2 the day after seeding. If the seeding takes on, it will reach confluence in 1-2 weeks.

Subculturing.

The cells are detached by treatment with trypsin 0.05%/EDTA 0.02% and subcultured. Seeding takes place with a concentration of 10^4/cm^2 on adhesive protein coated dishes in the complete culture medium. The subcultures reach confluence in 5-7 days after splitting. When the established primary cultures reach the end of the second passage, they are cryopreserved. These cells exhibit characteristic properties: contact inhibition, cobblestone morphology, synthesis of von Willebrand factor and thrombomodulin activity.

Vascular EC easily withstands cryopreservation in early passages without loosing their essential characteristics after thawing.

Freezing.

The cells are trypsinized before confluency is reached, and the cell suspension is washed in 10-20 volumes of complete culture medium (M199/RPMI + 30% serum). After centrifugation (250g in 7 minutes at 20°C) the pellet of cells is resuspended in chilled freezing medium (50% complete culture medium, 10% DMSO) placed on crushed ice and distributed in cryotubes. The cell suspension is distributed with 1 ml per tube containing $1-2 \times 10^6$ cells per ml. The cells are then cooled to -80°C (dry ice or deep freezer for at least 2 hours) and subsequently stored in liquid nitrogen until required for use.

Thawing.

Thawing is performed by warming the cells rapidly in 37°C water bath and immediately transferring them into 20 volumes of prewarmed complete culture medium. After centrifugation at 250g in 7 minutes, the cells are resuspended in culture medium and seeded under the desired conditions. The viability of the thawed cells (evaluated by Trypan blue dye exclusion) is 90 - 95%. Thawed cells can be subcultured for several passages without loosing their essential characteristic properties.

Seeding of endothelial cells on vascular graft material.

The material commonly used for vascular surgary is Dacron$^®$ (polyethylene terephtalate, PETF) and Gore-Tex$^®$ (expanded polytetrafluoroethylene, e-PTFE). The technique using these two materials is described.

Building of a test chamber.

Pieces of the biomaterial are cut from the grafts, stretched and mounted between two close-fitting concentric rings. The inner ring is cut from a barrel of polypropylene syringe and the outer ring from a polystyrene tube. Thus a small "culture dish" is obtained with the floor consisting of the inner part of the graft material (*Fig. 1*) with a surface of 1.1 cm². The mounted chambers can be sterilized by ethylene oxide.

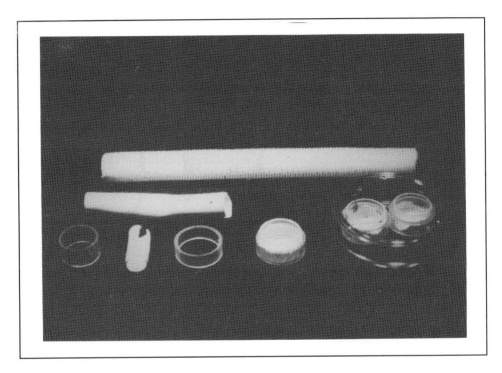

Fig. 1

Test chamber. The piece of biomaterial is immobolized between the two concentric rings.

e-PTFE (Gore-Tex®)

Prior to EC seeding on e-PTFE it should be coated with different adhesive protein mixtures: crude fibronectin (*van Wachem et al., 1987*), Factor VIII concentrate and Transglutin®. Coating is performed by incubating the graft material with a solution of proteins in phosphate buffer solution (PBS containing Ca^{2+} and Mg^{2+}) containing a concentration of FN of 0.5-3 mg/ml. The solutions with cFN, FN and Factor VIII concentrate solutions are adsorbed for approximately 60 min. at room temperature. TGL is mixed with dilute bovine thrombin (0.1 IU/ml). The mixture is immediately spread on the graft surface. The excess solution is removed in order to leave a thin film coating the graft surface. The TGL mixture is then left to polymerize at room temperature for one hour after which it is rinsed with PBS. An alternative method for coating of e-PTFE grafts is to apply the adhesive proteins with a syringe piston of the same diameter as the inner part of the test chamber until the solution starts to leak out through the material.

PETP (Dacron®)
Due to the porosity of PETP grafts, the coating procedure is different. TGL is applied in large amounts in order to soak the fabric and allow to polymerize for one hour. Microscopic monitoring indicates that most of the fabric is included in the gel which is formed by the polymerized glue. After this treatment the culture medium does not permeate through the PETP disc.

Seeding of the vascular grafts.
Seeding of EC is performed in principle as described above. Approximately 6×10^3 cells/cm^2 are used. The graft material is always tested in comparison to TCP dishes coated with the same adhesive proteins. Treatment of vascular prosthesis with endothelialization has not attained clinical use. The same method should be applied as described above. Special equipment must be designed to hold the items suspended and to reduce the amount of culturing medium necessary.

Safety aspects and biohazards.

Normal laboratory safety rules should be observed in the handling of glassware, sharp items and chemical substances. Cell culture needs in addition observation of strict sterility rules. Protection against biological hazards is achieved by using sterile working procedures which minimize direct contact with cells or reagents , using by heat-inactivated human serum from donors screened for hepatitis and HIV-viruses, and by working under biohazard laminar air flow hood. Waste materials should be discarded in biohazard containers which are incinerated. Work areas should be decontaminated with 70% alcohol solution, hypochloride solution and disinfectant sprays.

RESULTS

It is important to handle the saphenous vein individual fragments with care. Up to now 64% of the vessel fragments give rise to primary cultures. The cell density at confluence in primary culture is about 8.10^4 cells/cm^2, estimated by direct numeration on photographs (*Klein-Soyer et al., 1984*) or by image analysis. Over a period of one year, 123 saphenous vein individual fragments were processed in two ways:

Group A (74 fragments) were treated according to the procedure of veins undergoing treatment prior to aorto-coronary bypass, i.e. perfusion and distention with isotone sodium chloride and preservation in 9% sodium solution containing xylocaine.

Group B (49 fragments) were handled in order to maintain the integrity of the endothelial layer and prevent it from undergoing any traumatism.

In **group A** 56.7% of the samples failed to grow in primary culture, either due to contamination by smooth muscle cells (9.4%), to total absence of cells (39.1%) or to culture degeneration with presence of multinucleated cells (8.1%).

In **group B** 6.1 % failed to grow in primary culture, either due to contamination of smooth cells (2%), to total absence of cells (2%) or to other reasons (2%). The confluence speed was also different. In a 75 cm^2 flask confluence was obtained in **group A** after 21 ± 8.2 days and in **group B** after 16 ± 4.3 days. The final number of cells in the two groups was $\pm 2.95 \pm 0.9 \times 10^6$ cells/cm^2.

In practice approximately 2/3 of the vessel fragments may give rise to primary cultures.

Growth characteristics of endothelial cells on graft biomaterials.

Thawed or subcultured EC seeded on precoated e-PTFE.
Cell adhesion necessitates pre-treatment with adhesive proteins. The nature or the type of protein pretreatment seems to have no significant effect, and the adhesion is similar to that on TCP (tissue culture polystyrene) dishes precoated with the same proteins (see *Table I*).

Table I

Growth characteristics of human saphenous vein endothelial cells on tissue culture polystyrene or e-PTFE precoated with various proteins.

	% adherent cells 24 h after seeding	Doubling time (h)	Confluent cell density (cells/cm^{2}x10^{-3})
TCP/cFN	34.7 ± 10.3	48.6 ± 6.1	51.2 ± 7.8
TCP/FVIII	35.2 ± 10.2	48.6 ± 5.1	56.5 ± 6.7
TCP/TGL	30.6 ± 8.0	47.3 ± 4.0	66.4 ± 8.7
e-PTFEplain	0.96 ± 0.4	no growth	no growth
e-PTFE/cFN	22.3 ± 3.0	no growth	no growth
e-PTFE/FVIII	19.8 ± 3.0	no growth	no growth
e-PTFE/TGL	37.9 ± 16.3	47.1 ± 5.8	59.2 ± 4.9

TCP:	Tissue culture polystyrene	e-PTFE:	expanded PTFE
cFN:	crude fibronectin	FVIII:	factor VIII concentrate
TGL:	Transglutine®		

Results are the mean ± s.e.m. of 3 experiments using cells from different donors in which all conditions were assayed simultaneously. The results were analysed by 2 way variance analysis followed by a Duncan test.

Cell proliferation is not seen on e-PTFE coated with cFN or Factor VIII. Most of the cells adherent 24 hours after seeding are washed away during change of medium. The doubling time of EC seeded on e-PTFE coated with TGL is similar to that of EC growing on TCP dishes coated with any of the three types of adhesive proteins. Normally the final cell density reached after 12 days of culture is not significantly different in any of the conditions (see *Table I*).

Adsorption of adhesive proteins under pressure.

The EC will adhere to e-PTFE coated with cFN or TGL (see *Table II*) in a similar manner as is seen on TCP dishes. However, cell proliferation does not take place if e-PTFE is coated with cFN without pressure although the EC adhere. Contrary-wise, pressure application will allow both cell adhesion and cell proliferation comparable to that observed for TGL coated e-PTFE (see *Table II*). Likewise at confluence the cell densities are similar and comparable to that obtained in standard conditions as described.

Table II

Growth characteristics of human saphenous vein endothelial cells on e-PTFE coated with crude fibronectin or Transglutine® applied with or without pressure. Comparison with tissue culture polystyrene coated with the same protein mixtures.

	% adherent cells 24 h after seeding	Confluent cell density (cells/cm^{2}x10^{-3}
TCP/TGL	78.0	52.2
TCP/cFN	90.0	37.0
e-PTFE/TGL	93.3	47.1
e-PTFE/cFN	96.6	no growth
e-PTFE/TGLP	65.0	51.3
e-PTFE/cFNP	58.3	48.1

TGLP and cFNP: TGL and cFN applied under pressure.

PETP.
EC do not adhere to uncoated woven PETP. When the fabric is soaked with polymerized TGL the amount of cells adhering after 24 hours after seeding is identical to that observed on TCP coated with TGL (*Table III*).

Cell proliferation pretreated with TGL induces EC growth on PETP. The doubling time, however, is significantly higher than on TGL coated TCP ($p < 0.05$). Similarly, the confluent cell density as determined by image analysis is lower (see *Table III*) compared to other substrates.

Table III

Growth characteristics of human saphenous vein endothelial cells on tissue culture
polystyrene or PETP coated with TGL.

	% adherent cells 24 h after seeding	Doubling time (h)	Confluent cell density (cells/cm^{2}x10^{-3})
TCP/TGL	93.8 ± 1.0	31.4 ± 5.0	39.6 ± 1.3
PETP/TGL	101.0 ± 6.1	88.9 ± 18.4	23.5 ± 2.6

*Results are the mean ± s.e.m. of 3 experiments using cells from different donors in which
all conditions were assayed simultaneously. The results were analysed by variance analysis
followed by a Newman-Keuls test.*

DISCUSSION

Interpretation of results.

Improvement of human EC culture by harvesting non traumatized vein fragments.
Improved human EC culture by harvesting non-traumatized vein fragments seems
to be essential. Although the culture conditions for human EC are improved with
the use of adhesive proteins and high concentrations of human serum or growth
factors associated or not with heparin, the quality and the issue of the culture is
significantly dependent on the conditions of vein sample collection. Thus some
precautions are required during surgical collection of the vessels:

- avoid exposure to substances such as xylocaine or isotonic sodium
  chloride,
- no dilation of the vessel,
- preservation should be carried out in a buffer containing glucose and
  proteins like human albumin rather than just saline solution,
- minimize the time between collecting and processing of the vessel
  fragment.

Endothelialization of vascular prostheses.

The presented results makes it clear that neither PETP nor e-PTFE will support adhesion or growth of human endothelial cells unless they are precoated with adhesive proteins (*Callow, 1987*). But even if EC adhere to e-PTFE coated with cFN or Factor VIII concentrate no growth occurs. This can be explained by the fact that with time during medium exchange the adhesive proteins are desorbed and washed out. Thus the adhesive proteins are not only required to improve the initial adhesion of EC, but they also have to be cohesive with the graft material in order to mimic a basement membrane on which the new synthetized extracellular matrix produced by proliferating EC can attach. It has recently been shown that endothelialization of vascular e-PTFE grafts in the dog is greatly improved when fabric with increased microporosity (60 - 90 μm rather than 20 - 30 μm) is used (*Kogel et al., 1989*). This is in part due to a better mechanical anchorage of the cells to the biomaterial. The e-PTFE used in these results are of intermediate microporosity (30 μm). When they are coated under pressure, a better conservation of the protein layer is obtained on the graft (due in part to enlargement of the micropores), thus leading to attachment and proliferation af EC resulting in a confluent monolayer.

Coating of PETP grafts with the biological sealant TGL has two advantages: **1)** it impermeabilizes the PETP and therefore constitutes an alternative to conventional clotting of the graft with whole blood, and **2)** it provides a suitable surface for the EC attachment and growth. Finally it has been reported that there is better cohesion between the neointima and the biomaterial when PETP grafts are preclotted with a biological sealant due to the resorbability of the glue (*Jonas et al., 1987*).

When testing the stability of monolayers of EC seeded on pretreated biomaterials under steady conditions, the superior density is attained up to 5 x 10^4 cells/cm^2 allowing the establishment of a confluent monolayer within 24 hours. The monolayers will normally remain intact in vitro for at least 15 days without any microscopically detectable degradation. These monolyers exhibit a typical endothelial contact inhibiting cobblestone pattern and are non-thrombogenic. They express thrombomodulin activity and secrete vWF and tissue plasminogen activator following stimulation with thrombin.

Limitations of the method.

The described EC adhesion, growth and maintenance at confluence after precoating of vascular grafts with adhesive proteins take place under static conditions. Some authors (*Anderson et al., 1987*) have shown a rather poor stability of EC monolayers under static conditions. In that case the EC was seeded at high concentrations (5×10^5 cells/cm^2) leading to an adhesion density of over 3×10^5 cells/cm^2. The graft is probably overcrowded in such conditions and does not allow normal spreading of EC, thus resulting in detachment of part of the cells. When similar devices are maintained under flow conditions EC remain morphologically confluent for 72 hours without exposure of the matrix despite subsequent cell loss.

Preliminary experiments seeding Ec (5×10^4/cm^2) on e-PTFE vascular grafts of internal diameter 4 mm coated with TGL under pressure were as followed. The attachment period was 30 minutes after which the grafts were submitted to a culture medium flow equivalent to arterial shear rate ($700s^{-1}$) for 2 hours. Subsequent examination of the grafts by scanning election microscopy demonstrated that 88.6% of the seeded EC had adherred to the surface and demonstrated progressive spreading of EC with time which was almost total after only 120 minutes under flow without notable cell loss.

Such data suggest that EC available from small fragments of human veins when grown under optimal conditions can provide an ideal lining for small caliber (≤ 4 mm) vascular grafts. An adequate coating with adhesive proteins of these grafts allows good cohesion between the endothelial monolayers and the biomaterial and may facilitate its incorporation in perigraft tissue. The presence of endothelial lining on the inner surface is believed to improve the long term patency of such grafts.

REFERENCES

Amstein, C.F. & Hartman, P.A. (1975) "Adaptation of plastic surfaces for tissue culture by glow discharge". J. Clin. Microbiol., 2, 46-54.

Anderson, J.S., Price, T.M., Hanson, S.R. & Harker, L.A. (1987) "In vitro endothelialization of small-caliber vascular grafts". Surgery, 101, 577-586.

Berger, K., Sauvage, L.R., Rao, A.M. & Wood, S.J. (1972) "Healing of arterial prostheses in man: its incompleteness". Ann. Surg., 175, 118-127.

Callow, A.D. (1987) "Presidential address: The microcosm of the arterial wall. A plea for research". J. Vasc. Surg., 5, 1-18.

Camilleri, J.P., Phat, V.N., Bruneval, P., Tricottet, V., Balaton, A., Fiessinger, J.N. & Cormier, J.M. (1985) "Surface healing and histologic maturation of PTFE grafts implanted in patients for up to 60 months". Arch. Pathol. Lab. Med., 109, 833-837.

Chinn, J.A., Horbett, T.A., Ratner, B.D., Schway, M.B., Haque, Y. & Hauschka, S.D. (1989) "Enhancement of serum fibronectin adsorption and the cloned plating efficiencies of Swiss mouse 3T3 fibroblast and MM14 mouse myoblast cells on polymer substrates modified by radiofrequency plasma deposition". J. Colloid Interface Science, vol. 27, 67-87.

Gimbrone, M.A.Jr., Cotran, R. & Folkman, J. (1973) "Endothelial regeneration: Studies with human endothelial cells in culture", Ser Haemat. VI, 453-455.

Gospodarowicz, D., Vlodavsky, I., Greenburg, G. & Johnson, L.K. (1979) "Cellular shape is determined by extracellular matrix and is responsible for the control of cellular growth and function" in "Hormones and Cell Culture" (Sato, G.H. & Ross, R. eds.), vol.6A, 561-592, Cold Spring Harbor Conference on Cell Proliferation, Cold Spring Harbor Laboratory, USA.

Jaffe, E.A., Nachman, R.L., Becker, C.G. & Minick, C.R. (1973) "Culture of human endothelial cells derived from umbilical cord veins. Identification by morphologic and immunologic criteria". J. Clin. Invest., 52, 2745-2756.

Jonas, R.A., Schoen, F.J. Ziemer, G., Britton, L., & Castadeda, A.R. (1987) "Biological sealants and knitted Dacron Conduits: Comparison of collagen and fibrin glue pretreatment in circulatory models". Ann. Thoracic Surg., 44, 283-290.

Klein-Soyer, C. & Cazenave, J.-P., (1986) "Culture de cellules endothéliales vasculaires humaines" in "Physiopathologie de l'hémostase et de la thrombose" (Sultan, Y. & Fisher A.M., eds.) Paris: Doin Editeurs. Progrès en Hématologie, 8, 83-93.

Klein-Soyer, C., Stierle, A., Bouderbala, B. & Cazenave, J.-P. (1984) "Effect of an extract of human brain containing growth factor activity on the proliferation of human vascular endothelial cells in primary culture". Biol. Cell, 54, 9-20.

Kogel, H., Amselgruber, W., Frasch, D., Mohr, W. & Cyba-Altunbay, S. (1989) "New techniques of analysing the healing process of artificial vascular grafts, transmural vascularization and endothelialization". Res.Exp.Med., 189, 61-68.

Ruoslahti, E., Vuento, M. & Engvall, E. (1978) "Interaction of fibronectin with antibodies and collagen in radioimmunoassay". Biochem. Biophys. Acta, 534, 210-218.

Thornton, S.C., Mueller, S.N. & Levine, E.M. (1983) "Human endothelial cells. Use of heparin in cloning and long-term serial cultivation". Science, 222, 623-625.

Wachem, P.B.van, Beugeling, T., Feijen, J., Bantjes, A., Detmers, J.P. & Aken, W.G. van (1985) "Interactions of cultured human endothelial cells with polymeric surfaces of different wettabilities". Biomaterials, 6, 403-408.

Wachem, P.B.van, Mallens, B.W.L., Dekker, A., Beugeling, T., Feijen, J., Bantjes, A., Detmers, J.P. & Aken, W.G.van (1987) "The adsorption of (endothelial) fibronectin onto tissue culture polystyrene". J.Biomed.Mat.Res., 21, 1317-1327.

Wachem, P.B.van, Vreriks, C.M., Beugeling, T., Feijen, J., Bantjes, A., Detmers, J.P. & Aken, W.G.van (1987) "The influence of protein adsorption on interactions of cultures human endothelial cells with polymers". J.Biomed.Mat.Res., 21, 701-718.

Determination of Protein C in Plasma

Authors: L. Grunebaum, D. Gobaille, M. Grunert, M.-L. Wiesel & J.-M. Freyssinet

INSERM U311, Centre Régional de Transfusion Sanguine, F - 67085 Strassbourg Cedex

INTRODUCTION

Name of method:

Determination of protein C in plasma.

Aim of method:

The method is suitable for clinical, in vivo and ex vivo detection.

Biophysical background for the method:

The endothelium forms a thromboresistant and haemocompatible surface in the vasculature to which platelets normally do not adhere and coagulation is not activated. In contrast to this, biomaterial surfaces can play an important role in the contact activation of the coagulation. If clotting is triggered, it results in the generation of thrombin, the central enzyme (a serine proteinase) of the blood coagulation cascade. Under physiological conditions the generation of thrombin is controlled by two major regulatory systems namely **1)** the antithrombin system whose function is to inhibit directly the action of thrombin and other up-stream activated factors such as factor Xa, and **2)** the protein C pathway which inhibits the amplification of thrombin production (*Rosenberg, 1987*). Thus protein C is an essential natural anticoagulant in the blood. At a biomaterial interface, it can be expected that the antithrombin system will act at least as a regulator of the coagulation reactions. However, the involvement of the protein C pathway is more problematic since biomaterials lack endothelial cells which release thrombo-modulin, the essential thrombin cofactor for the activation of protein C (*Freyssinet & Cazenave, 1988*). However, it has now been established that protein C could also be activated by factor Xa (*Freyssinet et al., 1989*). Factor Xa is activated through a sequence of processes on the blood-biomaterial interface. Protein C in plasma

S. Dawids (ed.), Test Procedures for the Blood Compatibility of Biomaterials, 425–430.
© 1993 *Kluwer Academic Publishers. Printed in the Netherlands.*

could thus be a useful parameter to evaluate the haemocompatibility of biomaterials. This seems justified since activated protein C functions as a potent anticoagulant through its ability to degrade cofactors Va and VIIIa in the presence of protein S which acts as a specific cofactor together with negatively charged phospholipids. Once factors Va and VIIIa are inactivated, they can no longer accelerate thrombin generation.

Protein C and S are vitamin K-dependent proteins. They circulate in plasma as inactive proenzymes at concentration levels of 5 and 25 $\mu g/ml$ respectively. Several assays of protein C in plasma have been published (*Bertina, 1988*), but the finding that an enzyme from the venom of the snake, Agkistrodon contortrix contortrix can activate protein C has enabled the design of a simple functional assay. The purpose of this contribution is to describe an assay which is fast and reliable enough to be used in the evaluation of biomaterials.

Scope of method:

The analysis method is aimed at experimental procedures because of limited knowledge of its relevance in in vitro testing of biomaterials.

DETAILED DESCRIPTION OF METHOD

Equipment.

Several variants of the same method are now available from various manufacturers as kits such as:

ChromoTime System®, Behring, Marburg, Germany
Staclot® or Stachrom®, Diagnostica Stago, Asnières, France

Reagents.
Owren-Koller buffer

Outline of method.

The commercial kits for detecting protein C are complete with respect to the different compounds required. Furthermore they contain detailed specific procedures that will be summarized below. The common principle of the two chromogenic methods ChromoTime System® and Stachrom® Protein C is that protein C can be activated i plasma in the presence of venom from Agkistrodon contortrix contortrix and the amidolytic activity of the generated protein C can be measured using a chromogenic substrate. The principle of the chromometric method Staclot®Protein C is that in the presence of the activator from the venom of Agkistrodon contortrix contortrix, the partial thromboplastin time of plasma is prolonged due to the activation of protein C.

Description of procedure.

A careful introduction of a canule should be made into a peripheral vein. Blood is collected directly on 0.13 M trisodium citrate; 9 vol. of blood/1 vol. of anticoagulant. The blood is centrifuged and the plasma transferred to another test tube. As a "normal" reference, a plasma pool should be obtained from at least 20 healthy volunteers. This plasma pool can be considered to have 100% protein C activity.

The following sequence of steps in the analysis is given schematically below.

Chromogenic assays.

	ChromoTime System®	Stachrom®
Plasma	25 μl	50 μl
PC activator (venom)	250 μl	200 μl
mix, incubate	5 min. at 37°C	5 min. at 37°C
Substrate (37°C)	50 μl	500 μl
mix, incubate	-	5 min. at 37°C
mix, read OD/min	1 min.	-
(after lag phase of 5s)		
Add pure acetic acid, mix	-	200 μl
and read OD at 405 nm		

With the ChromoTime System® the amount of protein C can be determined automatically provided an appropriate instrument is used and provided a blank is performed using 0.154 M NaCl instead of plasma.

Using Stachrom® a calibration curve must be constructed using the same procedure. Pure normal plasma represents 100% protein C activity. The same plasma diluted 1:2 with 0.154 M NaCl represents 50% protein C activity while 0% is obtained in the presence of pure 0.154 M NaCl. The OD, measured at 405 nm, is plotted on the ordinate and the percent of protein C activity on the abscissa.

Chronometric assay.

Plasma dilution (sample or standard)	100 µl
Protein C deficient plasma	100 µl
Phospholipid suspension	100 µl
Protein C activator	100 µl
mix, incubate at 37°C	3 min.
0.025 M NaCl (37°C)	100 µl
Record clotting time	

The calibration curve is constructed by plotting the clotting time for standard plasma on a bilogarithmic scale with a percent of protein C activity on the abcissa and clotting time on the ordinate. Plasma dilutions are performed with Owren-Koller buffer. For the calibration curve a 1:10 dilution of normal plasma will represent 100% of protein C activity while a 1:80 dilution will represent 12.5% etc. The "normal" reference plasma is defined as a pool obtained from at least 20 healthy volunteers.

Safety aspects.

All the mentioned methods are safe provided all the blood samples have been screened in order to discard subjects having hepatitis B, HIV etc. Under all circumstances care should be taken to avoid direct contact with the blood. The environment should not be contaminated with blood.

RESULTS

Under physiological conditions the concentration of circulating protein C is at the level of 5 µg/ml corresponding to 100% activity. The normal range has been determined between 70% and 140%. Patients having protein C levels lower than 55% are at risk of recurrent thrombotic episodes. The limits of detection of these methods are approximately 10%, and the reproducibility is in the level of ± 10% of protein C activity. In case of the intravascular presence of biomaterials (or compounds made of biomaterials) this may lower the level of protein C. However, the presence of heparin will influence the detected level of protein C if the concentration is above 1 IU/ml.

DISCUSSION

In the absence of any apparent pathology the level of circulating activated protein C is approximately 5 ng/ml which is approximately 1/1000 of the protein C concentration in the blood. In an extreme pathological situation such as disseminated intravascular coagulation, the level of circulating activated protein C can reach levels of 200 ng/ml i.e. 2% of total protein C (*Bauer & Rosenberg, 1987*). With activation (i.e. consumption) the level of protein C in blood declines. None of the available methods for detection of protein C, neither functional nor immunological, is able to detect the differences between total protein C and activated protein C in pathology. This implies that protein C should only be assayed if a decrease of at least 10% of total is expected upon contact of blood with biomaterials. In the case of protein C activity reduction it should be established whether this is due to adsorption onto a biomaterial or to alteration of the protein C molecule. This can be achieved by comparing protein C to factor VII. These two vitamin K-dependent proteins of plasma show a high degree of structural homology. They have furthermore comparable half lives. Therefore, if the observed diminution of protein C is due to adsorption, one would expect that factor VII activity also would be diminished in a comparable proportion. The determination of protein S, the co-factor of the activated protein C molecule, is not more sensitive compared to that of protein C.

The implantation of biomaterial devices could lead to the development of an inflammatory process, a situation known to be favourable in leading to thrombosis. Cytokines such as interleukin-1 or tumor necrosis factor can similarly act on the endothelium to induce diminution of the expression of thrombomodulin, resulting in a diminished activation of protein C. However, in this situation it is unlikely that the global level of protein C could be sufficiently altered to be detected by these methods.

REFERENCES

Bauer, K.A. & Rosenberg, R.D. (1987) "The pathophysiology of the prethrombotic state in humans: insights gained from studies using markers of hemostatic system activation". Blood, 70, 343-350.

Bertina, R.M. (1988) "Assays for protein C". In "Protein C and related proteins" pp. 130-150, Churchill Livingstone, Edinburgh.

Freyssinet, J.-M. & Cazenave, J.-P. (1988) "Thrombomodulin" In "Protein C and related proteins" pp. 91-105, Churchill Livingstone, Edinburgh.

Freyssinet J.-M., Wiesel, M.-L., Grunebaum, L., Pereillo, J.-M., Gauchy, J., Schuhler, S., Freund, G. & Cazenave, J.-P. (1989) "Activation of human protein C by blood coagulation factor Xa in the presence of anionic phospholipids". Biochem.J., 261 (2), 341-348.

Rosenberg, R.D. (1987) "Regulation of the hemostatic mechanism". In "The molecular basis of blood diseases" (Stamatoyannopoulos, G., Nienhuis, A.W., Leder, P. & Majerus, P.W. eds.) pp. 534-574, W.B. Saunders Company, Philadelphia.

A Centrifugation Technique for the Preparation of Suspensions of Non Activated Washed Human Platelets

Authors: J.-P- Cazenave, J.N. Mulvihill, A. Sutter-Bay, C. Gachet, F. Toti, A. Beretz

INSERM U311, Centre Régional de Transfusion Sanguine, F - 67085 Strassbourg Cédex

INTRODUCTION

Name of method:

Centrifugation technique for the preparation of suspensions of non activated washed human platelets.

Aim of method:

A centrifugation technique has been developed for isolation and washing of platelets from human blood. The platelets are resuspended in a physiological buffer solution containing apyrase, thus enabling study of their intrinsic properties under well defined conditions, in particular physiological ionic calcium concentrations and the absence of coagulation factors or other plasma components.

Biophysical background for the method:

The method of isolation of human platelets by centrifugation and washing described by *Cazenave et al. (1983)* is directly derived from the technique of *Mustard et al. (1972)*. Blood collected into acid-citrate-dextrose is used to prepare platelet rich plasma, from which platelets are isolated by successive steps of centrifugation and resuspension in Tyrode's buffer, an iso-osmotic phosphate buffer at pH 7.30, containing glucose (0.1%), human serum albumin (0.35%) and physiological concentrations of Ca^{2+} (2mM) and Mg^{2+} (1mM). Prostacyclin (PGI_2) is employed to prevent transitory platelet activation during the preparation. Addition of apyrase (ATP diphosphohydrolase 3.6.1.5) to the final suspending medium prevents platelets from becoming refractory to ADP and enables them to maintain their discoid shape.

S. Dawids (ed.), Test Procedures for the Blood Compatibility of Biomaterials, 431–439.
© 1993 *Kluwer Academic Publishers. Printed in the Netherlands.*

Suspensions of washed platelets prepared by this method are stable for 5-8 hours at 37°C. Platelet aggregation studies may be carried out in the absence of plasma proteins or anticoagulants, under well defined conditions of pH, osmolarity and Ca^{2+} ion concentrtion. Furthermore, the platelets may be labelled with radioisotopes (^{51}Cr, ^{111}In-oxine, ^{14}C-or ^{3}H-serotonin) and hence used in vitro to study platelet secretion or accumulation on artificial or natural surfaces (*Kinlough-Rathbone et al., 1977; Mulvihill et al., 1987*) or in vitro to follow platelet survival and detect sites of sequestration or thrombus formation (*Eber et al., 1984*).

Scope of method:

The test may be used as in vitro screening procedure for biomaterials, always using standard reference material.

DETAILED DESCRIPTION OF METHOD

Equipment

Sorvall RC-3B centrifuge (Sorvall Instruments, DuPont Biomedical Division, Newtown, CO, USA) with rotor H-4000, using 15 or 50 ml conical bottom, polycarbonate centrifuge tubes (Corning Glass Works, New York, N.Y., USA). Automatic counter (Platelet Analyser 810, Baker Instruments, Allentown, PA, USA).

Wide bore Pasteur pipettes (Pastettes, Biolyon, Dardilly, France) and all other material in contact with blood, platelet rich plasma (PRP) or platelet suspensions are in plastic or siliconized.

Reagents.

All chemicals from commercial sources (Prolabo, Paris, France; Merck, Darmstadt, FRG; Sigma Chemicals, St. Louis, MO, USA) are of analytical grade.
^{111}In-oxine (1 mCi/ml) is obtained from the Commissariat à l'Energie Atomique, Saclay, France
^{14}C-serotonin (^{14}C-5-hydroxytrymptamine, 57 mCi/mmol) from Amersham, UK.

Acid-citrate-dextrose (ACD)

Trisodium citrate dihydrate 2.5 g, citric acid monohydrate 1.4g, anhydrous D(+)-glucose 2.0, in a final volume of 100 ml distilled water (pH 4.5, 450 mOsm).

Stock solutions for Tyrode's buffer (stored at 4°C)

Solution I: NaCl 16,0g, KCl 0.4g, $NaHCO_3$ 2.0g, $NaH_2PO_4.H_2O$
0.116g in a final volume of 100 ml distilled water.

Solution II: $MgCl_2.6H_2O$ 2.03g in 100 ml distilled water.

Solution III: $CaCl_2.6H_2O$ 2.19g in 100 ml distilled water.

Hepes buffer 0.5M: Hepes (N-2-hydroxyethylpiperazine-N'-2-ethanesulfonic acid, Sigma Chemicals)11.9g in 100 ml distilled water, pH adjusted to 7.5 with NaOH 2M.

Human serum albumin: Pasteurized human serum albumin for intravenous injection (200 g/l, purity >98%) is supplied by the Centre Régional de Transfusion Sanguine, Strasbourg, France.

Tyrode's-albumin 0.35% buffer

Solution I: 5ml, Solution II: 1ml, Solution III: 2ml, Hepes buffer: 1ml, Human Serum Albumin: 1.75 ml, Anhydrous D(+)-glucose: 0.1g, in a final volume of 100 ml distilled water, pH adjusted to 7.30 with HCl (1M) or NaOH (1M) and osmolarity to 295 mOsm with NaCl 30% or distilled water.

Washing solutions for platelet preparation

First wash: Tyrode's-albumin buffer (Roche, Neuilly-sur-Seine, France) containing 10 U/ml heparin

Second wash: Tyrode's-albumin buffer

Final suspension: Tyrode's-albumin buffer containing 2 µl/ml apyrase

Apyrase (ATP diphosphohydrolase 3.6.1.5)

Apyrase is prepared from potatoes by the method of Molnar and Lorand, modified by dialysing the final material against NaCl 0.9% (*Cazenave et al., 1983*). Protein concentrations is adjusted to 3 g/l and the solution is stored in 1 ml aliquots at -30°C. This preparation hydrolyses approximately 1.7 nmol ATP/min/µg of protein or 3.3 nmol ADP/min/µg of protein.

Prostacyclin (PGI_2) 1 mM

Prostacyclin as sodium salt (Upjohn Co., Kalamazoo, MI, USA) is dissolved at 4°C in 0.05 M tris buffer (pH 9.36 at 4°C): 10 mg PGI_2 for 23.4 ml buffer. The solution (PGI_2, 1mM) is stored in 100 µl aliquots at -30°C.

Outline of method.

From collected and anticoagulant treated blood the platelets are isolated after centrifugation. They are washed with Tyrode's solution containing containing PGI_2 (to prevent activation of the platelets during procedures) to remove all activating substances. To the last washing apyrase is added to prevent build-up of ADP in the suspension. These prepared platelets can be used for approximately five hours and can be used for platelet activation studies together with selected protein addditions and under well defined physical conditions. The platelets may further be radioisotoped tagged with [111]Indium during washing (removal of protein etc.). The washed platelets can now be utilized in measurement of platelet aggregation using microscopic and isotopic techniques. Likewise tagging with ^{14}C-serotonin during the first wash enables studies of serotonin release from platelets during activation.

Description of procedure.

Blood collection.
Blood is collected by venepuncture, using a wide bore (18/10) needle mounted on a short length (10-20 cm) of plastic tubing, directly into a conical 50 ml centrifuge tube containing 1 volume of anticoagulant ACD for 6 blood volumes (final pH 6.5 and citrate concentration 22 mM). The first few millilitres, contaminated by tissule thromboplastin, are discarded. Blood is then allowed to flow down the tube wall, thus minimizing air contact or bubble formation. Immediately after collection, the tubes are stoppered. Blood and anticoagulant are mixed gently and all tubes are placed in a water bath at 37°C for a maximum storage period of 30 min.

Washed platelet preparation.
Blood, Platelet Rich Plasma (PRP) and platelet suspensions are maintained at 37°C throughout the preparation procedure. The Sorvall centrifuge is prewarmed to 37°C and blood is centrifuged for 15 min at 1,000 rpm (175 x g) to obtain PRP. This portion is carefully removed and transferred to a stoppered 50 ml centrifuge tube. After incubation (15-30 min) at 37°C, PRP is centrifuged at 3,000 rpm (1,570 x g) for a period of time depending on the quantity of plasma:

Plasma quantity (ml)	15	20	25	30	35	40
Centrifugation time (min)	10	12	13	14	15	16

The supernatant platelet poor plasma is carefully removed by aspiration to eliminate all traces of plasma on the tube walls or near the platelet "button" to avoid the generation of thrombin traces during the subsequent washing steps. Platelets are then resuspended in the first wash solution, prewarmed to 37°C, with addition of 1 μl/ml PGI_2 1 mM. A wash volume of 10 ml is normally required for the platelet button corresponding to 50-100 ml of blood. This volume of suspension is transferred to a 15 ml centrifuge tube which is immediately stoppered and incubated for 10 min at 37°C. Then, after addition of 1 μl/ml PGI_2 1mM, it is centrifuged for 8 min at 2,600 rpm (1,100 x g). The platelet "button" is resuspended in the second wash solution (10 ml) also containing 1 μl/ml PGI_2 1mM. Incubation for 10 min at 37°C is followed by a second centrifugation for 8 min at 2,600 rpm. The platelets are then resuspended in the final suspension medium (20 ml), Tyrode's-albumin buffer containing 2 μl/ml apyrase. Platelet count is adjusted to 300,000/μl with the same medium. This washed platelet preparation is kept at 37°C in a stoppered tube under an atmosphere 5% CO_2/95% air in order to minimize pH changes. Under these conditions the suspension is stable for 5-8 hours.

When required, cytoplasmic labeling of platelets may be performed by incubation in the first wash solution with ^{111}In-oxine (5 μCi/ml), 15 min at 37°C as described by *Eber et al. (1984)*. Platelet dense granule contents may be labelled by incubation for 20 min in the first wash with ^{14}C-serotonin (0.2 μCi/ml) according to *Mustard et al. (1975)*.

Safety aspects.

Normal precautions are observed in the handling of radioisotopes, and radioactive waste materials are stored in lead shielded containers until disposal by a competent authority. Manupulation of blood samples is also subject to strict hygiene controls, all donors being screened for absence of infection by hepatitis or human immuno-deficiency virus.

RESULTS

Results are given in the subsequent tests in this chapter.

DISCUSSION

The use of washed platelet suspensions, separated from their plasma environment and in the absence of anticoagulants, is essential for the study of intrinsic platelet properties. Results reported by *Mustard et al. (1972)* and *Cazenave et al. (1983)* demonstrate that human platelets isolated by the centrifugation technique and resuspended in Tyrode's-albumin buffer containing apyrase retain their properties and ability to respond with aggregation and release of inducing agents as compared to platelets in citrated PRP. Suspensions are stable for 5-8 hours at 37°C. Examination by transmission electron microscopy confirms the discoid shape of the platelets, the absence of pseudopods and the presence of normal dense and α-granule contents. Under these conditions of Ca^{2+} ion concentration (2 mM), platelets aggregate in response to ADP (1-5 μM) in the presence of human fibrinogen (0.2 g/l) and disaggregate without appearance of a second aggregation wave accompanied by secretion of granule contents and generation of thromboxane A_2 from membrane arachidonate. Adrenaline does not aggregate platelets although it potentiates the effects of traces of any other aggregating agent (*Lanza et al., 1988*). After addition of collagen (1.25-2.5 x 10^{-3} g/l), thrombin (0.1-0.5 U/ml), arachidonic acid (0.1-0.25 mM) or ionophore A 23187 (1 μM), platelets aggregate and secrete their granule contents. In response to PAF-acether (0.1-1 μM) they aggregate without secretion. Addition of ristocetin (0.05-0.1 g/l) in the presence of EDTA (10 mM) and platelet poor plasma gives rise to a single wave of irreversible agglutination.

Citrate is the preferred anticoagulant for blood collection, as EDTA damages platelets and heparin modifies their function (*Ludlam, 1981*). However, human PRP prepared from blood collected into trisodium citrate has a depressed ionic calcium concentration which may cause platelet aggregation and release of substances during centrifugation (*Kinlough-Rathbone et al., 1983*). This problem is avoided by using ACD in which platelet activation is inhibited by low pH (6.5).

It is particularly important to prevent generation of traces of thrombin during preparation of the washed platelet suspension (*Cazenave et al., 1983*). Hence venepuncture for blood collection should be carried out with a minimum of vessel trauma and rapid blood flow while attention should be paid to eliminate all traces of plasma from the centrifuge tube before resuspending the platelet button after centrifugation of PRP. Platelets stimulated by exposure to low concentratins of thrombin may have reduced granule contents or be more sensitive to ADP or adrenaline induced aggregation (*Kinlough-Rathbone et al., 1983; Lanza et al., 1988*).

Suspending platelets in an artificial medium presents two main advantages:

a)   plasma enzymes are excluded
b)   it is possible to manipulate the inorganic ions, proteins and other constituents in the suspension.

When such a medium is employed, the pH should lie in the physiological range (7.2-7.4), to avoid affecting platelet function. The solution should be iso-osmotic, containing glucose as a source of metabolic energy and physiological concentrations of divalent cations (*Kinlough-Rathbone et al., 1977*). As shown by *Mustard et al. (1972)*, addition of apyrase to the final suspension to degrade released ADP prevents platelets from becoming refractory and maintain their discoid form and eliminates the formation of microaggregates with storage at 37°C. Transitory platelet activation during the preparation may be inhibited by use of PGI_2 in centrifugation and resuspension steps (*Cazenave et al., 1983*).

Limitations of method.

Platelet functions are not influenced by the presence of an anticoagulant. Provided the necessary precautions are taken to avoid thrombin generation and platelet activation in the course of the preparation, platelets resuspended in the final medium containing apyrase retain intact discoid shape and functional properties for storage periods of 5-8 hours at 37°C.

As compared to gel filtration, platelet yields from PRP are lower and more time is required to carry out the successive centrifugation and incubation steps. One further limitation of the method is the possibility that subpopulations of platelets may be selectedduring centrifugation, either in initial preparation of PRP from whole blood or in subsequent steps of the washing procedure.

Related testing methods.

Isolation of platelets from human blood may also be performed by density gradient centrifugation. In the albumin technique described by *Walsh et al. (1977)*, citrated PRP is layered onto a continuous albumin gradient at 300 mOsm in calcium free Tyrode's buffer containing apyrase at pH 6.5. After centrifugation, the platelets are resuspended and the washing procedure is then repeated in the absence of apyrase.

In order to avoid preparation of PRP by centrifugation, in which subgroups of platelet population may be selected, *Corash et al. (1977)* have developed a method whereby whole blood is centrifuges through an arabinogalactan (Stractan) gradient, platelets thus being separated in a single step from plasma and other cellular blood components. The Stractan is later removed on an isoosmolar albumin gradient.

Gel filtration of PRP on Sepharose 2B is another widely used alternative method (*Tangen et al., 1973; Lages et al., 1975*). By this technique, platelets appear in the void volume, while the majority of plasma proteins are retained in the gel and elute later. The separation is rapid and reproducible with minimal loss of platelets from the initial PRP, an advantage when only small blood volumes are available, but the platelet count in the resulting suspension is often low and the technique is not readily adapted to preparation of sufficient quantities of platelets for ex-perimental work. In addition, it does not enable elimination of larger plasma components, e.g. immunoglobulins or factors V and VIII-von Willebrand.

The principal advantages of the centrifugation technique for isolation of human plasma are that washed platelet suspensions may be prepared in large quantities with the required platelet count and in the absence of red blood cells and plasma proteins, including high molecular weight constituents such as multimetric von Willebrand factors. Use of an artificial buffer medium as the suspending fluid allows adjustment of pH, osmolarity and divalent cation concentrations to physiological levels.

REFERENCES

Cazenave, J.-P., Hemmingendinger, S., Beretz, A., Sutter-Bay, A. & Launay, J. (1983) L'aggrégation plaquettainre: outil d'investigation clinique et d'étude pharmacologique. Méthodologie. Ann. Biol. Clin. 41, 167-179.

Corash, L., Tan, H. & Gralnick, R. (1977) Heterogeneity of human whole blood platelet subpopulations. I: relationship between buoyant density, cell volume and ultrastructure. Blood 49, 71-87.

Eber, M., Cazenave J.-P., Grob, J.-C., Abecassis, J. & Methlin, G. (1984) [111]Indium labelling of human washed platelets; kinetics and in vivo sequestration sites. In "Blood Cells in Nuclear medicine, Part I. Cell Kinetics and Biodistribution". (Hardeman, M.R. & Najean, Y., eds.) pp.29-43, Martinus Nijhoff Publishers, Boston.

Kinlough-Rathbone, R.L., Mustard, J.F., Packhan, M.A., Perry, D.W., Reimers, H.-J. & Cazenave, J.-P. (1977) Properties of washed human platelets. Thromb. Haemostas 37, 291-308.

Kinlough-Rathbone, R.L., Packham, M.A. & Mustard, J.F. (1983) Platelet aggregation in "Methods in Hematology" (Harker, L.A. & Zimmerman, T.S. eds.) vol. 8: Measurements of Platelet Function, pp. 64-91, Churchill Livingstone, New York.

Lages, B., Scrutton, H.C. & Holmsen, H. (1975) Studies on gel filtrated human platelets: isolation and characterization in a medium containing no added Ca^{2+}, Mg^{2+} or K^+. J. Lab. Clin. Med. 85, 811-825.

Lanza, F., Beretz, A., Stierlé, A., Hanau, D., Kubina, M. & Cazenave, J.-P. (1988) Epinephrine potentiates human platelet activation but is not an aggregating agent. Amer. J. Physiol. 225, H1276-H1288.

Ludlam, C.A. (1981) Assessment of platelet function in "Haemostasis and Thrombosis" (Bloom, A.L. & Thomas, D.P. eds.) pp. 775-795, Churchill Livingstone, New York.

Mulvihill, J.N., Huisman, H.G., Cazenave, J.-P., van Mourik, J.A. & van Aken, W.G. (1987) The use of monoclonal antibodies to human platelet membrane glycoprotein IIb-IIIa to quantitate platelet deposition on artificial surfaces. Thromb. Haemostas. 58, 724-731.

Mustard, J.F., Perry, D.W., Ardlie, N.G. & Packham, M.A. (1972) Preparation of suspensions of washed platelets from humans. Brit. J. Haematol 22, 193-204.

Mustard, J.F., Perry, D.W., Kinlough-Rathbone, R.L. & Packham, M.A. (1975) Factors responsible for ADP-induced release reaction of human platelets. Amer. J. Physiol. 228, 1757-1765.

Tangen, O., McKinnon, E.L. & Berman, H.J. (1973) On the fine structure and aggregation requirements of gel filtered platelets (GFP). Scand. J. Haematol 10, 96-105.

Walsh, P.N. Mills, D.C.B. & White, J.G. (1977) Metabolism and function of human platelets washed by albumin density gradient separation. Brit. J. Haematol. 36, 281-296.

Measurement of Circulating Fibrinopeptide A in Plasma

Authors: J.-M. Freyssinet, C. Wagner, L. Grunebaum, M.-L. Wiesel

INSERM U311, Centre Régional de Transfusion Sanguine, F - 67085 Strassbourg Cedex

INTRODUCTION

Name of method:

Measurement of circulating fibrinopeptide A in plasma as indication of activation of blood coagulation.

Aim of method:

The method is useful for ex vivo and in vivo detection of activation of the coagulation system.

Biophysical background for the method:

Fibrinogen is a soluble plasma protein of 340.000 Dalton composed of three chains called Aα, Bβ and γ arranged in a covalent symmetrical structure (2-fold symmetry) of type $(Aα, Bβ, γ)_2$. Its transformation into an insoluble fibrin polymer gel is the last event of the coagulation cascade. Thrombin, the ultimate serine-proteinase which is generated during the culminating step of the coagulation cascade, hydrolyses two peptide bonds at the internal end of the Aα and Bβ chains. This results in liberation of two peptides: Fibrinopeptide A and B (FPA and FPB). Human FPA has 16 amino acid residues while FPB has 14. In each case the C-terminal residue is ARG. Such peptide bonds are highly suseptible to hydrolysis by thrombin, especially those of the Aα chains since the removal of FPA is sufficient to allow polymerization of the fibrin monomers (*Doolittle, 1987*). The concentration of circulating FPA appears as a sensitive and reliable parameter to assess the activation of the coagulation system. This does, however, require that the samples are specifically anticoagulated immediately in order to block any further generation of FPA after blood collection.

441

S. Dawids (ed.), Test Procedures for the Blood Compatibility of Biomaterials, 441–447.
© 1993 *Kluwer Academic Publishers. Printed in the Netherlands.*

The contact of blood with biomaterial devices is a possible source of activation of the coagulation reaction. To detect an activation due to in vivo or ex vivo contact of biomaterials requires an assay of high sensitivity. The described assay of fibrinopeptide A in plasma has been successfully used in the clinic and may be applied as a sensitive parameter in evaluation of designs made from biomaterials.

Scope of method:

The method is established as a clinical routine and may be used as general screening and for experimental purposes. The handling of blood samples is of major importance to avoid continous generation of FPA after the blood collection.

DETAILED DESCRIPTION OF METHOD

Equipment.

Asserachrom® FPA, Diagnostica Stago, Asnières-sur-Seine, France. This kit includes all reagents necessary to perform the determination of FPA (about 50 determinations):
- FPA coating agent: synthetic FPA in the appropriate coating buffer (freeze dried)
- anti-FPA serum: rabbit serum (freeze dried)
- anti-IgG peroxidase: goat anti-rabbit-IgG coupled with horseradish peroxidase
- orthophenylene diamine (OPD): substrate for peroxidase
- reference FPA
- special anticoagulant solution:
    - 0.11 M tri-sodium citrate
    - 1.000 IU/ml standard heparin
    - 1 TIU/ml aprotinin
    - 0.1% (^{w}/v) NaN$_3$
- bentonite suspension: 80 mg/ml
- tween 20: 2% (^{v}/v) solution
- sulphoric acid (H$_2$SO$_4$) 3M/l
- hydrogen peroxide (H$_2$O$_2$) 30%, free of inhibitors
- micro-Elisa plates, Dynatech M129B, Nunc, Roskilde, Denmark
- plate reader (492 nm wavelength) Roche-Kontron, Switzerland
- multi-channel pipette 200 μl, 100 μl and 50 μl

- washing equipment (for micro Elisa plates) Roche-Kontron, Switzerland; Nunc, Roskilde, Denmark

Outline of method.

The assay of FPA is a competitive enzyme-linked immunoassay (CELIA) modified in order to increase sensitivity and to improve reproducibility. FPA of control or test samples is first allowed to interact with a known amount of rabbit anti-FPA antibody. Excess antibody sites are then neutralized by exposure to synthetic FPA immobilized on a solid support. These rabbit anti-FPA antibodies thus fixed to the support are revealed by anti-rabbit IgG conjugated with horseradish peroxidase giving rise to a colored reaction upon exposure to an appropriate substrate. At this level it is important to emphasize that uncleaved FPA can be recognized on the fibrinogen molecule and it is thus essential that fibrinogen should be totally removed from plasma before starting the FPA assay. This can be achieved by adsorption on bentonite.

Description of procedure.

Sample collection and treatment.
Blood is collected by venipuncture or directly from the biomaterial device in a plastic container containing 1 Vol. of anticoagulant for 9 Vol. of blood. The first drops of the sample should be discarded and the collection should be as rapid as possible. Only a clean venipuncture should be accepted. The plasma is separated by centrifugation at 5000 x g for 10 min. at 4°C. Bentonite is then added to the plasma and the suspension is mixed by inversion for 10 min. at room temperature. The adsorbed plasma is collected by centrifugation at 3000 x g for 10 min. at room temperature. The treated samples can be used either for FPA determination immediately or stored frozen for later analysis. Before bentonite treatment the stability of the sample is approximately 2 hours. After bentonite treatment it is stable up to 8 hours.

Preparation of reagents.
From the kit the reagents are made ready. They are either freeze dried or in solution. It is important closely to follow the instructions of the manufacturer for reconstitution or dilution as the composition or presentation can vary.

Coating procedure.

In each well of the micro-Elisa plates there is added 200 μl. If tubes are used, 500 μl - 1 ml is applied. It is essential to use the same volume throughout the entire procedure. Incubation is carried out overnight at room temperature. The next day 5 successive washings are performed just before subsequent use. Storage of the coated support is possible according to the manufacturers instruction.

Preparation of the assay.

Samples are normally ready for use after adsorption has taken place on bentonite. However, if the expected FPA content is higher than 25 ng/ml (see *RESULTS*) an appropriate dilution can be performed using the dilution buffer. First the construction of the standard curve for the FPA reference solution is made by serially diluting from 1:1 up to 1:32. The 1:1 dilution represents a FPA level of 25 ng/ml and the 1:32 corresponds to 0.78 ng/ml.

In a series of plastic tubes 1 ml of the standard or test sample and 0.1 ml of anti FPA rabbit serum are incubated either for 1 hour at 37°C or overnight at 2-8°C. The assay can then be performed according to the following scheme without any interruption between the different steps:

1. The competition stage. From the incubation mixture of FPA-antiFPA 200 μl is added to each well, and incubation is allowed to proceed for 1 hour at room temperature. The plate is then washed 5 times with the washing solution, taking care not to touch the coated support.

2. Fixation of immunoconjugate. Anti-IgG peroxide complex, 200 ml, is added to each well and incubation is allowed to proceed for 1 hour at room temperature (20°C). The wells are then washed 5 times with the washing solution.

3. Colour development. The ODP/H_2O_2 solution, 200 μl (prepared following the manufacturer's instructions) is then added to each well, and the reaction is allowed to proceed for exactly 3 min. at room temperature (20°C). The reaction is stopped by the addition of 50 μl of 3M H_2SO_4. The colour is then allowed to stabilize for 10 min. at room temperature, preferably in the dark. Absorbance at 492 nm is measured using the plate reader.

Calculations.

FPA levels are expressed on the abscissa on a logarithmic scale, and the absorbance at 492 nm is plotted on the ordinate on a linear scale. There is inverse proportionality between the absorbance (A_{492} nm) and the FPA levels. The concentration of substrate (OPD) and other conditions are such that a linearity cannot be verified above 1.5 absorbance unit.

Safety aspects.

No special precautions should be taken apart from the necessity of having meticulously clean equipment and avoiding touching the coated surfaces.

RESULTS

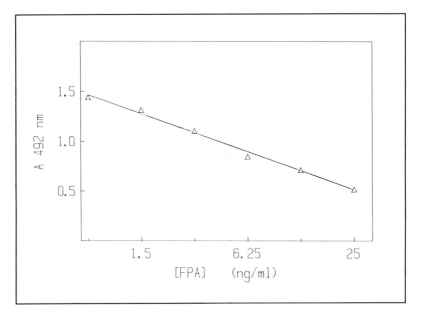

Fig. 1

A typical calibration curve for the determination of FPA in plasma by a modified competitive enzyme-linked immunoassay. (Note that the concentration of FPA on the abcissa is expressed on a log scale whereas absorbance on the ordinate is linear).

In *Fig. 1* a typical standard curve is given for FPA determination by CELIA. The minimum limit of detection with such a kit is 1 ng/ml. Under normal physiological conditions the levels of circulating FPA are generally lower than 3 ng/ml. In phatological conditions the values are greater than 3 ng/ml. During extreme pathological conditions like disseminated intervascular coagulation the concentration of FPA can reach 500 ng/ml. As the physiological concentration of fibrinogen is approximately 3 mg/ml and the molecular weight of FPA is approximately 1% of that of fibrinogen, the highest theoretical value of FPA should not exceed more than 30 μg/ml in the absence of any inflammatory syndrome. During inflammatory conditions it could reach 100 μg/ml.

DISCUSSION

The CELIA assay determines reliably and sensitively FPA. The interpretation of results is easy, especially in the case of blood-biomaterials interaction, because any elevation of FPA will reflect activation of the coagulation cascade resulting in the generation of thrombin. One characteristic of fibrinogen is the presence of phosphorylated serine residues, particularly SER 3 of the Aα chain and consequently of FPA. The degree of phosphorylation of this residue is approximately 23% in the normal population. In case of an inflammatory condition this parameter can be increased up to 50%. The level of fibrinogen can also be increased up to 3-4 fold under similar circumstances, but well after its degree of phosphorylation. The latter thus appears as a suitable signal to monitor the early events of an inflammatory process (*Seydewitz & Witt, 1985*).

The circulation of blood through biomaterials can induce an inflammatory response. Therefore the determination of the degree of phosphorylation FPA could also be very a very useful indicant for the assessment of such a state. However, due to differences in sensitivity the degree of phosphorylation can only be determined on total FPA when using a HPLC technique (*Seydewitz & Witt, 1985*). This implies that fibrinogen has to be exposed to a relatively high concentration of thrombin to release maximal amount of FPA. The minimum quantity of FPA detectable upon separation by HPLC on a Lichrosorb RP-18 (5 μm) is about 5-10 μg. Such an amount is considerably greater than conventional concentrations of FPA even in the case of an extreme pathological situation such as disseminated intravascular coagulation.

Other sensitive methods for the determination of activation of coagulation have been proposed. They are based on the measurement of protein C activation peptide or prothrombin fragment 1 + 2 (*Bauer & Rosenberg, 1987*) and are described elsewhere in this chapter.

An increase of FPA level testifies to the activation of coagulation while an increase of the degree of phosphorylation signals the development of an inflammatory process. It should be noted that although the determination of the degree of phosphorylation seems more sophisticated than that of circulating FPA, both would certainly be relevant since an acute inflammatory reaction constitutes a favourable circumstance for the occurrence of a prethrombotic state (*Nawroth & Stern, 1986*). This is probably as well true in the case of blood-biomaterial interactions.

REFERENCES

Bauer, K.A. & Rosenberg, R.D. (1987) "The pathophysiology of the prethrombotic state in humans: insights gained from studies using markers of hemostatic system activation". Blood, 70, 343-350.

Doolittle. R.F. (1987) Fibrinogen and fibrin. Haemostasis and Thrombosis. (Bloom, A.L. & Thomas, D.P. eds.) pp.192-215, Churchill Livingstone, Edinburgh.

Nawroth. P.P. & Stern, D.M. (1986) "Endothelial cells as active participants in procoagulant reactions". In "Vascular endothelium in hemostasis and thrombosis" (Gimbrone, M.A., ed.) pp.14-39, Churchill Livingstone, Edinburgh.

Seydewitz, H.H. & Witt, I. (1985) "Increased phosphorylation of human fibrinopeptide A under acute phase conditions". Thrombosis Res., 40, 29-39.

Assessment of the Fibrinolytic System

Authors: M.-L. Wiesel, J. Dambach, L. Grunebaum & J.-M. Freyssinet

INSERM U311, Centre Régional de Transfusion Sanguine, F - 67085 Strassbourg Cedex

INTRODUCTION

Name of method:

Assessment of the fibrinolytic system.

Aim of method:

The method can determine in vivo, ex vivo and in vitro an activation of the fibrinolytic system.

Biophysical background of the method:

Fibrinolysis is the physiological extension of blood coagulation. It represents the enzymatic system capable of dissolving blood clots (*Collen et al., 1985*). This system comprises a central proenzyme, plasminogen (Plg) which is activated by different types of activators (PA) to a plasmin which hydrolyses fibrin clots. Inhibition of this system may occur either at the level of activators by means of plasminogen activator inhibitors (PAI) or at the level of plasmin, essentially by a specific inhibitor, α_2-antiplasmin. PA and PAI are both synthetized by endothelial cells.

As for the coagulation system, surface phenomenon play an important catalytic role in fibrinolysis. As soon as fibrin is formed by the clotting system, tissue plasminogen activator (t-PA), which is one of the physiological plasminogen activators, and plasminogen (Plg) are both adsorbed onto the fibrin clot. Thus fibrin focalises the action of t-PA on Plg accelerating the activation exactly where it is necessary. In the presence of fibrin the effect of a specific plasmatic inhibitor of prourokinase (PUK), which is another PA, is neutralised and Plg absorbed on the fibrin can be rapidly activated by PUK. Plasmin formed on the surface of fibrin clots is protected from rapid inactivation by α_2-antiplasmin and can continue to digest fibrin. For

449

S. Dawids (ed.), Test Procedures for the Blood Compatibility of Biomaterials, 449–458.

digest fibrin. For potent fibrinolysis, a constant flux of circulating Plg molecules will assure the replenishment of inactivated plasmin molecules on the clot.

Physiological fibrinolysis is thus triggered by the fibrin itself and will normally remain strictly localised to this fibrin surface. There is no systemic fibrinolysis or fibrinogenolysis. Disorders of fibrinolysis can predispose to bleeding in the case of hyperactivity of the system or, on the other hand, to thrombotic events in the case of hypofibrinolysis.

Fibrinolytic parameters are of fundamental importance for the evaluation of haemocompatible properties of biomaterials which could activate or depress normal fibrinolysis by modulating PA or PAI or by acting directly on Plg.

The aim of this paper is to describe relevant tests for the exploration of fibrinolysis within the field of biomaterial study.

The analysis of four indicants is described:

1. Plasminogen (Plg) in plasma giving an estimate of the fibronolytic potential,
2. Plasma euglobulin clot lysis time giving an estimate of the activity of plasma PA,
3. Tissue plasminogen activator (t-Pa) in plasma
4. Plasminogen activator inhibitor (PAI) in plasma secreted by endothelial cells.

Scope of methods:

The methods cover measurements of plasminogen (Plg), plasma euglobulin clot lysis time, tissue plasminogen activator (tPA) and plasminogen activator inhibitor (PAI) in blood. The tests are available for routine in vitro screening, but of limited relevance in materials testing.

DETAILED DESCRIPTION OF METHODS

Equipment

Standard colorimeter for reading at 405 nM and assessories.

Reagents
Several kits exist on the market. They contain almost all necessary reagents. The described tests outline the principles. The instructions of the kit should be followed closely.

For measurement of plasminogen (Plg) in plasma:
S 2251, 3 mMol, Kabi, Sweden
Streptokinase solution, 10.000 U/ml
Tris NaCl buffer (Tris 0.05 mMol, NaCl 0.012 mMol) pH 7.4
Pool of normal plasma from 20 healthy persons
Acetic acid 0.25%

For measurement of plasma euglobulin clot lysis time:
Barbital buffer 0.01 Mol
Acetic acid 50%
Thrombin solution at 10 U NIH/ml
Melting ice

For measurement of tissue plasminogen activator (t-PA) in plasma:
CBS 10-65 substrate, 0.015 Mol, Pentapharm, France
Purified plasminogen 3 U/ml
Phosphate buffer pH 6.8
Phosphate buffer with 0.1% Bovine serum albumin pH 7.5
ELISA plate, covered with insolubilised and stabilized fibrin
Purified t-PA: 2 U/ml
Acetic acid 0.25%

For measurement of plasminogen activator inhibitor (PAI) in plasma:
Plasminogen activator reagent (PAR)
Desofib-x: des AA fibrinogen partially digested with plasmin
Standard t-PA from human melanoma 6000 IU/ml
t-PA-PAI free plasma
Tris buffer

Acetate buffer 1 Mol pH 3.9
Stop/fibrin

Outline of method.

For plasminogen (Plg) assay.
A number of different methods generally called chromogenic amidolytic assays
relate to the following: A synthetic peptide with a C-grafted paranitroanilide (pNA)
group behaves as a substrate for plasmin. If Plg is to be measured, it is first
activated by addition of a bacterial potent plasminogen activator, streptokinase
(SK) in excess in order to have no residual free Plg. The generated plasmin induces
release of pNA from the synthetic peptide. The free pNA is detected by measuring
OD at 405 nm (*Soria et al., 1976*). Improvement of sensitivity can be obtained by
replacing the chromogenic substrates by fluorogenic substrates, but the principle of
the method remains the same. The most widely used chromogenic substrates of
plasmin and SK-activated Plg are:

(S2251) H-D-Val-Leu-Lys-pNA, Kabi, Sweden
(CBS 30-41) H-D-But-CHA-Lys-pNA, Pentapharm, France
H-D-Nva-CHA-Lys-pNA, Behring, Germany

Description of procedure.

For plasminogen (Plg) assay.
The plasma sample to be assayed is usually diluted to 1:40 or 1:50. Activation of
Plg is allowed to proceed for 10 min. at 25°C, 30°C or 37°C after addition of SK
in excess, approximately 1000 - 5000 U/ml final concentration. Under these
conditions there is no residual free Plg, and the enzymatic activity of SK-Plg
complexes cannot be inhibited by plasma inhibitors.

After exactly 10 min. activation the chromogenic substrate is added at a final
concentration of 0.3 - 0.6 mMol. After 2 - 3 min. the reaction is either blocked with
acetic acid (5 - 10% final concentration) and the developed colour is read at 405
nm, or the linear variation of OD is recorded at 405 nM for 1 or 2 min. to obtain
stable values. A standard curve previously constructed using known amounts of
reference Plg or, for preference, a curve worked out on a pool of normal plasma
then allows the exact determinination of Plg concentration.

The pool of normal plasma is prepared from 20 healthy blood donors. Blood is collected on tri-sodium citrate (0.13 Mol), 1 volume citrate/9 volumes of blood. Plasma is obtained by centrifugation for 20 min. at 2000 g. This reference plasma can be kept frozen for 2 months at -80°C.

S 2251 is diluted in Tris NaCl buffer, 2 volumes of S 2251/5 volumes of buffer. Preparation of standard dilutions in Tris NaCl buffer is the following: dilution of 1:41 which corresponds to 1 U/ml. Further dilutions can be made corresponding to 0.75, 0.50 and 0.25 U/ml.

Plasma samples to be tested should likewise be diluted in Tris NaCl buffer 1:41.

For kinetic assay 200 µl of diluted solution, either standards or samples are prewarmed at 37°C for 2 - 6 min. SK (100 µl at 10.000 U/ml) is then added and incubated for 10 min. to allow the complete assembly of SK-Plg complexes. After exactly 10 min. 700 µl of the substrate buffer mixture kept at 37°C is added and the OD is read at 405 nm at t0 and t1. OD variations per minute are determined.

The standard curve is traced on linear coordinates and Plg activity in plasma is deduced from this standard curve. Normal values should be between 0.8 and 1.2 U/ml.

Outline of method.

For plasma euglobulin lysis time assay.
This is a more sensitive test than that of the clot lysis time of whole or diluted blood. It represents a good estimate of the activity of plasma PA. In this procedure a separation is made of the fraction containing PA and that containing PAI by acidification which caused the euglobulin fraction to precipitate.

Citrated plasma is diluted to 1:10 at 4°C and acidified using acetic acid to pH = 5.9 for 10 min. This pH value must be respected to obtain maximal precipitation of the euglobulin fraction. The euglobulin fraction is then centrifuged and the precipitant is redissolved in Tris buffer. The precipitate contains fibrinogen, Plg and PA. The supernatant contains most of the PAI except C1 inactivator and fast t-PA inhibitors (*Kluft & Brakman, 1975*). The redissolved euglobulin precipitate is clotted with a solution of thrombin or a mixture of thrombin and calcium. The time required for the clot to hydrolyse is recorded. If there is not enough fibrinogen in the euglobulin fraction to allow clotting, fibrinogen at a concentration of 3 g/l must be added before precipitation. The normal clot lysis time is above 3 hours. If there is a high fibrinolytic activity in the plasma, the clot lysis time should be shortened proportionally. Likewise a condition of hypofibrinolysis will be responsible for an abnormally long clot lysis time.

Description of procedure.

For plasma euglobulin lysis time assay.
Citrated plasma for testing is diluted 1:10 in distilled water on melting ice. The pH is adjusted to 5.9 with acetic acid. This mixture is kept 10 min. on melting ice to allow precipitation of the euglobulin fraction. The precipitate is obtained after centrifugation at 150 g for 5 min. at 4°C. The supernatant is removed and the walls of the test tubes are carefully dried. Precipitant is then redissolved in the same volume of the Barbital buffer as that of the starting plasma. This mixture is clotted with half its volume of thrombin solution at 10 U NIH/ml. The time for lysis is then measured.

Outline of method.

For tissue plasminogen activator (t-PA) assay.
The amidolytic assay of t-PA is based on the high affinity of t-PA to bind on fibrin. In the first step t-PA either in plasma or euglobulin solution is adsorbed to an insoluble fibrin film. Other PA are eliminated by washing. In the second step Plg in excess is added to the t-PA adsorbed on the fibrin leading to plasmin production. In the last step a chromogenic substrate is added and the variation of OD is read at 405 nm. A standard curve is established using known amounts of purified t-PA. This technique detects only free t-PA (*Angles-Cano et al., 1987*).

Description of procedure.

For tissue plasminogen activator (t-PA) assay.
Blood is collected by a clean venepuncture in tri-sodium citrate 0.13 Mol, 1:9 vol of blood, on melting ice. Plasma is then rapidly obtained by centrifugation for 10 min. at 3000 g/min at 4°C.

A standard curve is established by diluting t-Pa from 2 to 0.2 U/ml.

A sample of 100 µl plasma or, as reference, a dilution of t-PA to be assayed is incubated for 60 min. at 37°C in the wells of the ELISA plate pre-covered with fibrin. After one hour this plate is washed 3 times with 300 μl phosphate-albumin buffer.

A mixture of 1 ml purified Plg and 1 ml chromogenic substrate in 10 ml phosphate albumin buffer is prepared. 100 μl of this mixture is added to each well on the ELISA plate and incubated at 37°C for 30 min. OD is then read at 405 min. The standard curve is traced on linear coordinates and the t-PA activity is calculated from the OD obtained for each aliquot.

Outline of method.

For plasminogen activator inhibitor (PAI) assay.
The endothelial cells secret PAI_1 is present in plasma and platelets. It forms rapidly a complex with t-PA. PAI_2 is produced by placenta and is normally only present in plasma during pregnancy. PAI_3 is present in plasma and urine and inhibits preferentially urokinase. PAI can be measured by immunological methods, but it is preferable to use an assay that measures PAI activity.

The amidolytic assay is based on the principle that plasma t-PA is rapidly neutralized by PAI. Three steps are necessary:

1. plasma to be assayed is diluted in plasma selectively depleted in PAI and fibrinopeptide-A, but containing all other inhibitors of fibrinolysis,
2. dilution is then added to solution of a known quantity of t-PA. Thus the t-PA-PAI complex can form and residual t-PA remains in solution,
3. residual t-PA is determined as above in the presence of fibrin and Plg by measuring the amidolytic activity of generated plasmin.

Several kits are available and details are given of one of the available techniques (Streptolyse™/fibrin®, Biopool, Diamed).

Description of procedure.

For plasminogen activator inhibitor (PAI) assay.
Initially two standards are prepared: PAI "40" (24 μl t-PA-PAI free plasma + 25 μl Tris buffer) and PAI "0" (25 μl t-Pa-PAI free plasma + 25 μl t-PA at 40 IU/ml). Furthermore intermediate dilutions are prepared containing 10, 20 and 30 IU/ml. For each plasma to be tested a volume of 25 μl plasma is mixed with 25 IU t-PA (of standard 40 IU/ml). All tubes are incubated for 15 - 20 min. at 25°C. Then 50

µl of acetate is added to each tube and carefully mixed. Antiplasmin activity is destroyed by incubating for 20 min. at 37°C. Then 2 ml NaCl 0.15 Mol is added and mixed. Formation of t-PA-PAI complex is achieved and residual free t-PA can be measured. Thereafter 20 µl of each sample previously prepared is added to the microtest plate in the kit. To each well is added 200 µl PAR and the microtest plate is placed in the refrigerator (4°C) for 15 min. To each well 10 µl of Desofib-X is then added and mixed. The plate is incubated at 37°C for 15 min. Then 25 µl of Stop/fibrin is added. Absorbance is measured at 405 nm and 492 nm for each well. A 492 is substracted from A 405 to obtain the value of A 405/492. A standard curve is constructed by plotting A 405/492 for the standards against the corresponding theoretical PAI activity on a linear scale.

Safety aspects.

No special precautions ar necessary. Care should be taken to avoid contact with blood especially if it has not been tested for hepatitis, HIV etc.

RESULTS

Normal ranges vary from kit to kit, but are quoted in the instructions. Unfortunately results often cannot be compared from kit to kit as the processes differ.

DISCUSSION

Physiological fibrinolysis is induced by fibrin and remains strictly localized to it. The vessel wall, particularly endothelial cells, play a fundamental role in the regulation of normal fibrinolysis. Biomaterials can exhibit poor or good haemocompatibility, but when used ex vivo an inflammatory syndrome can be triggered which will be difficult to evaluate and can stimulate or depress the fibrinolytic system. The determination of Plg is important because it is the key enzyme of fibrinolysis. Euglobulin lysis time allows an estimation of plasma PA activity and gives a rapid overview of the status of the fibrinolytic system. The determination of t-PA or PAI enables localization of the principal abnormalities of fibrinolysis, i.e. excessive or depressed liberation of either activator or inhibitor.

Interpretation of results.

Activation of fibrinolysis cannot directly be related to the property of the biomaterial. Percutaneous inwelling of catheters, operative incorporation of a device or ex vivo connection of biomaterials can easily cause an activation of the coagulation and fibrinolysis through the tissue damage. Therefore a very careful evaluation of the causes is mandatory. In many cases the activation of the fibrinolytic system induced by the biomaterial may be minor compared to the physiological reactions. Due to the lack of t-PA on polymer surfaces one can expect a low fibrinolysis.

The euglobulin clot test is considered as a global test and is generally of limited value unless pronounced fibrinolysis is present.

Limitations of method.

Normal or pathological values are strongly dependent on the concentration of the various components used in the individual kits or assay systems and in this regard it is important closely to follow the manufacturer's guidelines. As mentioned the changes induced by biomaterials may be minor compared to those elicited by the normal response from damaged vascular tissue. Some of the kits are difficult to use when reproducible results are desirable.

Related relevant testing methods.

Immunological assay of plasminogen,
using radial immunodiffusion or immunoelectrophoresis with specific anti-plasminogen antiserum is still a frequently used technique to determine the plasminogen concentration. It gives a good evaluation of plasminogen, but does not evaluate the ability of plasminogen to be converted to active plasmin. This explains why functional techniques measuring the enzymatic properties of generated plasmin might be preferred.

Immunological assay of tissue plasminogen activator,
using IRMA or ELISA is currently applied to measure t-PA antigen. As for other coagulation or fibrinolysis proteins an assay measuring t-PA activity is preferable to an immunological assay since the latter detects both active t-PA and t-PA complexed to its specific inhibitor PAI.

REFERENCES

Angles-Canoe, E., Arnoux, D., Boutiere, B., Masson, C., Contant, G., Benchimol, P., & Sampol, J. (1987) "Release pattern of the vascular plasminogen activator and its inhibitor in human plasma as assessed by a spectrophotometric solid-phase fibrin-t-PA activity assay". Thromb. Haem., 58, 843-849.

Collen, D. & Lijnen, H.R. (1985) "The fibrinolytic system in man: an overview". In "Thrombolysis: Biological and terapeutic properties of new thrombolytic agents. Contemporary issues in haemostasis and thrombosis. 1 (Collen, D., Lijnen, H.R. & Verstraete, M. eds.) pp.1-14, Churchill Livingstone, Edinburgh, London, Melbourne and New York

Kluft, C. & Brakman, P. (1975) "Effect on euglobulin fibrinolysis: involvement of C_1 inactivator". In "Progress in chemical fibrinolysis and thrombolysis. Vol. 1 (Davidson, J.F., Samama, M.M. & Desnoyer, P.C. eds.) pp. 375-381, Raven Press, New York.

Soria, J., Soria, C. & Samama, M. (1976) "Dosage du plasminogène à l'aide d'un substrat chromogène tripeptidique". Pathol. Biol. Fr., 24, 725-729.

Chapter VII

TEST METHODS FOR THE DETECTION OF
THROMBOSIS, FIBRINOLYSIS AND EMBOLI FORMATION

Chapter eds.: R. Eloy[1] & J.-P. Cazenave[2]

1)   INSERM U 37, Chirurgie Vasculaire et Transplantations d'Organes, F - 69500 Bron
2)   Centre Régional de Transfusion Sanguine, F - 67085 Strassbourg, Cedex

INTRODUCTION

When blood makes contact with an artificial surface a complex series of reactions occurs including deposition of plasma components, activation of intrinsic clotting mechanism deposition, aggregation of platelets and other blood cells, leading to the formation, growth and/or embolization of thrombin. These phenomena present serious drawbacks to the use of artificial organs, devices, cardiovascular prostheses and catheters.

Blood-material interactions are known to be regulated by at least three factors: material surface (physicochemical nature and structure), blood composition and eventual non-physiological blood interfaces (e.g. blood/air interface) and blood flow conditions.

Analysis of blood-material interactions is used to determine the blood compatibility of material and/or to examine the mechanism of thrombogenesis. Many experimental procedures have been developed. No one of the existing analytical techniques can provide by itself all the data. Further, as outlined for other chapters in this book, the combination of many analytical techniques, each characterized by its specificity and limitations is therefore required to measure blood-material interactions. Most of the systems reported in the literature are studied in vitro, when contact between blood and an artificial surface occurs outside the body. In this chapter dealing with biomaterials implanted in or connected to the cardiovascular system, both in vivo and ex vivo methods will be analyzed. This description is limited to the most widely used procedures:

S. Dawids (ed.), Test Procedures for the Blood Compatibility of Biomaterials, 459–460.
© 1993 *Kluwer Academic Publishers. Printed in the Netherlands.*

I.   The ex vivo evaluation of thrombogenicity (Dudley test).

II.   The ex vivo acute arteriovenous shunt.

III.   The chronic artificial external arteriovenous shunt.

IV.   The implantation of vascular prostheses.

The effects of anticoagulant, temperature and animal species on blood-material interactions will be specifically discussed. There is not up till now a "best" way to conduct these investigations, given that multiple variables characterize artificial surface-induced thrombogenesis. Nevertheless, a multi-testing approach focused on the final use of the material is suggested.

Dynamic Ex Vivo Thrombogenicity Test in Modified Dudley Test

Authors: M.-C. Rissoan, R. Eloy & J. Baguet

INSERM U 37, Chirurgie Vasculaire et Transplantation d'Organes, F - 69500 Bron

INTRODUCTION

Name of method:

Dynamic ex vivo thrombogenicity test: Modified Dudley (MD) test.

Aim of method:

This test allows a study of the blood-material interactions under dynamic conditions. Through a catheter inserted into e.g. the cephalic vein of a dog, blood in the native state without anticoagulant is bled at low flow rates. The time related blood flow and the time for the blood to clot in and obstruct the catheter is measured. It has been demonstrated (*Dudley et al., 1976*) that the blood flow and the flow time is strongly related to properties of the material.

Background of method:

The MD test is an ex vivo dynamic test which allows the study of haemocompatibility of biomaterials in conditions of low venous blood flow. This method has the following characteristics:

1. The blood only encounters the surface of the foreign material to be tested and after insertion there is normally no opportunity for the clotting sequence to be initiated by any other foreign material.

2. No surgical procedure is involved and thus the animal has essentially neither a recovery period nor any significant tissue damage which could influence the results. Normally the test is carried out on the animal during anaesthesia.

S. Dawids (ed.), Test Procedures for the Blood Compatibility of Biomaterials, 461–480.
© 1993 *Kluwer Academic Publishers. Printed in the Netherlands.*

3. The procedure can be used for a statistically valid number of test and control studies in the same animal under identical conditions.

4. An extended range of haemodynamic flow conditions can be simulated by controlling the pressure at the outflow of the catheter.

5. The simple experimental set up eliminates the need for highly trained personnel or specialized facilities.

Contact between blood and foreign material sets off a sequence of protein adsorption/desorption processes which subsequently trigger the clotting cascade and enables adhesion and activation of thrombocytes and leucocytes.

Adhesion of activated platelets to the catheter wall also depends on haemodynamic conditions (function of shear rate) which, as well, influence the speed of protein adsorption (*Turitto et al., 1977, 1979*). The adsorption pattern of proteins to polymer surfaces is a primary event of importance since it determines the subsequent thromboembolic events. These can be considered to be fully accomplished within thirty minutes of blood flow and a constant equilibrium of adsorption of proteins and platelets is present on the surface (*Ihlenfeld et al., 1978*). The conditions of a MD test could be, a priori, considered as representative of blood-material interactions.

Dogs are preferentially used because they have a highly responsive blood coagulation system. It is possible to relate the observation in dogs' blood to the behavior in man (*Mason & Reads, 1971; Eloy et al., 1987*) although the test is applied on other animals: sheep (*Hecker, 1979*), baboons (*Eloy et al., 1988*).

Using an open ended system without recirculation, it is possible to detect biochemical changes generated by the material by regular sampling. The continuous blood sampling allows sequential determination of the blood contents of biological markers, i.e.: FpA, platelet factors, heparin (released from the surface).

The described modification of the original Dudley test encompasses simultaneous introduction of catheters in both cephalic veins of the upper limbs of a dog or human volunteer in order to:

-   use long catheters (30 cm length, 1 mm inner diameter) to increase the surface of the biomaterial,

- compare the relative time of coagulation and other parameters of clotting of one material to a reference material,
- analyse the first interaction between blood and biomaterials,
- associate the study of patency time with

    a)   continuous blood flow recording through the catheter,

    b)   kinetics of generation of activated coagulation and platelet factors

    c)   the rate of release of biologically active materials grafted on the polymer surface,

    d)   analysis of the surface by scanning microscopy. For the rate of release can be detected.

Scope of method:

The MD test allows a short term study of the interactions between blood and biomaterial in conditions that are comparable to those obtained in vivo in man.

Besides patency time, the test provides insight into the study of more complex phenomena such as thrombosis by fractional analysis of the outflowing blood for e.g.: protein adsorption, platelet adhesion, clot formation, early markers of thrombin activity, platelet activation, release of surface grafted compounds, release of material additives (plasticizers etc.). It has been particularly useful for studying the relative behavior of polymer surface-blood interactions in the comparison between treated and non treated surfaces.

DESCRIPTION OF THE METHOD

Equipment required.

a.   Surgical requirements should envisage that the test can be performed on adult mongrel dogs weighing 15/20 kg with respiratory assistance.

The blood flow rate is continuously monitored with an ultrasound microflow velocimeter (Microflow MF20, Eden Eme, D - 770 Überlingen).

b.   Biochemical studies: 5 ml polystyrene tubes (Fumouze Laboratories, Distriphar, F - 92541 Montrouge Cedex).

For FpA determination.
Equipment for FpA determination is similar to that described in the chapter on *"Test methods for the activation of haemostasis ..."* with the following specificities:

-   Kit EIA-Asserachrom FpA (Stago, F - 92602 Asnieres Cedex) and dog serum with 25 µg/flacon of dog FpA (equivalent fibrinogen which serves as reference curve going from 25 ng to 250 ng/ml),
-   anticoagulant solution: Sodium citrate 0.11 M, Heparin, Aprotinin and Sodium Azide,
-   Kinetic microplate reader (V max, Vietech Sarl, F - 69720 Saint Laurent de Mure),
-   NUNC plates with certificate, 96 flat bottomed wells (ref.454 Poly-Labo Bloc, F - 67028 Strassbourg Cedex)

For heparin release.
Heparin is determined by its anti-IIa and anti-Xa activities on chromogenic substrates, respectively CBJ 34.47 and CBS 3.39 (Stago: Stachrom heparin and Hepachrom). No heparin is present in the anticoagulant solution.

Preparation for scanning microscopy.
The rinsing solution is a buffered cacodylate 0.4 M solution. The fixative solution is buffer cacodylate and a solution of glutaraldehyde mixed as:

A - Buffer cacodylate 0.4 M,
      21.4 g of cacodylate in 250 ml H_2O, adjust to the pH to 7.4 with HCL RP, PO = 400 mosmol/l.
      cacodylate: dimethylarsinic acid Sodium salt (Ref. 820 670 Merck, Darmstadt, FRG).
B - Glutaraldehyde solution at 4% (final concentration).
      16 ml glutaraldehyde 50% + 184 ml distilled water.
      (Biological grade - Polyscience INC., Washington PA 18576, USA).

Solution of fixation is mixed before assay consisting of roughly:
      2 volumes of glutaraldehyde 4%
      1 volume of cacodylate buffer 0.4 M
      1 volume of water

This solution can be used within 24 hours.

Outline of method.

The catheters or tubings are placed symmetrically after the dog has been anaesthetized. The intravascular portion should extend a few centimeters above the insertion site of the catheter. Flow from both catheters should start simultaneously. The catheter (tubing) does not require insulation, but the environmental temperature should be 20°- 25°C (room temperature). Blood samples should be collected in test tubes placed in ice.

After the test has been fulfilled, the catheter should be withdrawn, and the veneous lesion should be carefully closed if it has been cut. If insertion has been made with a stylet, careful compression is sufficient. The distal ligature on the vein can be removed if it is planned to run the test on the animal again.

Description of procedure.

For dog.
The dog (15-20 kg) is anaesthetized by administration of Sodium Thiopental (25 mg/kg). Respiratory assistance is ensured via intratracheal intubation. Before surgery, catheters are flushed with saline and closed at one end with a plug. Catheters of the polymer 30 cm long, 1 mm ID are simultaneously implanted via a short cutaneous incision and phlebotomy in the cephalic veins over a length of 2 cm (*Fig. 1*) and secured.

The plug is removed and the chronometer is started. The blood flow rate is set to 4 ml/min by adjusting the heigth of the distal end of the catheter and continuously monitored by ultrasound microflow velocimeter. The blood effluents (1.5-2 ml) are collected in 5 ml polystyrene tubes (with anticoagulant solution) immersed in ice at time intervals of 1, 2, 3, 5, 8, 10 min and then collected uninterrupted until clotting occurs within the catheter.

For catheters or tubes with heparinized surfaces the blood effluents are collected on anticoagulant solution containing no heparin (Sodium Citrate, 0.11 M, pH = 7.3). The sequential sampling is performed at: 1, 2, 3, 5, 8, 10, 15, 30, 45, 60, 75, 90, 120 min.

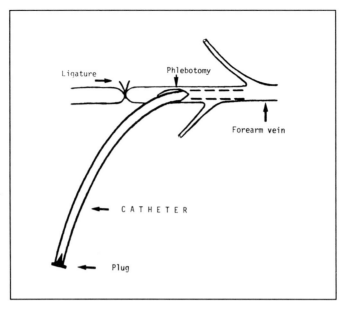

Fig. 1

Diagram of implantation of catheter.

After 15 minutes of continuous flow a plug is placed in the distal end of the catheter. Every 15 min. it is removed to evaluate blood flow and the sample is analysed for FpA and heparin.

Blood samples are immediately centrifuged at 2800 rpm at 4°C for 15 min. Plasma is transferred to another vial and frozen at -20°C until analysis can take place.

The catheter is removed after clotting. Buffer cacodylate is gently injected retrograde through the catheter to remove non-adherent thrombi. Immediately after, the catheter is cut and split along its length and cautiously placed in the fixative solution before preparation for and examination by scanning electron microscopy.

For human volunteers.
A short skin insertion for indwelling is performed. Catheters are implanted trans-cutaneously into the left and right cephalic vein. Reference catheters and test catheters to evaluate are compared in each individual using both arms. Blood flow through each catheter is monitored by ultrasound microflow velocimetry. Blood is

collected continuously into three plastic vials for determination on appropriate media of:

- Fibrinpeptide A (FpA)
- Heparin
- ß-Thromboglobulin (ßTG) and Platelet factor 4 (PF4) markers of platelet activation (see chapter on *"Test methods for the activation of haemostasis ...")*

RESULTS

Typical materials tested were:

1. a copolymer polyurethane (PU), polyvinyl chloride (PVC) and heparinized PU-PVC (ionic bonding of heparin 10% of the bulk "Rhone-Poulenc", CRC, Rhone Poulenc, F - 69190 Saint Fons),

2. aliphatic non heparinized polyurethane (Tecoflex-Thermidics, Lab. Vygon, F - 95540 Ecouen).

Table I

The relationship between shear rate and diameter of the catheter.

Diameter in cm	Shear Rate sec^{-1}		
	1ml/min	4ml/min	6ml/min
0.1	163.00	672.00	1019.00
0.2	20.04	84.40	128.00
0.3	6.00	24.82	37.64
0.4	2.55	10.55	16.00

A representative recording is given for PU and heparinized PU in *Fig. 2.*

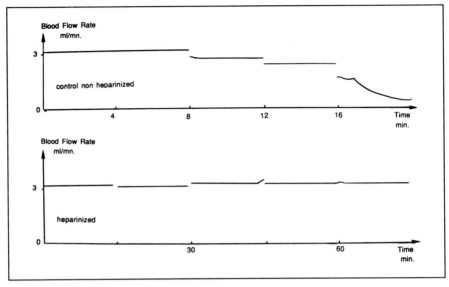

Fig. 2

*Blood flow rate versus time for **above**: non heparinized catheter PU. Thrombosis started at 16.30 min, for **below**: heparinized catheter PU. After 60 min. the heparinized PU catheter is still patent.*

Patency time and FpA generation.

The results (mean ± SD) obtained with silicone catheters versus polyurethane (PU) (*Fig. 3*) and silicone compared with polyethylene (PE) (*Fig. 4*) are summarized as follows:

Table II

Material	Silicone n = 5	PU n = 5	Silicone n = 5	PE n = 5
Patency time in minutes	8 ± 1.2	23.6 ± 7.5	11.6 ± 4.1	14.7 ± 7.2

In *Figs. 3 & 4* the values of the single experiments are shown illustrating considerable variations.

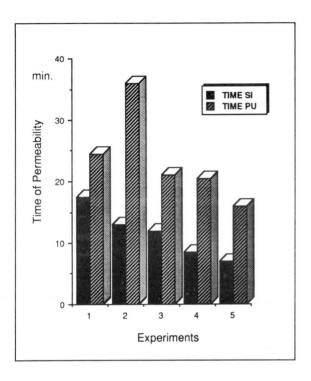

Fig. 3

Paired times of patency of silicone and polyurethane in 5 different experiments.

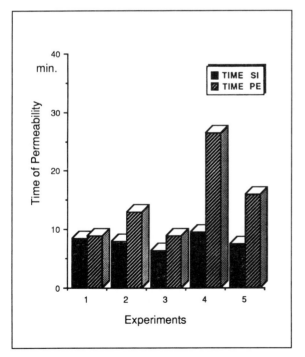

Fig. 4

Paired times of patency of silicone and polyethylene in 5 different experiments.

470

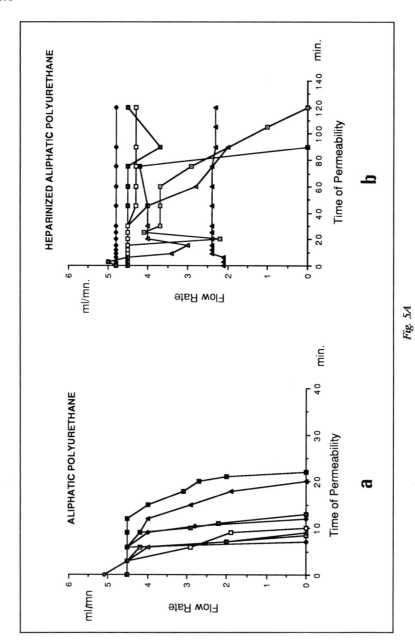

Fig. 5A

Comparison in individual experiments of clotting time (a & b) and FpA generation (c & d, see following page) of aliphatic polyurethane (a & c) to heparinized aliphatic polyurethane (b & d).

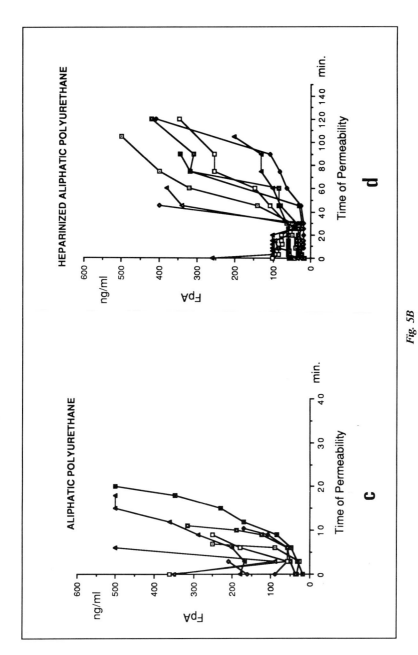

Fig. 5B

Comparison in individual experiments of clotting time (a & b, see previous page) and FpA generation (c & d) of aliphatic polyurethane (a & c) to heparinized aliphatic polyurethane (b & d).

The amounts of FpA generated at the 6th minute following catheter insertion are of 8 ± 6 (ng/ml) with silicone and 3 ± 1.4 with polyurethane, likewise showing large variations. In *Fig. 5* the time dependent flow (= time of permeability) and time dependent generation of fibrinopeptide A (FpA) is given. It can be seen that the decrease of flow (a & b) is preceded by a rapid rise in liberation of FpA (c & d).

Heparinized materials.

In dogs.
Testing in dogs of heparinized materials can be expected to show correlations as indicated in *Fig. 5*. The relationship between the catheter flow time and the kinetics of generation of FpA is evident with this method: a rapid increase of FpA generation is followed by an almost exponential decrease of blood flow leading to total clotting of the tube. In heparinized aliphatic polyurethane, heparin significantly prolongs the flow time to more than two hours, despite the plug.

In humans.
Testing in humans of non-heparinized and heparinized catheters showed thrombosis of non-heparinized polyurethane occurring in the testing conditions at the following times: 15 min., 17 min., 35 min., 23 1/2 min. and 26 1/2 min. No thrombosis occurred for at least 65 min. in the case of heparinized catheters.

Heparin release is illustrated in *Fig. 6* showing that detectable amounts of heparin activity (anti-Xa and anti-IIa) were present throughout the period of testing. The value of heparin presented a high initial level which gradually decreased. By comparing the concentration and cumulative amounts of both activities, since blood flow and sequence of blood sampling were identical in each case, it appears that anti-Xa activity is generally released at a higher level than anti-IIa activity.

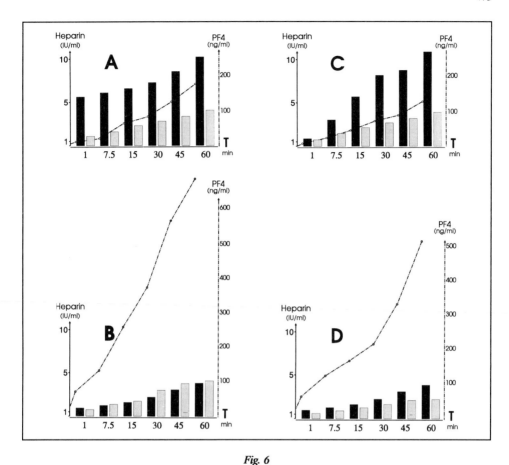

Fig. 6

Anti-Xa (dark columns), anti IIa (open columns) activities (IU/ml) from blood samples collected via heparinized catheters in four healthy volunteers (A, B, C, D).

The levels of FpA in non heparinized catheters rose exponentially to over 50 ng/ml as shown in *Fig. 7*. On the other hand, levels of FpA in heparinized catheters remained below 10 ng/ml even at the 60th minute and despite the plug.

Generation of PF4 and ßTG indicating activation of thrombocytes are shown in *Figs. 8 and 9*. The levels of both parameters appear higher in non-heparinized versus heparinized catheters. This is already present in the first sampling tube which corresponds to the first blood effluent in contact with the material (0.3 minutes). The mean course of PF4 and ßTG for both materials show that the pattern of liberation is significantly different in heparinized and non-heparinized materials for PF4 ($p < 0.025$) and for ßTG ($p < 0.01$) during the first twenty minutes of catheter implantation.

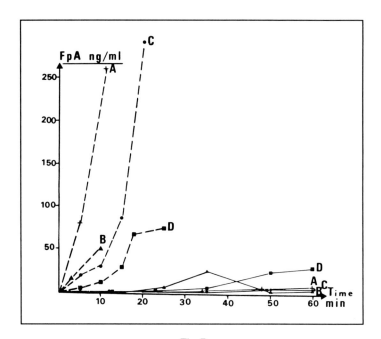

Fig. 7

Generation of fibrinopeptide A (FpA) in blood samples collected via non heparinized (dotted lines) and heparinized (full lines) catheters in four healthy volunteers (A,B,C,D).

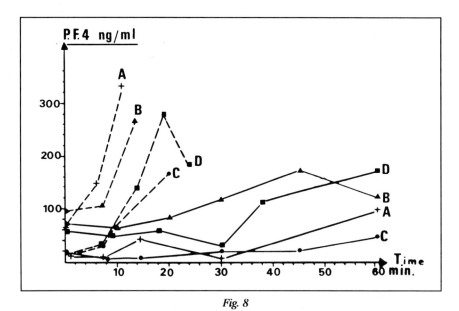

Fig. 8

Generation of platelet factor 4 (PF4) in blood samples collected via non-heparinized (dotted lines) and heparinized (full lines) catheters in the same individuals (A,B,C,D).

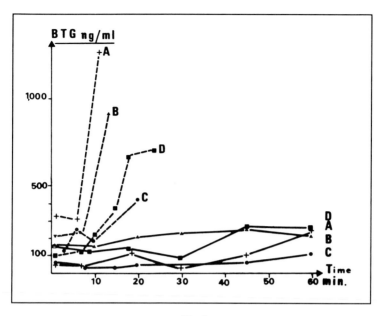

Fig. 9

Generation of β-thromboglobulin (βTG) in blood samples collected via non-heparinized (dotted lines) and heparinized (full lines) catheters in the same individuals (A,B,C,D).

Scanning electron microscopic examination.

Sequential examination ranging from initial exposure to cellular deposition of the blood material interactions is given in *Fig. 10* to compare polyurethane and heparinized polyurethane interactions with blood. The sequences illustrate clearly protein adhesion, platelet adhesion followed by spreading, activation, aggregation and fibrin formation, leading to clotting in non-heparinized materials whereas the sequence is stopped at the level of limited platelet adhesion in the case of heparinized polyurethane. Heparin released from this material, although inhibiting thrombin formation, does not prevent platelet adhesion.

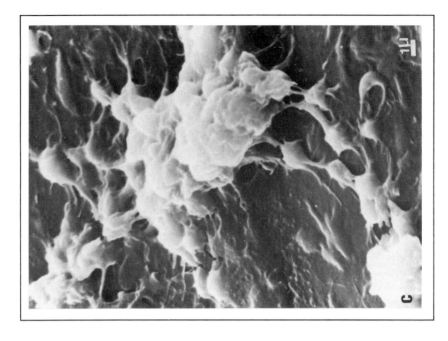

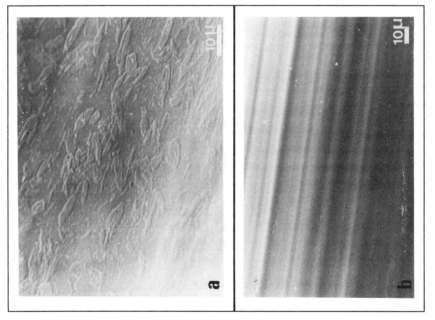

Fig. 10 A

*SEM examination of the PUs before and after implantation in dogs. **Before** implantation: a) ultrastructure of non-heparinized PU catheters. b) ultrastructure of heparinized PU catheters. **After** implantation: c) typical platelet aggregate on a layer of adherent proteins at the time of thrombosis (21 minutes) of the catheter (a).*

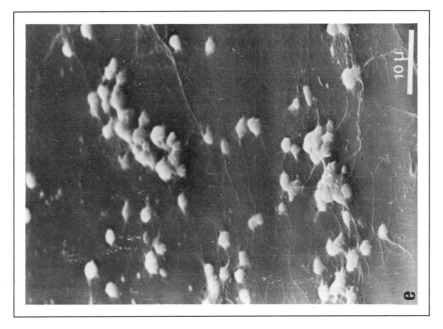

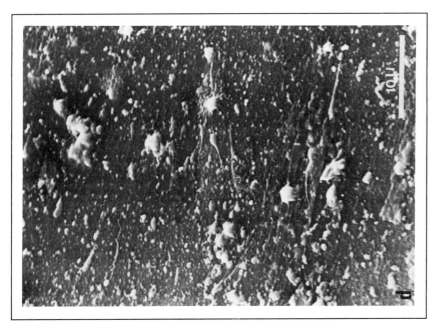

Fig 10 B

SEM examination of the PUs after implantation: d) isolated adherent platelets on a heparinized PU surface (catheter b) after 120 min. e) adherent platelets without activation of heparinized PU surface (catheter b). No fibrin is formed after 75 minutes.

DISCUSSION

While in vitro experiments are usually easier to control and results more quantifiable, blood is usually anticoagulated in these situations and not tested under conditions similar to clinical conditions. Moreover, the test material is not exposed to the same clinical rheological conditions. Admittedly the relevance of in vitro tests to in vivo blood material responses has not been regularly assessed.

The above described MD test belongs to ex vivo testing methods where a blood flow from the body runs via the biomaterial to the open with no recirculation. There is thus no additional interference.

- Standard catheters and experimental catheter are studied simultaneously.
- A flow rate of 1 to 6 ml/mn is easily obtained.
- The methods are rapid and reproducible with 4 to 6 animals according to the desired accuracy of comparison and differences between the materials.
- The animals are anaesthetized for a short time and thereby not exposed to a large trauma.

In addition to the the classical clotting time information obtained with the original Dudley test, this procedure allows good control of flow conditions and monitoring of important indicants of blood clotting activation, i.e. the rate of thrombin generation and of platelet activation.

The relevance of the test has been established on human volunteers with small tubings of 1-1.5 mm ID as it was performed in dogs, using peripheral vein catheterisation. A good relationship of data and ranking of materials can be established in both canine and human situations for the same material. In *Fig. 11* the measurement of FpA in human and canine plasma is shown using human anti-FpA. Closely aligned graphs show that the molecular structure in the two species are very resembling.

By using an ex vivo model (*Didisheim et al., 1983*) it has been shown that the relative ranking of human platelet and fibrinogen levels measured on five materials was similar to that observed in dogs although haemostatic mechanisms of most, if not all animals differs significantly from that of man.

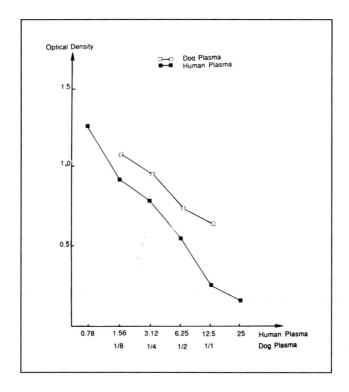

Fig. 11

Standard curve of FpA immunoenzymatic assay using anti-human FpA as antibody.

The MD test only requires that it must be performed in a manner that avoids blood-air interface since it may by itself activate haemostasis. Variability of the results will occur if careful execution of the test is not respected, Only 4 to 6 experiments are generally necessary to discriminate between relatively similar materials, and the simultaneous control test of the contralateral vein is essential to provide quality check of the results.

The test can be expanded by temporary measurements. At the end of the experiment the extent of platelet adhesion and of specific protein adsorption can be measured by using radiolabelling procedures, by applying desorption of adsorbed cell and plasma constituents or by using immunoenzyme assay of adsorbed proteins (see chapter on *"Test methods for the detection of protein adsorption on polymers"*).

480

Results observed are directly related to two orders of parameters:

1. The status of the experimental animal in terms of biochemical parameters, general and local haemodynamic conditions. Particular attention should be given to the initial flow through the material in order to standardize rheological conditions.

2. The physiochemical nature of the material. These are directly related not only to the chemistry of the polymer but also to the surface ultrastructure and thus to the technical conditions of transformation of the polymer.

REFERENCES

Didisheim, P., Stropp, J.Q., Dewanjee, M.K. & Wahner, H.W. (1983) Thromb. Hemostasis, 50, 60.

Dudley, B., Williams, J.L., Able, K. & Muller, B. (1976) Trans. Amer. Soc. Artif. Int. Organs, XXII, 538-544.

Eloy, R., Belleville, J., Paul, J., Pusineri, Ch., Baguet, J., Rissoan, M.C., Cathignol, D., French, P., Wille, D. & Tartullier, M. (1987) Thrombosis Research, 45, 223-233.

Eloy, R., Belleville, J., Rissoan, M.C. & Baguet, J. (1988) Journal of Biomaterials Application, 2, 4, 475-519.

Hecker, J.F. (1979) Trans. Amer. Soc. Artif. Organs, XXV, 294-298.
Ihlenfeld, J.V., Mathis, T.R., Barber, T.A., Mosher, D.F., Riddle, L.M., Hart, A.P., Updike, S.H. & Cooper, S.L. (1978) Trans. Amer. Soc. Artif. Intern. Organs, XXIV, 722-733.

Mason, R.G. & Reads, M.S. (1971) J. Biomed. Mat. Res., 5, 121.

Turitto, V.F., Muggli, R. & Baumgartner H.R. (1977) Annals New York Academy of Sciences, 283, 284-292.

Turitto, V.F & Weiss, H.J. (1979) Journal of Rheology, 23, 6, 735-749.

Ex Vivo Acute Arteriovenous Shunt

Authors: E. Chignier & R. Eloy

INSERM U 37, Chirurgie Vasculaire et Transplantations d'Organes, F - 69500 Bron

INTRODUCTION

Name of method:

Ex vivo acute arteriovenous shunt.

Aim of method:

The method can be applied to monitor the short time pattern of blood-material interactions on polymeric surfaces in an extracorporeal blood circuit.

Biochemical and biophysical background of the method:

Ex vivo experiments using a test section or chamber located outside the body can be used to test blood-material interactions with flowing unaltered blood. The performance of blood contacting materials can thus be assessed under near normal conditions. Arteriovenous shunts were introduced by *Rowntree et Shionaya, 1927* for studies on experimental thrombosis and the mechanism of thrombus formation.

The basic principle is that the blood, after flowing through the test section, then returns to the vascular system (closed system). Several parameters of the blood-material interactions may be analyzed (patency time, amount of thrombus formation, kinetics of haemostasis activation ...).

The monitoring of the haemodynamic conditions are especially important using this method, since they qualitatively and quantitatively influence blood-material interactions. Wall shear rates should correspond to those encountered in the cardiovascular system as reviewed recently by *Goldsmith & Turitto (1986)*. Briefly, it is accepted that (*Anderson & Kotke-Marchant, 1985*):

481

S. Dawids (ed.), Test Procedures for the Blood Compatibility of Biomaterials, 481–493.
© 1993 *Kluwer Academic Publishers. Printed in the Netherlands.*

1. platelet adhesion will increase with platelet concentration,
2. an increase of blood flow or a decrease in vessel diameter will lead to increased deposition. Also, platelet adhesion will increase as shear rate increases,
3. as the axial distance from the entry increases, platelet deposition decreases. This implies greater platelet depositions at the proximal end of tubings (*Adams & Feuerstein, 1981*),
4. platelet deposition increases with time.

Although the ex vivo model is representative of the clinical use of a vascular device and thus allows an evaluation of thrombus formation, more acute criteria are needed to establish reliable comparisons between new polymers and/or chemically treated surfaces. In this way we can hopefully determine the basic mechanism of blood-material interactions. Such criteria include:

1. evaluation of platelet activation,
2. evaluation of protein adsorption/activation or inactivation in at least three major systems: coagulation, fibrinolysis and complement,
3. evaluation of granulocyte and monocyte deposition and their possible activation.

The described method is designed to allow repeated access to measure several of these parameters coupled with the advantages of an ex vivo shunt test and follows mainly the description made by *Ihlenfeld et al., 1979*. Originally, Ihlenfeld based his material haemocompatibility discrimination on the study of the amount of labelled platelets and fibrinogen material deposition. This author recognized PVC as having poor blood compatibility, because it rapidly activates platelet and fibrin. This results in a transient maximum thrombus formation time after 10 to 15 minutes of blood contact.

The ex vivo canine femoral AV shunt as employed by Ihlenfeld provides blood flow rates of 385.3 ± 26.0 ml/min with 3 mm ID PVC tubes and of 173.5 ± 13.2 ml/min with 2.5 mm ID PVC tubes. According to the author's description, these flow rates were controlled only by the animal's cardiac output.

Scope of the method:

This method is employed to study short time (up to 48 hours) blood-material interactions for general comparative screening of haemocompatibility of native and modified polymer surfaces.

The ultimate criterion used in this method is the time of clotting. However, supplementary measurements are generally performed on platelet adhesion-activation, protein adsorption and activation of e.g. coagulation factors.

DETAILED DESCRIPTION OF METHOD

Equipment/materials.

Preparation of the tubing test samples.
Prior to implantation, the ends of each segment are cut clean at 45°. The segments are washed with detergent soap solution in distilled water, thoroughly rinsed with sterile distilled water and left overnight filled with Tyrode's solution (pH = 7.35).

Blood samples are obtained through a pair of Y connectors made from heparinized small diameter polyethylene tubes, connected at a 45° angle with the tubing section on the arterial and venous sides. This allows continuous or sequential blood sampling at both sides of the shunt for the determination of the amount of the biochemical marker generated along the shunt. Blood velocity is monitored continuously in the middle part of the shunt by means of a Microflow Doppler ultrasound flowmeter.

Polymer tubings (1 mm ID) used for haemocompatibility testing are prepared as follows:

A 50 cm long segment is cut and receives at 3.5 cm from each end a 10 cm long heparinized tubing (1 mm ID). These Y-connectors are mounted with glue in the main tube section at an angle of 45° and are used as sampling lines. The whole assembly is cleaned with sterile distilled water in an ultrasonic bath and sterilized by ethylene oxide gas and then aerated adequately prior to use.

Preparation of the animals.

The test is generally performed on adult mongrel dogs, weighing 18-27 kg, but other mammals with good congruence between femoral vessel size and the diameter of tubings to be tested may be used.

Before surgery it is wise to check the haemostatic parameters of the animals (blood differential cell and platelet count, fibrinogen (or factor II, VII and X), hematocrit analysis and platelet aggregation in the presence of ADP and epinephrine).

Outline of the method.

Under general (intravenous) anaesthesia, the femoral artery and vein in one leg of a mongrel dog are exposed and ligated as shown in *Fig. 1*. Both vessels are cannulated and the segment of tubing is filled with fresh Tyrode's solution.

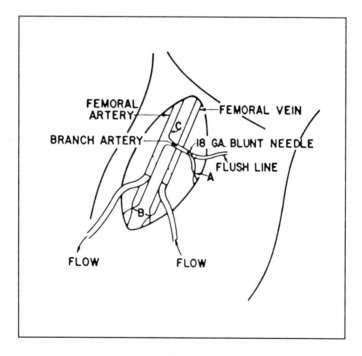

Fig. 1

Shunt cannulation procedure using the technique described by Ihlenfeld et al. (1979). The branch artery is cannulated proximal to the shunt entrance to permit introduction of Tyrode's solution into the shunt to remove the volume of blood contained within the shunt which is 1 m 77 cm long and 3 mm ID.

The shunt is sampled for platelet and fibrin deposition after 2 minutes of exposure to blood flow and then every minute between 5 and 60 minutes, and then at 90 and 120 minutes by means of radiolabelled platelets and fibrinogen.

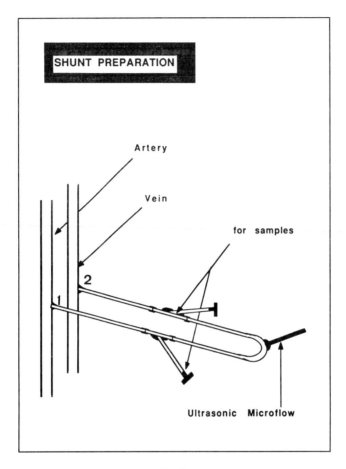

Fig. 2

Shunt cannulation following our procedure: the shunt is 50 cm long and 1 mm ID. It is inserted into the arterial and venous collateral branches (see text). A pair of Y-connectors allows sequential blood sampling at both (arterial and venous)sides of the shunt.

In this description the Ihlenfeld's method (*Fig. 2*) has been slightly modified. In order to avoid vessel trauma and subsequent interaction with coagulation parameters, the femoral vessels are not ligated. The temporary haemostasis is achieved by applying a lace (following the Blalock technique) placed around the femoral artery and vein, above and below a small branch of each vessel. Secondly, to avoid inequalities between the recipient artery and the shunt section and consequent turbulent blood flow, it is advisable to use the small vessel branches to insert the shunt section,

Description of the procedure.

Surgical procedure.

Healthy mongrel dogs are used for the acute ex vivo AV shunts. The animal has fasted overnight and is anaesthethized with Sodium Thiopental (30 mg/kg) tracheal intubated and ventilated with Bird® respirator. The abdominal skin area and the tights are shaved and surgically prepared with an antiseptic solution (Betadine, Sarget). The animal is placed on its back on a mattress containing warm water and draped with sterile sheets. Only the upper thigh is uncovered. Body temperature is recorded continuously with a rectal probe. Both legs are used simultaneously for testing shunts, one for the assay material and the other for reference material.

The Femoral artery and vein are isolated in the femoral triangle, and each vessel is secured with two laces, above and below a small collateral branch, respectively the lateral circumflex femoral artery and the medial saphenous vein. The two laces around the artery and the vein are tightened up to provide temporary haemostasis without causing trauma of the vessel walls. The circumflex femoral artery (*1 in Fig. 2*) and the medial saphenous vein (*2 in Fig. 2*) are distally ligated. A veinotomy is performed, using microscissors, and the shunt tubing, previously filled with saline solution to avoid blood-air contact, is introduced into the vein branch until it is flushed with the main trunk and is then secured. The same procedure is then carried out for the circumflex femoral artery. The two laces are then released, and the blood streams through the catheter, establishing the AV shunt. The blood flow is measured with the Doppler ultrasound device.

Blood samples are taken sequentially, at first every minute and then every 6 or 12 hours through the Y-connector to measure the difference between the arterial and venous concentrations of markers, i.e. Fibrinopeptide A (FpA) which results from thrombin activation. The experiment ends when clotting occurs as detected by the blood flow device. The animal is then sacrificed.

In some cases, by using thromboresistant polymers, the test sections remain patent up to 6 hours. In these circumstances the procedure is carried out as follows: The assay tubing is gently positioned under the skin of the thigh and the wound is closed, leaving the sampling tubing outside, occluded by plugs and protected with a dressing. The animals are awakened and returned to their cages. The blood flow is regularly controlled by the same Doppler ultrasound device. Every 12 hours, the plugs are removed and blood samples are taken for FpA measurements.

At the end of the experiment the test tubing is removed from the vessels, gently flushed with saline, and several fragments are processed for scanning electron microscopy or for analysis of the stabilized blood-material interface.

RESULTS

Following this protocol, two main groups of parameters of thrombogenicity , i.e. haemodynamic and hematologic, can be analysed.

Haemodynamic parameters.

The results given here are obtained with different polymer tubing tested in this model (1 mm ID):

1. silicone (n = 3),
2. polyurethane, bulk heparinized, containing 10% ionically bonded heparin (n = 5),
3. polyurethane as above, but without heparin (n = 5).

The course of blood flow and the time of thrombosis for each material are illustrated in *Fig. 3*. It demonstrates that the release of small amounts of interfacial heparin inhibits thrombus formation even in the low flow blood conditions, where thrombosis develops in polyurethane catheters in a mean time of 7 minutes (from 6 to 10 minutes) which is close to the thrombosis time (from 4 to 15 minutes) for the silicone tubes.

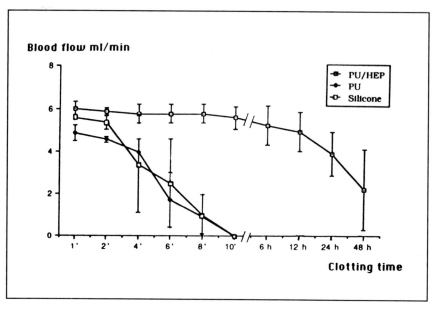

Fig. 3

Blood flow rates versus thrombosis observed in a series of experiments using 1 mm ID catheters. Mean and SD values of PU bulk heparinized catheters (-⊙-). Two of them were stopped respectively 48 and 72 hours after the beginning of the experiment. All silicone catheters (-○-) as well as control PU catheters (-●-) thrombosed between 4 and 10 minutes,

Hematologic parameters.

Using immunoenzyme linked assay for FpA (Stago-Asnière, France) and an antibody against human FpA which develops cross-reactivity towards dog FpA, normal values for dogs range between 20 - 30 ng/ml.

Fig. 4 illustrates the course of this marker in the arteriovenous shunts in dogs.

These results demonstrate the early rise of FpA marker in catheters before thrombosis occurs already after the fourth minute of blood flow through the shunt.

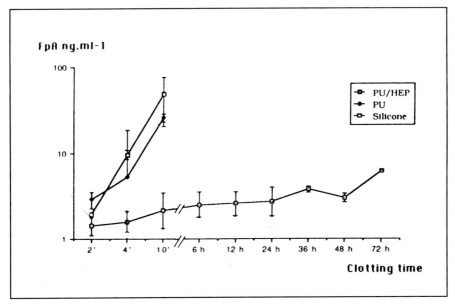

Fig. 4

FpA measurements in blood samples were correlated with clotting time in the same series of experiments indicated in Fig. 3. The initial levels of FpA were higher in PU (-•-) and silicone (-○-) than in PU bulk heparinized (-⊙-) catheters, and the variation of FpA amounts always preceeded the clotting.

DISCUSSION

The recirculation of the blood into the vascular system of the dog is one limit to the interpretation of the results, since the values measured are the combined result of:

1) generation of FpA at the blood-material interface,
2) dilution in total blood volume
3) rapid catabolism of the radioactive marker which has, in this case, a half-life time of 2 - 3 minutes.

On the other hand, the model allows the animal's homeostatic compensatory mechanisms such as inflammatory, haemostatic hormonal systems to function normally.

As outlined by *Anderson & Kotke-Marchant, 1985,* the systems whose action and feedback are largely or completely inhibited during in vivo testing may have an influence on the interactions of platelets with materials tested in ex vivo systems.

Interpretation of results.

The correlation between the well-defined clotting time of a catheter preceeded by a decrease in flow and the rise in FpA generation confirms that this sequence of events is a good indicator for the study of haemocompatibility. Although clotting time is the final step relating to the performance of the tube, it cannot be used alone to discriminate between properties of related materials.

The monitoring of blood flow provides information concerning the thrombus formation process and its growth. In addition, the amount of thrombin generated per unit of surface of a biomaterial in given rheological conditions can be determined by the assay of thrombin on its physiological substrate, i.e. fibrinogen.

Markers of platelet function or activation are highly indicative such as PF4 (see chapter on *"Test methods for the activation of haemostasis and other activation systems"*). Quantitative assays of platelet adhesion and/or release can be applied provided that cross reacting markers are available. In addition, by varying the rheological conditions (by increasing the tubing diameters) and using the same polymer interface, the direct effect of flow in the same animal can be compared and analysed.

Limitations of the method.

- The testing is limited to small diameter tubes (1 to 3 mm ID).
- Only short time periods of blood flow contact are used, ranging from minutes to 72 hours.
- Only two materials can be tested simultaneously in the same animal.
- Activated coagulation factors or cells are recirculating and/or consumed.

Related relevant testing methods.

Canine ex vivo series shunt *(Lelah et al., 1984).*
The experimental technique utilized by Lelah et al. allows for the simultaneous investigation of up to ten different materials in an acute canine ex vivo series shunt.

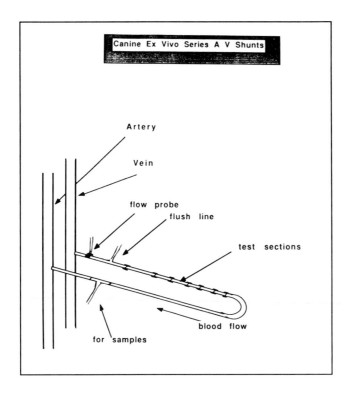

Fig. 5

Schematic representation of the canine ex vivo series shunt described by Lelah et al.

The test sections are connected in series (*Fig. 5*), thus reducing the time and expense associated with animal testing. The anaesthetized animal has its femoral artery and vein exposed and ligated, then cannulated with polyethylene entrance and exit sections connected to the shunt with the test tubing. The shunt section is initially filled with degassed divalent cation-free Tyrode's solution. A small branch artery, proximal to the shunt cannulation site, is cannulated and serves to introduce Tyrode's solution at a rate of about 1 ml/s to avoid blood dilution. Immediately following flushing, the 4 cm long test section is removed and fixed with freshly prepared glutaraldehyde/phosphate buffered solution. Each segment is then sectioned under buffer, and a 3.2 cm fragment of each test surface is tested for radioactivity in an automated dual channel gamma counter (Beckman 5500). A 6 mm segment is examined by scanning electron microscopy. A new set of identical test surfaces joined as a shunt section is inserted for each period of time (from 30 sec. to 60 min. of blood contact). The blood contacting and flushing procedures are repeated, and blood flow is continuously monitored. Blood samples are periodically taken from the jugular vein to assess radioactivity level and hematological taken from the jugular vein to assess radioactivity level and hematological parameters. Platelet aggregability and fibrinogen are monitored at the start and at the end of experiments. No animals are used for repetitive experiments, each animal is sacrified at the end of testing.

This protocol is used by Lelah and coworkers to examine platelet and fibrinogen deposition on polyethylene, silastic, polyvinylchloride and oxidized polyethylene surfaces. Peak levels of platelets and fibrinogen were found to be inversely related to flow rate over the range studied.

Baboon flow regulated shunt.

Mackey and coworkers (*Mackey et al., 1984*) employed an ex vivo system for the study of small caliber vascular grafts implanted in baboon femoral artery and vein. Synthetic graft materials of 4 mm ID were placed into the shunt circuit and studied for ^{111}In-labelled platelet uptake associated with subsequent morphological analysis of the graft segments. ^{111}In-labelled platelet interactions with the test materials are monitored using gamma camera scanning.

Throughout the period of scanning, a constant flow rate is maintained. Flow rates of 25, 100 and 200 ml/min. were chosen for this study. Grafts were removed from the shunt circuit after 1 hour and processed for morphological analysis by scanning electron microscopy.

Mackey based his study on three baboons, in rotation, and experiments at each flow rate were carried out in each animal in random sequence. The order of materials in the circuit (arterial or venous limb of the shunt) does not influence platelet deposition, even if up to four materials are placed in the circuit simultaneously.

Advantages of the baboon model are: 1) conservation of the animals, 2) possibility of interrupting shunt studies at specified intervals to remove segments of the test materials for morphologic analysis and 3) an animal can be studied every two weeks without significant morbidity. The placements of the percutaneous shunt tend to eliminate the problems of infection presented by implanted AV shunts.

REFERENCES

Adams, G.A. & Feuerstein, I.A. (1981) Am. J. Physiol., 240, XX 99.

Anderson, J.M. & Kotke-Marchant, K. (1985) CRC Critical Reviews in Biocompatibility, 1, 2, 11-203.

Goldsmith, H.L. & Turitto, V.T. (1986) Thromb. Hemostas., 55, 3, 415-435.

Ihlenfeld, J.V., Mathis, T.R., Riddle, L.M. & Cooper, S.L. (1979) Thromb. Res., 14, 953-967.

Lelah, M.D., Lambrecht, L.K & Cooper, S.L. (1984) J. Biomed. Mat. Res., 18, 475-496.

Makey, W.C., Keough, E.M., Connolly, R.J., Mc Cullogh, J.L., Ramberg-Laskaris, K., O'Donnell, T.F., Fonall, T. & Callow, A.D. (1984) J. Surg. Res., 37, 112-118.

Rowntree, L.G. & Shionaya, T. (1927) Exp. Med. 46, 7-17.

Chronic Artificial External Arterio-Venous Shunt

Authors: E. Chignier & R. Eloy

INSERM U 37, Chirurgie Vasculaire et Transplantations d'Organes, F - 69500 Bron

INTRODUCTION

Name of method:

Chronic artificial external arterio-venous shunt.

Aim of method:

Chronic external arterio-veneous shunts (CAVS) represent a chronic experimental model in which many materials with inner diameter (ID) of 3mm or less, of variable length and/or design can be tested successively. The CAVS can be used for several months twice weekly, providing a well-controlled and reproducible biological and rheological test system at a relatively low cost.

Biochemical, biophysical, physical background of the method:

The first author who employed this method was *Frasher et al. (1967)*, who published an adaptation for beagle dogs of the artificial arteriovenous shunt technique for haemodialysis in human patients described by *Schribner et al. (1960)*.

Upon exposure to flowing blood, a test material is rapidly coated with a layer of blood proteins and, simultaneously or consecutively, platelets and other blood cells are implicated in the formation and growth of thrombi, whereby the lumen can be partially or totally blocked.

Dynamic conditions of blood flow in CAVS permit the measurement of cell deposition on the polymer surfaces of interposed tubing during the blood flow contact period.

Standard large diameter (3-4 mm) CAVS (*Fig. 1a, 1b*) results in relatively high blood flow rates between 250-400 ml/min and wall shear rates of 1500-4000 s^{-1}.

495

S. Dawids (ed.), Test Procedures for the Blood Compatibility of Biomaterials, 495–507.
© 1993 *Kluwer Academic Publishers. Printed in the Netherlands.*

Distinctions are made between low flow conditions, where fibrin formation and red thrombus formation predominate, and high flow conditions, where platelet deposition and white thrombus formation prevail.

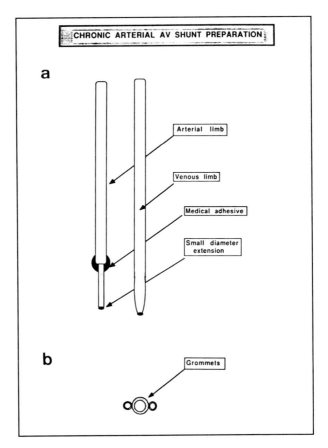

Fig. 1

Main AV shunt preparation

The occurence of platelet adhesion phenomena is largely depending on flow rate and shear stress (*Turitto et al., 1977*). At shear rates above $1000\ s^{-1}$ there is an increase in fibrin deposition and a decrease in platelet adhesion and platelet thrombi formation rates. Under the conditions of high wall shear stress ($1500 - 4000\ s^{-1}$) existing in CAVS, platelet-material surface interactions dominate.

When antithrombogenic surfaces are tested, it is presumed that they exert a dominant effect on the intrinsic coagulation system. They should therefore be examined under low flow conditions where fibrin formation (red thrombus formation) is greater than platelet activation (white thrombus formation).

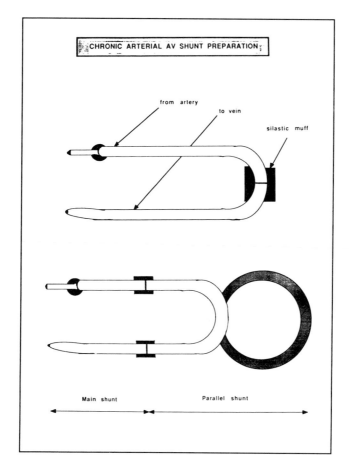

Fig. 2

Parallel flow test section preparation.

A parallel flow shunt system (*Fong et al., 1985*) which enables testing of heparinized or other materials at low flow rate (*Fig. 2*) seems to be a good tool. This system is connected in parallel to the CAVS. The flow rate through the test branch of 1.14 mm ID is 2.5 ml/min with a corresponding shear stress rate of 200-600s^{-1}.

Using Poiseuille's law it is possible closely to calculate the expected flow in the tubing, ignoring the pressure loss in the Y connections which divide the flow. It should be noted that the radius of the tubing plays a large role in flow resistance. Shunt flow rates which initially are 250-400 ml/min will gradually reduce to 130-

170 ml/min when chronic conditions are obtained i.e. two weeks after the initial procedure. Flow rates may vary in an experiment by approximately ± 5 ml/min, while in the parallel shunt system they can vary by less than 0.5 ml.

Scope of method:

The method is used in general screening for haemocompatibility concerning:

- blood-material interactions
- thrombus formation rate
- blood protein-material interaction
- blood cell activation
- blood flow mechanics of thrombus formation

DETAILED DESCRIPTION OF METHOD

Equipment/Material.

Equipment required.
Silastic medical grade tubing from Dow Corning, Midland, MI are used with the following diameters:

    3.18 mm ID x 6.35 mm OD
    2.64 mm ID x 4.88 mm OD
    1.98 mm ID x 3.18 mm OD

Outline of method.

Establishment of main CAVS (Chronic external arterio venous shunt).
Two lengths of Silastic® medical grade tubing (Dow Corning) of 3.18 mm ID per 6.35 mm OD (each one 50 cm long) are prepared beforehand, serving as the venous and the arterial cannulae. On the latter, a 10 cm extension Silastic® medical grade tubing (Dow Corning) is mounted measuring 1.98 mm ID per 3.18 mm OD, joined with Silastic® medical adhesive Silicone Type A (*Fig. 1a*).

Both prepared tubings are cured at 100°C for 2 hours to ensure a good bond.

The tip of each tube is tapered by grinding the external wall of the tubes to facilitate cannulation of the iliac vessels (*Fig. 1a*). A larger diameter Silastic® tubing (3.18 mm ID) is used for the central piece. The two side rings are prepared from smaller Silastic® tubing (1.98 mm ID, 3.18 mm OD) which are attached to the center ring by Silastic® adhesive (*Fig. 1b*). The tubes and grommets are placed in an ultrasonic bath to remove the debris resulting from the grinding and adhesive process and subsequently steam autoclaved for 1 hour at 120°C.

Establishment of parallel flow system.
The test section corresponds to the small diameter tube to be tested. In the present report polyethylene (Leadercath, Vygon, France) was used. The parallel flow system is made for each experiment as follows: Two holes are drilled at a distance of 2.5 cm from each end of a 20 cm long Silastic® tube of 3.18 mm ID. The holes are punched with a stainless-steel needle with a diameter corresponding to that of a small tube which is twisted through the Silastic® tube wall. The Silastic® tube is held in position in a specially prepared support to ensure that the angle between the tube and the drilling needle is 30°. The drilling tube is cleaned in an ultrasonic bath. A short segment (0.5 cm) of medical grade Silastic® tubing with an internal diameter corresponding to the test tubing external diameter is glued with medical grade Silastic® adhesive to the outside of the drilled tube at each hole at an angle of 30°, forming Y-connectors. Each end of the smaller diameter tubing to be tested can be inserted through these Y-connectors into the larger diameter Silastic® tube. To minimize turbulence the tips of the test tubing are cut flush with the lumen of the larger tubing at an angle of 30°.

The complete assembly is sterilized in ethylene oxide gas. The material must be properly degassed prior to use. The Silastic® main tube support can be reused several times (5 times) after cleaning by flushing the tubing with 3.5 ml of saline followed by ultrasonic bath cleaning for 20 min. without modification of test results. During the period of testing the Silastic® main tube support is clamped in its central segment close to the Y connectors to direct blood flow into the parallel shunt which is formed by the material to test.

Surgical procedure.
Male healthy mongrel dogs weighing between 15 and 20 kg are preferred for CAVS. Haematocrit, activated coagulation time, platelet count and fibrinogen should be checked for each animal before and after CAVS implantation.

The dog is fasted overnight and anaesthetized with Sodium Thiopental (30 mg/kg IV), tracheally intubated and ventilated with a BIRD® respirator. The skin of the abdomen area plus the back are shaved and surgically prepared with antiseptic solution (Betadine, Sarget Lab.) while the animal is lying supine on a mattress inflated with warm water. The animal is draped with a sterile sheet, only the abdomen is uncovered. Body temperture of the dog is controlled with a rectal thermometer, until it awakens. A laparotomy is performed to expose the common iliac vessels (artey and vein) bilateral.

The external iliac artery on the left side and the external iliac vein on the right side are tightened with a ligature as the corresponding limbs of the shunt are introduced into the vessels. The Silastic® tubing with the smaller diameter is used for arterial cannulation, the other for the venous side. The artery and vein are ligated distally to the shunt. Before insertion into the vessels, each limb of the shunt is filled with heparinized saline (15 IU/ml saline) to prevent blood-air contact. The filled tubings are clamped (Craford Vascular clamps) distally as the tapered ends are introduced, first into the venous limb and then into the arterial limb. Each tubing is inserted 10 cm into the blood vessel and anchored with grommets using 4.0 non resorbable polypropylene monothread (Ethnor, France) sutures. Then each arm of the shunt is positioned ventrally with the two clamped shunt limbs on each side of the body. Tunneling is made for each shunt limb. The skin is retracted and two stab wounds are made, not more than 6 cm apart, to permit the exit of the shunt limb. The manoeuvers are performed with special care so as to maintain the shunt limb clamped. One or two further stab wounds are performed following the same procedure until the shunt limb can exit at the shoulder plate level. Identical manoeuvres are done with the other shunt limb. The extra lengths of the external portions are then cut to provide a good adjustment of both tips (almost 10 cm free on each side). A Silastic® muff is slipped over the cut ends of each limb to guarantee the contact between both limbs and restore the AV blood flow through the shunt.

An elastic bandage (Elastoplast) 7.5 cm in width is used to cover the exteriorized shunt to prevent damage (e.g. caused by rubbing against the cage) and at the wound sites that could lead to infection. This bandage is changed as required in order to clean the stab wounds, and a snug jacket is worn for further protection. A collar is put on the dog's neck to keep it from biting and scratching its wounds and the shunt.

During the first post-opertive week the dog receives 1) Penicillin therapy 1 mill. IU, and 2) anticoagulant (Coumadin) treatment. The dose of Coumadin is calculated according to the conventional prothrombin time and adjusted to obtain 50% of the initial prothrombin level. CAVS flow rate is measured daily using a Doppler ultrasound device. This must beforehand be calibrated with timerelated collection of blood (or saline) in an in vitro-peristaltic pump flow circuit.

After one week of "maturation", the anticoagulant and antibiotic therapies are stopped. At that time the haematologic parameters are normalized and the shunt is ready to be used as a testing system.

Description of procedure.

Testing protocol.
At the start of an experiment, the animal is restrained by hand and placed comfortably in a hammock system where its limbs can move freeely to maintain cardiac output. No anaesthesia is necessary. Tubes of 3mm ID or less are tested successively every 2 or 3 day as follows:

The sterilized test section is filled with sterile saline and the animal's chronic shunt is clamped with tubing protected clamps (Craford vascular clamps) at both arterial and venous limbs to stop the blood flow temporarily. The shunt is then disconnected and a ring probe of the Doppler device is mounted around the arterial limb of the shunt. The test section is quickly fitted into each limb of the shunt with the aid of two Silastic® muffs identical to the one employed to connect the limbs of CAVS. After connections are secured, tubing clamps are removed to expose the test section to the flowing blood. The total occlusion time is measured and may be less than 5 minutes. When the test is finished, the test section is quickly removed from the AV shunt after clamping the two limbs of the main shunt which are reconnected by means of the Silastic® muff. After releasing the clamp, the blood flow rate of the shunt is monitored as above, and the animal is returned to its cage.

Thrombosis of the main shunt is assessed manually twice daily by evaluating the temperature of the shunt tubing. When occluded, the temperature of the Silastic® tube decreases rapidly within 5 min. and it feels cold. Within 30 min. the thrombus retracts and is seen as a column in the tube with serum around.

Haematological parameters (haematocrit, activated coagulation time, platelet count and fibrinogen) are recorded between each protocol testing.

502

RESULTS

Typical results.

These are given for haemodynamic, haematological and morphological parameters

Haemodynamic parameters.
The patency rate of a CAVS lasts from several weeks to several months.

Table I

Parallel flow testing system obtained at different p.o. days. The blood flow rate obtained in the main shunt and in the parallel section system is related to the shunt internal diameter (ID) and to its length. The different thrombosis times depend on the blood/material interface.

Dog	Blood Flow Rate		Tubing	Length	Thrombosis of the //
N°	AV shunt	Parallel shunt	ID		shunt*
	n = 12	n = 12	(mm)	(cm)	(min.sec)
816	180 ± 43.58	3.9 ± 0.35	1	20	10' ± 2
821	167.5 ± 45.96	2.3 ± 0.50	1	20	10' ± 2
854	248.33 ± 42.15	4.20 ± 0.75	1.9	20	10'.22 ± 2

*) Polyethylene tubes

Typical values are summarized in *Table I*, corresponding to chronic canine external AV iliac shunt and the parallel flow testing system. The flow rate should be recorded before and after the utilization of the parallel flow testing system. The blood flow rate in the latter is recorded directly on the parallel branch at each assay.

Haematological parameters.

The chronic external AV iliac shunt does not alter significantly the haematological parameters recorded before the shunt implantation and after its maturation. The mean of the results obtained in four experiments and the standard deviations are given in *Table II*.

Table II

Haematological parameters recorded before AV chronic iliac shunt implantation and after its "maturation" (n = 4).

PARAMETER	BEFORE AV SHUNT	AFTER 2 WEEKS AV SHUNT
Activated coagulation time (sec)	14.58 (± 1.87)	18.80 (± 13.15)
Platelet count (x 10^9/l)	198.75 (± 73.68)	183.75 (± 39.74)
White blood cells (x 10^9/l)	11.67 (± 5.17)	38.92 (± 22.79)
Red blood cells (x 10^{12}/l)	5.57 (± 0.91)	6.02 (± 1.45)
Haematocrit	43.00 (± 1.87)	41.48 (± 6.82)
Haemoglobin (g/l)	146.25 (± 9.25)	142.60 (± 29.52)
Fibrinogen (g/l)	3.23 (± 1.18)	3.61 (± 0.98)

The same haematological parameters should be measured before and after material testing using the parallel flow testing system. The results which are summarized in *Table III* corresponded to 6 experiments performed on the same animal with a mean contact time of test system of 10 min 25 sec (± 2 min 9 sec.).

Table III

*Haematologic parameters recorded before and after parallel flow system testing (n = 6)
performed on one animal (Dog N° 854), contact time = 10 min. 25 se. (± 2 min. 9 sec.)*

PARAMETER	BEFORE	AFTER
Activated coagulation time (sec)	13.15 (± 0.15)	14.75 (± 2.05)
Platelet count (x 10^9/l)	240.00 (± 10.00)	-
White blood cells (x 10^9/l)	21.50 (± 10.00)	31.15 (± 20.15)
Red blood cells (x 10^{12}/l)	5.32 (± 0.28)	5.19 (± 0.99)
Haematocrit	40.00 (± 2.00)	35.00 (± 5.00)
Haemoglobin (g/l)	132.50 (± 10.50)	113.50 (± 21.50)
Fibrinogen (g/l)	3.65 (± 1.35)	3.06 (± 0.90)

Morphological parameters.
After the blood flow contact one can study the tested material means of scanning electron microscopy which provides an evaluation of the number and morphology of the blood cells, particularly platelets at the material surfaces.

DISCUSSION

Interpretation of results.

Haemodynamic parameters.
A relatively high flow rate (135-330 ml/min) is an indication of the quality of the shunt. However, the blood flow rate generally shows a tendency to diminish after two weeks regardless of whether or not a device was placed in the external shunt

limbs. *Galletti et al. (1981)* has described the same phenomenon with external iliac shunts which remained patent for an averge of 48 days (7 - 63 days).

Reasons for failure.
With such a protocol the reasons for failure can be: 1) thrombosis of the iliac vessels, most commonly the iliac artery, 2) accidental AV shunt disconnection provoked by scratching of the dog, 3) small bowel volvulus. All these incidents will occur during the first postimplantation week. Other causes of failure which occur later are e.g. 1) skin infection around the zones where the shunt limbs emerge, 2) accidental AV shunt disconnection provoked by scratching, 3) late venous wall hypertrophy with impaired flow and venous limb thrombosis.

Haematological parameters.
No noticeable changes are generally observed with the main shunt. However, when it is used as a system for 3 mm internal diameter tubing testing (*Fong et al. 1985*), the contact time of blood with specific fabrics (heparinized materials) can affect the haematological parameters. In such cases a limited (20%) reduction in platelet count has been observed after 100 min. of blood exposure to heparinized tubing without modification of the initial clotting time.

Limitations of the method.

1. Continuous and meticulous care of the animal and shunt is required.
2. Cutaneous infection along the shunt site occurs frequently.
3. Only tubes of 3 mm ID or less can be used when applying the parallel flow method.
4. Long term use is limited to less than 8-10 months in each animal.
5. Possible cardiac overload is a risk caused by the fistulous high flow rate.
6. Progressive thickening of the venous wall with increased collagen content and gradually decreasing venous flow and thrombosis of shunt.

Comprehensive list of relevant testing methods and their value relative to this method.

Flow chamber testing system.
Flow chambers have been used for the analysis of factors determining thrombus formation on foreign surfaces. *Leonard et al. (1970)* and *Benis et al. (1975)* among other authors utilized a flow chamber for varying times of exposure (in minutes) and regulated blood flow rates (100-400 ml/min.).

The geometry of the flow chamber which was introduced into an arterio-venous shunt varied with the authors. Experiments fall into different categories:

For *Leonard et al.* three categories of results were obtained with this system: 1) thrombosis time, 2) platelet and fibrinogen turnover rates and 3) thrombogenesis when a normal haematological response has been pharmacologically altered to suppress either platelet activity or enzyme reactions.

For *Benis et al.* the flow chamber testing system allowed study of the thrombogenesis of a material by measuring the dry weight of the thrombus inside the flow chamber device according to the blood flow rate.

Testing in vivo for hybrid artificial organs by means of a cell culture device. *(Galletti et al., 1981).*

Paediatric dialysis shunts approximately 50 cm long and 2.6 mm ID were attached to the external iliac vessels of beagle dogs, providing a blood flow rate of between 160 and 200 ml/min. The shunt flow was used as the perfusion medium by insertion of a tissue culture device, the heart of the animal acting as the perfusion pump. The AV shunt allowed a high enough blood flow rate which progressively diminished when the shunts had been in place over 6 weeks. Nevertheless, this blood flow rate was enough to ensure a good flow testing without anticoagulation as long as it exceeded 140 ml/min., but thrombosis was common below 100 ml/min.

REFERENCES

Benis, A.M., Nossel, H.L., Aledort, L.M., Koffsky, R.M., Stevenson, J.F., Leonard, E.F., Shiang, H. & Litwak, R.S. (1975) "Extracorporeal model for study of factors affecting thrombus formation", Thrombos. Diathes. Haemorrh. 34, 127-144.

Fong, W.I., Zingg, W. & Sefton, M.V. (1985) "Parallel flow arteriovenous shunt for the ex-vivo evaluation of heparinized animals", J. Biomed. Mat. Res. 19, 161-178.

Frasher Jr., W.G. (1967) "Blood sampling by a chronic artificial external arteriovenous shunt in dogs", J. Appl. Physiol. 22, 2, 348-351.

Galletti, P.M., Trudel, L.A., Panol, G., Richardson, P.D. & Whittlemore, A. (1981) "Feasibility of small bore AV shunts for hybrid artificial organs in non heparinized beagle dogs", Trans. Am. Soc. Artif. Int. Organs, XXVII, 185-187.

Leonard, E.F. & Friedman, L.I. (1970) "Thrombogenesis on artificial surfaces. A flow reactor problem", Chemical Engineering Symposium Series, 66, 99, 59-71.

Schribner, B.H., Caner, J.E.Z., Buri, R. & Quinton, W. (1960) "The technique of continuous hemodialysis", Trans. Am. Soc. Art. Int. Org., 6, 88-103.

Turitto, V.T., Muggly, R. & Baumgartner, H.R. (1977) "Physical factors influencing platelet deposition on subendothelium: Importance of blood shear rate", Ann. N.Y. Acad. Sci., 283, 284-292.

Intravascular Implantation Test

Authors: **E. Chignier, J. Guidollet & R. Eloy**

INSERM U 37, Chirurgie Vasculaire et Transplantation d'Organes, F 69500 Bron

INTRODUCTION

Name of method:

In vivo chronic implantation for testing vascular prosthesis devices and intravascular testing for haemocompatible materials.

Aim of method:

The described procedures:

- simulate functional clinical conditions of an implanted prosthesis. This represents normally the last step of the necessary testing procedure before application in man. Patency of the vascular prosthesis related to the cellular ingrowth, in various anatomical condtitons, including peripheral arterial and venous locations on the one hand and the course of the blood-material interaction in acute and chronic follow-up on the other hand, represents the major information available through this method,

- measure, in haemodynamic conditions, blood clotting at the contact with a material and its possible consequence: distal embolization,

- measure the mechanical behaviour of the prosthesis after various periods of implantation, thereby providing further knowledge regarding the stability of the material and of the structure of the prosthesis.

S. Dawids (ed.), Test Procedures for the Blood Compatibility of Biomaterials, 509–533.
© 1993 *Kluwer Academic Publishers. Printed in the Netherlands.*

Background of the method:

The functions of the normal arterial wall are mechanical in nature, including viscocity, plasticity and elasticity and are related to a highly organized macromolecular elastomeric structure, where collagen and elastin fibers coact in a two stage visco-elastic response to stress. The activating systems of the blood are major components in the regulation of normal haemostasis and fibrinolysis and are locally regulated by the normal endothelial lining.

The ability of a vascular substitute to perform adequately in this environment can only be explored in vivo when normal rheologic conditions occur such as stasis, turbulence, flow separation, high shear rates etc.

Unlike thrombosis which is an early event that can be evaluated within minutes to days following exposure to blood, the evaluation of gradual excessive pseudointimal proliferation as well as subsequent calcification requires observation for weeks to months. The potential for pseudointimal growth appears to continue for a prolonged period while calcification of the pseudointima occurs preferentially in young growing subjects.

Ingrowth in a vascular prosthesis is a complicated array of processes occurring at the blood-prosthesis interface and which depend upon the nature of the tissue developed at the inner surface of the prosthesis. The exact nature of this pseudo-intimal tissue is not fully determined. Glycosyltransferases which are implicated in the formation of glycoproteins and glycosaminoglycans are believed to be responsible for the macromolecular construction in the intimal tissue of normal vessels.

The extracellular matrix of vascular tissues is a composite of collagen, non-collagenous glycoproteins, proteoglycans (PGs) and glycosaminoglycans (GAGs). PGs are large carbohydrate-protein macromolecules that consist of a protein core covalently linked to several chains of one or more GAGs. Glycosyltransferases are involved in the biosynthesis of GAGs, by addition of successive oses and their derivatives to the growing carbohydrate chains. Extraction of the normal arterial wall GAG is shown in *Fig. 1*.

The enzymes responsible for the generation of glycan chains are secreted by endothelial and smooth muscle cells and are of either cytoplasmic (xylosyl and galactosyl transferases) or microsomal (sialyl transferase) origin. The former act by placing the oses initiating the chain i.e. xylose and galactose, and the latter by placing the terminal ose, i.e. sialic acid. They have a capacity to determine the GAG composition. The augmentation of sialyl transferase activities provokes an increase in sialic acid which is coincident with an elevation of the dermatane sulfate and implies an augmentation of anionic charges, binding cations such as Ca^{2+}. GAGs are known to influence cell adhesion, migration, proliferation (*Radhakrishnamurthy et al., 1982*). They are also involved in cellular metabolic changes in haemostasis, calcification and lipoprotein binding (*Wight, 1985*), and thus play a role in vascular repair and atherosclerotic processes. Analysis of these provide a possible approach to quantify abnormalities in the tissue overgrowing vascular prostheses.

Despite major advances in manufacturing practices, user experience and consumer awareness, modern vascular surgery remains complex. Failures and complications are not uncommon. Material failure leading to graft dilation, compliance mismatch, disruption of anastomoses, perigraft seroma, neo-intimal hyperplasia, thrombus and/or embolization, infection, may account for the substantial proportion of precocious or late thromboses.

Mechanical testing of devices and materials using non-biological systems has provided essential information about their useful lifespans, but does not suffice. Vascular prostheses implanted in the organism for prolonged periods undergo chemical degradation which influences compliance and other important mechanical properties gradually leading to failure.

Scope of the method.

To test in vivo implantable vascular devices for:

- suitability of properties for implantation (suturability, flexibility etc.),
- mechanical properties during long term implantation and pattern of failure,
- formation of endothelial coverage (pseudointima and calcification),
- pattern of failure and embolization in relation to caliber of vessel.

Numerous analyses are subsequently performed in order to characterize the vascular healing in contact with the prosthetic material. They allow the patency of the graft to be tested as well as qualitative and quantitative parameters of the newly developed tissue (amount of tissue developed, nature and functions of cells present, macromolecular charateristics of extracellular matrix etc.) Thus, only indicative methods are reported here illustrting the information that can be gathered from this testing method. Depending on the procedure adopted, the equipment will be different.

DETAILED DESCRIPTION OF THE METHOD

Equipment/materials.

Healthy adult mongrel dogs weighing 25-30 kg should be preferred. Before surgery, hematological parameters of the animals are checked to rule out hypercoaguability. No anticoagulants should be given neither before, during nor after surgery. Sterile vascular prostheses of clinical quality are used and prepared according to manufacturers' guidelines. Implantation periods range from hours to years.

For GAG identification all chemicals (GAG standards, mucopolysaccharidase, chondroitinase ABC, AC, testicular hyaluronidase) are obtained from Sigma.

Bouin's solution composition for histological procedures:

Picric acid (Sigma)	15.2 g
Glacial acetic acid (Sigma)	95.2 ml
Formaldehyde (Sigma)	643.0 ml
Distilled water qsp	2000.0 ml

Fixation fluid for graft/vessel specimens: Glutaraldehyde 2% solution buffered with 0.2 M sodium cacodylate (pH 7.45, 400 mOs).

Specific equipment is mentioned under each description of procedure.

Outline of the method.

Identification of endothelial cells at the blood-material interface.
A longitudinal fragment of the prosthesis including both anastomosis sites and adjacent host vessel is gently rinsed with saline to remove excess of blood and fixed overnight in a hypertone glutaraldehyde solution. The fragment is then divided into four parts:

a. 2 mm^2 samples are taken from the adjacent host vessel, proximal and distal anastomoses and the middle of the prosthesis. They are processed for scanning electron microscopy.

b. Identical samples are processed for transmission electron microscopy including X-ray microanalyses.

c. Samples from the same areas are processed for conventional histological evaluation.

d. A final group of samples is frozen and processed for immunocytochemical analysis and/or lipid inclusion identification.

Identification of glycoaminoglycans (GAGs).
The identification of GAGs is accomplished by electrophoresis with the following procedures:

a. Identification of GACs by overloading with GAG standards (Sigma).
Samples (5-10 μl) containing 10 - 25 μg of GAG from the aortic wall or from pseudointima and of standard GAG are applied on the same electrophoretic strip. The increasing metachromatic staining intensity of one of the different GAGs from the sample is verified by overloading in the standard GAG. Thereby, the similarity between the loading of GAG from the arterial wall and standards can be assessed. The overloading of arterial wall heparan sulfate reveals precisely the differences in the electrophoretic mobilities of the two samples.

b. Identification of GAGs by enzymatic hydrolysis with specific mucopolysaccharidases (Sigma).
Mucopolysaccharidases allows one to identify several glycosaminoglycans. Chondroitin sulfate A and C and dermatan sulfate are hydrolysed by chondroitinase ABC and AC. Hyaluronic acid which is broken down by testicular hyaluronidase is only partly hydrolysed by chondroitinase AC at pH 4.8 in normal conditions. Heparan sulfate is not split by these two enzymes. The enzymatic assays can be used only with pure GAG from the aortic wall. The action of chondroitinase AC is inhibited by an excessive quantity of dermatan sulfate or heparan sulfate.

514

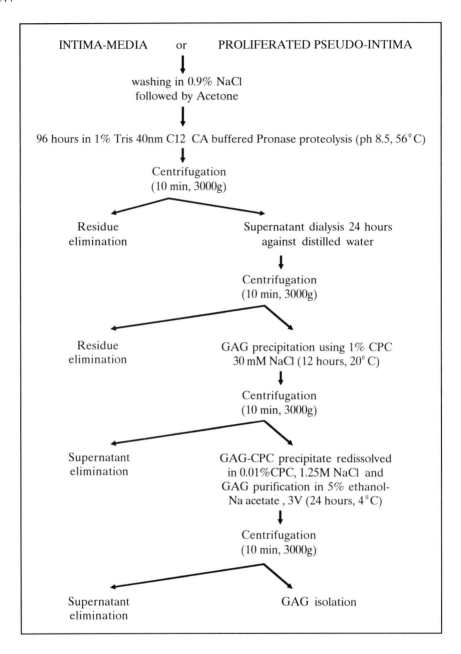

INTIMA-MEDIA   or   PROLIFERATED PSEUDO-INTIMA

washing in 0.9% NaCl
followed by Acetone

96 hours in 1% Tris 40nm C12 CA buffered Pronase proteolysis (ph 8.5, 56°C)

Centrifugation
(10 min, 3000g)

Residue
elimination

Supernatant dialysis 24 hours
against distilled water

Centrifugation
(10 min, 3000g)

Residue
elimination

GAG precipitation using 1% CPC
30 mM NaCl (12 hours, 20° C)

Centrifugation
(10 min, 3000g)

Supernatant
elimination

GAG-CPC precipitate redissolved
in 0.01%CPC, 1.25M NaCl and
GAG purification in 5% ethanol-
Na acetate , 3V (24 hours, 4°C)

Centrifugation
(10 min, 3000g)

Supernatant
elimination

GAG isolation

Fig. 1

Procedure of extraction of glycosaminoglycans (GAG), following Picard's technique (Gardais et al., 1969).

GAGs from intima are separated by electrophoresis on cellulose acetate strips (Cellogel-Sebia). Identification by electrophoresis analysis is carried out by comparing electrophoresis before and after hydrolysis of GAG using specific enzymes.

Identification of four metachromatic zones which appear when electrophoresis of GAG from the aortic wall is performed in pyridine formate buffer is thus obtained. Loss of the metachromatic spots on the electrophoresis proves the enzymatic sensitivity of the GAG studied and provides its identification by enzymatic hydrolysis characteristics. The GAG which has the slowest mobility does not disappear with chondroitinase ABC, but does with chondroitinase AC and hyaluronidase. The GAG which gives the second metachromatic spot does not disappear with chondroitinase AC but disappears with chondroitinase ABC. The fastest moving metachromatic spot disappears with chondroitinase ABC and AC. These results are consistent with the nature of the different GAGs separated by electrophoresis and are, according to their increasing mobilities: hyaluronic acid, heparan sulfate, dermatan sulfate and chondroitinsulfate C or A.

Description of the procedure.

Surgical procedure.
Animals are given general anaesthesia by e.g. intravenous injection of initial 0.25 g/kg body weight of Sodium Thiopental (Nesdonal®).

Endotracheal intubation is used for assisted respiration. Replentishing injections of Nesdonal® are given according to the duration of the operation. The abdominal aorta is exposed through a median xipho-pubic incision and its infrarenal segment is dissected and replaced by a prosthesis. The tube whose diameter depends on the aortic diameter (generally 6 or 8 mm) is placed with end-to-end anastomoses with continuous sutures using Monobrin 4.0 thread (Ethicon®).

Knitted Dacron® grafts are preclotted under the animal's blood pressure as follows: after proximal anastomosis has been performed, the prosthesis is compressed between two fingers to avoid yarn trauma occurring with normal clamping. The proximal vascular clamp on aorta is released allowing the blood to ooze through the prosthesis until clot formation in the knit has been formed (generally less than 1 min.). The proximal vascular clamp is closed again. Excess of clots are removed by gentle aspiration in the prosthesis. The distal anastomosis is then performed.

Expanded polytetrafluoroethylene may be implanted without previous blood clotting of the graft wall.

The abdominal cavity is closed by the usual surgical techniques with minimal trauma, and the animals are left in normal laboratory conditions. Patency of grafts is assessed periodically by examination of the femoral pulsation and using doppler ultrasound recording (8 MHz directional doppler device from Mira Electronique, France).

At the time of sacrifice (between 15 days and 1 year) the animals are anaesthetized as before and endotracheally intubated. Angiograms are taken in order to visualize and determine the quality of suture zones (*Fig. 2*).

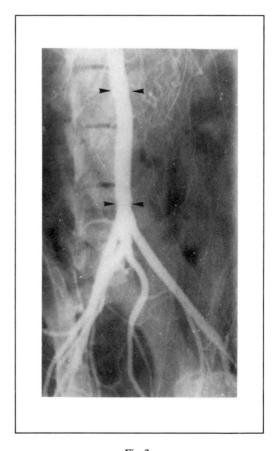

Fig. 2

Angiogram showing good quality of sutures zones (▶ ◀) 1 year after a Dacron® prosthesis implantation replacing dog aorta.

The abdominal aorta with the graft is dissected free, the prosthesis outer diameter and length are measured and the animal is given an overdose Sodium Thiopental (Nesdornal®) to pass away. The aorta between the renal and the iliac arteries is removed, gently flushed with saline to remove blood and is immediately placed on crushed ice and divided longitudinally including the host vessel and suture zones into two unequal parts:

a.   the dorsal or posterior fragment of 2 mm is processed for morphological studies,
b.   the remaining anterior fragment is immediately frozen and processed for biochemical studies.

Procedure for histological, immunocytochemical and ultrastructural characterization.

Histological studies. Once removed, the prosthesis and the adjacent arterial wall including suture zones are stored at +4°C. The whole length of the prosthesis is taken for morphological investigations and divided into two parts: one part fixed in Bouin's solution, embedded in paraffin for histological and immunocytochemical investigations; the other part immersed in 2% glutaraldehyde solution buffered with 0.2 M cacodylate. Both hematoxylin-eosin and Weigert's staining are used. The internal surface of the prosthesis in cross section and the lumen narrowing are measured at four sites before splitting the graft. Histomorphometrical analysis is done with the Kontron Mop/AMO2, an opto-manual electronic planimater. By Weibel's technique (*Weibel 1969*) the area and perimeter of each histological structure may be measured.

Immunocytochemical staining. Specific rabbit antisera against purified dog factor VIII related antigen (VIII. Rag) provided by dr. Benson, New York State Department of Health, is used. A first layer (dilution 1/8) provides an indirect immunofluorescent staining. A section of the tissue is incubated at room temperature for 2 hours in a humidified chamber and then rinsed three times for 5 min. in phosphate buffered saline (PBS). The section is then exposed to Fluorescein isothiocynate (FITC)-conjugated goat antirabbit globulin (Sigma) for 90 min. at room temperature using a 1/20 dilution, and the washing procedure is repeated. A drop of mounting fluid consisting of 10% glycerol and 90% PBS is added. The sections covered by a coverslip are examined on a Poyvar (Reichert Jung) epifluorescence microscope using an HBO-200 high pressure mercury vapor burner, 330-380 exciter filter and 418 LP barrier filter (*Fig. 3*).

518

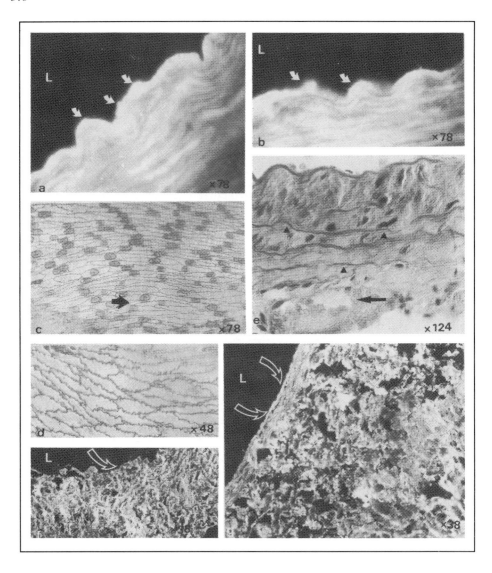

Fig. 3

a *(x78˙)* + **b** *(x78˙) Normal vessel intima specifically visualised with immunocytochemical staining using FV III RAG Antibodies (→)*

c *(x78˙)* + **d** *(x48˙) Silverstaining of normal endothelial lining with cell nuclei (→)*

e *(x124˙) Media and adventitia of normal vessel containing elastic fibers (▲), collagen and vasa vasorum (←). Normanski technique.*

f *(x 38˙)* + **g** *(x38˙) Section of GORETEX^R prosthesis. A smooth glistening coating replaces endothelial lining (→).*

˙ *original magnification*

Fig. 4 A

a *(10μ) SEM of normal endothelial lining showing nuclear prominence and contiguous cells.*
b *(400μ) SEM of GORETEX^R prosthesis lumen showing morphological heterogeneous surface after 9 months*
c *(40μ) SEM of Dacron^R prosthesis with fibrosis. The lumen is adjacent to a zone of fibrin deposition.*
d *(20μ) SEM of the fibrotic tissue containing fibroplasts (→).*

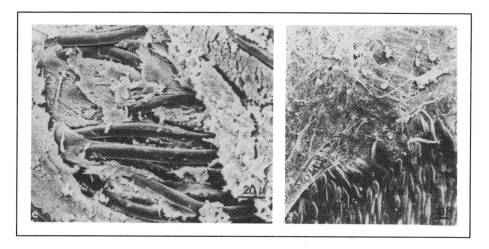

Fig. 4 B

e (20µ) SEM of a typical uncovered Dacron^R fiber at lumen as they are always seen.
f (10µ) Suture zone after 3 months with endothelial cells grown from the adjacent vessel wall
(not shown).

Ultrastructural studies.

A. Scanning electron microscopy (SEM) (*Fig. 4A & 4B*): After washing, the specimens are first dehydrated in progressive acetone vapor diffusion. They are the further dried by the critical point method using acetone and liquid carbon dioxide. Finally they are mounted on aluminium stubs coated with gold palladium and observed on a scanning electron microscope (Jeol JS M 35 CF) at a voltage of 15 KV.

B. Transmission electron microscopy (TEM) (*Fig. 5*): After washing, two specimens 1 mm in diameter are post-fixed for 1 hour at room temperature in a 1% osmium tetroxide buffered cacodylate 0.1 M solution. They are dehydrated in successive concentrations of ethanol to absolute alcohol and embedded in epoxy resin with routine techniques. Finally, ultrathin sections are obtained from the blocks. They are mounted on copper grids, stained with uranyl acetate and lead citrate for 5 min. at room temperture and observed on a Philips 300 microscope at a voltage of 75 KV.

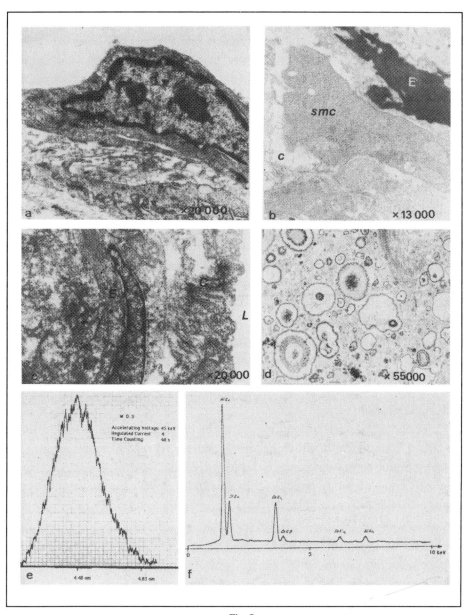

Fig. 5

a *(x20,000*) Normal ultrastructure of vessel intima.* **b** *(x13,000*) Normal ultrastructure of vessel media, smc = smooth muscle cell, E = elastic lumina, c = collagen.* **c** *(x 20,000*) Fibrocellular tissue covering a vascular prosthesis lacking the normal ultrastructural characteristics, L = lumen, E = Fibroplasts, C = collagen fibers* **d** *(x55,000*) Calcified tissue of collagenous prosthesis.* **e** *+* **f** *Microanalysis by EDS corresponding to carbon particle migration (WDS) and crystallin deposition (calcium).*

* *original magnification*

C. Wavelength dispersive and X-ray microanalyses are carried out on ultrathin sections (150 nm thickness) processed as above (excluding osmium post fixation and uranyl and lead citrate staining) and are then mounted on aluminium grids to be examined with a Camebax microprobe equipped with two kinds of X-ray detectors:

-   for Energy Dispersive Spectrometry (EDS): A solid-state Si (Li) detector coupled to a Tractor TN 2000 electronic is applied for calcium and silicon detection. The area of the Ca K_α X-ray from 3.880 to 3.520 keV and the area of the Si K_α X-ray from 1.860 to 1.630 keV are measured.

-   for Wavelength Dispersive Spectrometry (WDS): Four monocrystals and a proportional gaseous fluxcounter (PGFC, argon and methane) are employed for C detection and the monocrystal concerned is the ODPb one (2d = 100.7Å). The C X-ray heigth is measured at λ 1 = 4.48 nm for the peak and at λ 2 = 4.83 nm for the background. Analyses are performed under an accelerating potential of 45 KV with a regulated beam current of 100 nA on the specimen for a counting of 40 sec.

Biochemical procedure.
GAG electrophoresis: Electrophoresis is run on cellulose acetate strips (Cellogel from Sebia) 2.5 x 17 cm in a 0.15 M acetate buffer, pH 6.0 (90 min., 150 V), and in 0.1 M pyridine formate buffer, pH 3.0 (60 min., 150 V). This buffer is currently used for its high resolution quality. Strips should be equilibrated in the buffer for 30 min. before applying samples of GAG.

Samples of GAG standards and aortic wall GAGs which should not contain more than 20 μg in 10 μl are positioned on the moist strips. This procedure can also be used in a preparative manner for a more complete identification.

After electrophoresis of GAG, the strips are immersed in methanol. A small band (0.5 cm) is cut off along the strip. The bands of GAG are stained by Alcian blue, and the picture is used to locate the zones which must be cut on the main strip. Each fragment obtained is eluted in water (yield 90 %). Eluates are concentrated under vacuum, then GAGs are precipitated by 1% CPC dissolved in 1.25 M NaCl and precipitated again by a 5% Sodium acetate ethanol 100% solution.

<u>Identification and isolation of GAG bands</u>: Alcian blue staining is carried out by immersion for 2 minutes in a 0.5% acetic acid/water/methanol (10/40/50) solution of Alcian blue (Alcian blue 8 GX). The strips are then successively treated by an acetic acid/water/methanol (50/45/50) solution.

The zones corresponding to the stained bands containing GAG are carefully cut out and eluted in 1 ml of an 80% acetic acid solution and examined in a Beckman UV spectrophotometer at 675 nm. The quantity of each GAG is calculated in relation to the elaute of standards.

<u>Enzymatic hydrolysis of GAG</u>: Electrophoresis is completed by enzymatic and acid hydrolysis. This is performed by chrondroitinase ABC and AC according to the method of *Yamagata et al.(1968)*. The incubation is done in an enriched Tris buffer solution (3g Tris, 2.4 g 0.13 NHCl, pH 8.0).

The reaction mixtures containing in 50 μl of Tris buffer 25-30 μg of purified arterial wall GAG and either 0.1 unit of chondroitinase ABC or 0.1 unit of chondroitinase AC, are incubated at 37°C for 30 min. The blank mixture without the enzyme is incubated under the same conditions. The enzymatic reactions is stopped by heating (1 minute at ~ 100°C). After centrifugation, 10 μl of supernatant is applied on to the electrophoresis strip as described above.

RESULTS

Normal GAG values for reference materials.

In normal arterial tissue the GAG content represents about 0.2% of dry degreased tissue. A similar amount of GAG is present in the carotic artery, in aorta and endocardium.

Typical experimental values.

The distribution of GAG values in the pseudointimal tissue proliferated into Dacron® prostheses is presented in *Table I*. Control groups are tissue of intima from recipient aortae.

Table I

Level of Uronic Acids, Osamines and Glucosaminoglycans (HA, HS, DS, CS) of control vessel wall juxtaprosthesis (A) and tissue developed within the prostheses (B).

	Control	A	B
Uronic acids	512 ± 10	238a ± 4	460b ± 6
Osamines	100 ± 2	102 ± 1	80a ± 1
HA	36 ± 4	17b ± 1	124a ± 2
HS	183 ± 6	31a ± 2	81a ± 1
DS	134 ± 4	129 ± 3	200a ± 2
CS 4,6-C	257 ± 8	163a ± 3	135a ± 3

The results are expressed in µg/g of tissue and given as mean value of 6 to 8 experiments and standard deviation for n determinations. The values of each group are compared to those of control using Student's t-test. (a) $p < 0.001$, (b) $p < 0.005$.

Table II

Enzymatic activity (mean ± SD) of particles and soluble glycosyltransferases in: Tissues obtained from the aortic wall contiguous to the prosthesis (group A), Tissues grown on the inside (group B) and on the outside (group C) of the prosthesis.

	Sianyl Transferase	N-Acetyl Glucosaminyl- Transferase	Xylosyl Transferase	Galactosyl Transferase
Control Group	440 ± 10	350 ± 10	670 ± 10	340 ± 3
Group A	470 ± 10	430 ± 10	280 ± 10	80 ± 2
Group B	2.400 ± 40	2.100 ± 20	280 ± 10	140 ± 5
Group C	750 ± 10	740 ± 10	150 ± 5	70 ± 2

The results are expressed in picomoles of ^{14}C-carbohydrate per gram of protein.

Microsomal transferases, i.e. N-acetyl-glucosaminyltransferase and sialyltransferase were significantly increased (5-6 times) in the tissue which developed within the prosthesis as compared to both control aortic intimal tissue and the contiguous aortic wall. A smaller increase in microsomial transferases is also located to the tissue around the prosthesis with a 1.7-2.1 increase as compared to the normal intimal aortic control tissue. Two glycosyltransferases of the soluble cytoplasmic phase, i.e. xylosyl- and galactosyltransferases is generally very reduced in the newly formed tissues (inside and outside control aortic layer). This decrease is even more marked in the newly developed tissue on the outside of the prosthesis.

Table II shows the enzymatic activity analysis of proliferated pseudo-intimas on Dacron® prostheses.

Morphological experimental results.

Conventional histological examinations.
Normal vessel sections (*Fig. 3*) show three different tissue layers:

1. The intima or internal layer, composed of a monocellular layer illustrated with immunocytochemical techniques using specific Factor VIII Rag antibodies (*Fig. 3a-b*) or by in situ Silver staining (*Fig. 3c-d*).
2) The media or medial layer composed of smooth muscle cells and extracellular matrix including collagen and elastin (*Fig. 3e*).
3) The adventitia or external layer which provides nutrition and innerva-tion to the vessel. Graft sections studied by conventional or immunocy-tochemical techniques (*Fig. 3f*) serve to identify the inner surface which is frequently covered by a smooth glistening coating (*Fig. 3g*) or a thin cellular layer. Fibrin deposition can also be identified relating to the degree of microscopic thrombosis. Using morphometric data, the internal layer (whether fibrotic or thrombotic in nature) may be analysed to evaluate the degree of available vascular section of the graft, exclusively on complete graft sections.

Scanning electron microscopy.
The normal endothelial surface consist of typical nuclear proeminent and continuously contiguous cells (*Fig. 4Aa*). On the other hand, the prosthesis blood interface remains a morphologically heterogeneous surface (*Fig. 4Ab*), even after long term implantation. This morphology is common to synthetic grafts and requires a very careful analysis of the samples to avoid errors of interpretation. As

a general rule (*Fig. 4Ac*) the prosthetic implants are covered with fibrinous deposits or fibrous tissue and fibroblastic cells (*Fig. 4Ad*) lacking the morphological characteristics of normal endothelial vascular cells (nuclear and cytoplasmic shape, microvillous projections and cell junctions). Most commonly, the prosthetic fibers appear uncovered (*Fig. 4Be*). Only in the suture zones of long term (up to 3 months) implanted synthetic prostheses endothelial cells are identifiable (*Fig. 4Bf*). In contrast, when biological prostheses are implanted, an endothelialization, even though incomplete appears very quickly.

Ultrastructural and X-ray microanalysis data.
Normal vascular ultrastructural characteristics corresponding to intimal endothelial lining (*Fig. 5a*) is seen, associated with typical smooth muscle cells and intracellular matrix of the medial layer (*Fig. 5b*). Implanted vascular prostheses are covered with typical fibroblastic cells and collagen matrix (*Fig. 5c*). In cases of calcification or particle migration (*Fig. 5d*) the origin and nature of crystalline deposition can be determined by X-ray microanalysis (*Figs. 5e, 5f*).

DISCUSSION

Interpretation of results.

Biochemical data.
Xylosyl- and galactosyltransferases are the initiators of the biosynthesis of the glucidic linkage of glycosaminoglycuronoglycans. The decreased enzymatic activity of xylosyltransferase corresponds to a decrease in the enzyme involved in the branching of the initial sugars (xylose and galactose). Since glycosaminoglycuronoglycans are anionically charged, if not counterbalanced, this phenomenon is believed to lead to a large decrease in the anionic charges of the arterial wall.

Nevertheless, observation of the second category (N-acetyl-glucosaminyltransferase and sialyltransferase) shows that these microsomal enzymes present a greatly increased activity particularly within the prosthesis and, to a lesser degree, in fibrous tissues grown outside the prosthesis. Thus it cannot be excluded that the loss of electric negative charge is counterbalanced by the placement of a terminal sialic acid which is very anionic, as suggested by the increase in sialyltransferases. Moreover, the increased activity of the enzymes modifying the terminal sugars of the glycoproteic chain may be related to the altered behavior of these macromolecules towards cellular or biochenmical blood components, i.e. positive cations such as Ca^{2+}.

The high level of hyaluronic acid measured in the pseudo-intimal layer suggests a loss of differentiation of more specialized cells in favour of fibroblasts at the expense of the growth of endothelial cells. Also the decrease in heparan sulfate and chondroitin-sulfate and the increase in dermatan-sulfate seems to imply a marked impairment of the biosynthesis of the macromolecular components of a normal vessel wall.

The de novo pseudo-intima shows the pattern of an altered functional healing tissue unable, inter alia, to synthesize structural glycoproteins normally present on the cell surface membrane at the blood interface. It is suspected that these macromolecular modifications may be the first steps toward the pseudo-intimal hyperproliferation and subsequent calcification or thrombosis.

These investigations performed on two different polymers with different fabrics and one carbonaceous material demonstrates that in no case does a typical macromolecular matrix develop.

Morphological data.
Historical and morphometric results confirm that, even after long periods of implantation in the dog model, synthetic grafts are poorly invaded by tissue in the case of PTFE® grafts. Similarly Dacron® prostheses have either hypertrophic pannus ingrowth zones or are devoid of tissue covering on the threads. But in any case it has been possible to demonstrate an endothelial cell layering on the middle portion of the prosthesis.

These histological findings are confirmed by SEM and TEM and demonstrate that the graft blood interface can generally still be observed even after 3 months of implanation. This surface is characterized in places by a fibrous superficial tissue which never exhibits an endothelial cellular structure. In addition, ultrastructural microanalysis data reveal the migration of particles from the material as well as calcifications representing the major side effects in long term prosthesis implantation.

Limitations of the method.

- Specialized facilities and training in accurate vascular surgery is needed.
- Only one or two vascular prostheses can be tested on the same animal.
- Large size prostheses (>6 mm ID) require larger and costly animals.
- Reliable testing requires a test length of the prosthesis that exceeds the diameter by a factor of 10 (i.e. a vascular prosthesis of 6 mm ID must be implanted over a total length of up to 60 mm).
- The biological processes described in healthy animals at the contact with vascular prostheses can not always be directly related to the behavior of these prostheses in arteriosclerotic human patients.

Related testing methods.

Testing of in vivo small caliber vascular prostheses (≤ 3mm ID), (*Chignier et al., 1983***).**
The distal site of implantation, i.e. carotid, femoral vessels, is chosen in order to evaluate a given material and its function with distal haemodynamic conditions. Since differences in anastomotic diameters can create a disturbed flow and thrombosis, carotid and/or femoral implantations are more suitable. End-to-end anastomoses using procedures similar to those described above are performed. This model allows acute investigations on the graft itself (patency time, labelled cell and/or protein uptake) and long term investigations as described in the aortic graft model.

Patch angioplasty testing (*Planche et al., 1987***).**
A variety of synthetic or biological vascular devices are utilized for patch angioplastic stenosis correction, either on congenital or acquired defects. The in vivo model requires a growing animal or an adult one, depending on what the utilization of the material is intended for. Patching is performed in the aortic or pulmonary artery according to its utilization in high or low flow conditions respectively (*Fig. 6*).

After intraveneous anaesthesia and endotracheal intubation as described above, a left thoracotomy performed in the 4th intercostal space exposes the left pulmonary artery or the aorta by incision of the pericardium. The artery is then mobilized from its size to the hilar region (in the case of pulmonary artery) and a 2 cm long segment is isolated by the application of a U-shape vascular clamp.

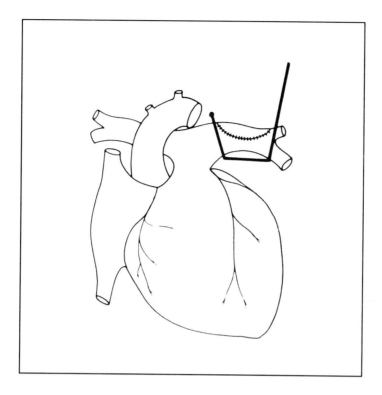

Fig. 6

Schematic representation of the patch angioplasty testing.

The wall of the artery is resected in part and an oval-shaped patch measuring 15 x 8 mm is inserted using running everted edge sutures of polypropylene 6/0 thread to bridge the defect. The patch extends either from the main pulmonary left trunk to the first lobar artery or is placed on the areas of the aorta (see *Fig. 6*). Positioning the patch on the upper edge of the artery facilitates angiographic observation. After releasing the vascular clamp, the thorax is closed anatomically over a suction drain tube which is removed 2 hours later. The animals should receive antibiotic therapy for 7 days and their weight be monitored.

The evaluation methods combine: 1) in vivo studies which are designed for the determination of the haemodynamic pattern of the patch. These evaluations encompass cineangiography and intra-arterial pressure measurements above and below the angioplasty thereby allowing an extended analysis of the vessel function at the angioplasty zone, 2) histological and ultrastructural or immunocytochemical studies including calcification and lipid deposit investigations.

Microsurgical vascular techniques for vascular prosthesis testing (*Chignier et al., 1985, Fonegra et al., 1982*).

The need for replacement of small vessels (smaller than 3 mm ID) is evident in many clinical situations such as reimplantation of amputated fingers or limbs, free intestine, skin flap transfers, blood flow restoration to occluded coronary, cerebral vessels or to ischemic distal parts of both the upper and lower extremities. Until now, more than three decades after the successful application of synthetic vascular grafts, it has not been possible to make a vascular substitute for small vessels with a diameter below 1.5 mm ID. Partial or total obliterations is the usual outcome mainly due to local thickening of the pseudo-intima. Pannus ingrowth from the two ends of the prosthesis leads to increasing reduction of the prosthetic internal diameter. Thus this experimental model has a very important field of application. Again to attain the principle of the in vivo model, i.e. test the function in relation to the site of its application, these prosthetic tubes of synthetic or biological origin are implanted in rats or rabbits using microvascular techniques.

A section of aorta or femoral arteries is replaced under intraperitoneal anaesthesia (Sodium Pentobarbital, 30 mg/kg body weight) using an end-to-end anastomosis between the prosthetic tube and the vessel. Suturing is made with microsurgical thread and techniques. It should be performed by a trained microvascular surgeon. The test material is 6 mm long, and the internal diameter should correspond to the internal diameter of the vessel to avoid haemodynamic irregularities. After clamp release, a patency test is run for at least 10 minutes to ensure the patency of the graft, thus avoiding that thrombosis due to technical errors occurs and is included in the evaluation of the material. When the material under study is not available in tubular form, it may be analysed as a vascular patch (*Chignier et al., 1985*). In this case, under intraperitoneal anaesthesia, a window 6 mm long and 1 mm wide is created in the anterior wall of the infrarenal aorta of 2-months old rats. The aorta is isolated and clamped. The defect is bridged with a patch using interrupted sutures (10/0 microvascular thread) under an operating microscope. The abdominal cavity is closed anatomically and the rat is left for a period ranging from minutes to 2 years.

This rat model is very useful when a great number of animals are needed to test sequentially a vascular device, as rats are cheaper and easier to use. However, it is difficult to perform repetitive blood analyses in rats, and morphological studies are thus the only ones which are feasible.

Vena Cava Ring Test (*Gott et al., 1969*).

The advantage of the vena cava ring test is its relative simplicity. Rings are fabricated from the material to test measuring 9.5 mm long and 6.3 mm ID with both edges streamlined to minimize blood flow turbulence. The wall thickness ranges from 0.5 to 3 mm. They are implanted in the inferior vena cava of prescreened mongrel dogs weighing 7 to 10 kilograms, under intravenous general anaesthesia using Sodium pentobarbital (25 to 30 mg/kg) and breathing is maintained by a mechanical respirator.

Under aseptic conditions a thoracotomy is performed in the fifth right intercostal space, thus giving access to the right atrium, the superior and the inferior vena cava. The vena azygos is ligated and tapes are placed around both superior and inferior vena cava as far from the atrium as possible. A "purse string" suture is placed in the right atrial wall which is opened to give access to the inferior vena cava.

During the short period necessary for inserting the test ring, the blood flow is stopped by traction of the tapes previously placed around the vena cava. The "purse string" atrial suture is tied and the tapes are removed. A wrap is placed tightly around the vena cava at the height of the ring to prevent pouch formation between the ring and the vein wall. The animal can only be used once. No systemic anticoagulant is used. Sometimes two test rings are inserted in both the superior and the inferior vena cava of the same animal with no difference in the degree of thrombi formation. The in vivo clot formation inside the test ring after periods of time ranging from 1 hour to 2 weeks of implantation is studied.

The procedure is considered outdated and should be used with caution in conjunction with other tests because the rigid and non-compliant rings promote significant mechanical changes in the vena cava properties. On the other hand, the thrombi formed inside the ring may be "washed" by the blood stream or be destroyed by fibrinolysis pathways, leading to false results. Another limitaion of this system is that the small testing device leads to a reduced blood surface area, and the measures are an end-product blood clotting version, only studied in a venous flowing system.

Renal embolus test.

Kusserow et al. (1976) described a test to measure the entrapped emboli within the kidney resulting from thrombi generated at the contact of test ring materials implanted in the abdominal aorta immediately above the origin of the renal arteries of anaesthetized mongrel dogs. The authors added a subtotal aortic constriction placed immediately below the origin of the renal arteries, inducing over 90% of the blood flowing through the test device to pass through the kidneys.

Test ring devices are 8 mm long, 6 mm ID and 8 mm OD, i.e. they are roughly isodiametric with the inner aorta diameter. They are made with rounded edges to reduce intimal injury and flow disturbances and are inserted using a special Teflon device (*Kusserow et al., 1976*). The ring is implanted for a period of 3 to 5 days, then the animals are heparinized immediately before sacrifice. This permits the removal of both kidneys with their renal arteries and attached aorta containing the test device without further clot formation.

The published results are expressed in relation to the extent of ring thrombosis which is assessed visually. A scale from zero (no thrombosis) to four (total occlusion) is accompanied by a schematic topographic map showing the thrombotic deposition on the ring and the extent of embolization in the kidneys. Six or seven experiments are considered necessary for a given material.

The advantages of the method are the possibility of testing arterial circulation, observing organ distal embolization (dog kidneys are especially sensitive to embolization in comparison to human ones) in addition to thrombus deposition analysis, and the severe character of this test system. These advantages are counterbalanced by the fact that a clean material surface after the testing is not necessarily a blood compatible or a thromboresistant one. This has sometimes led to a premature and optimistic prediction of blood compatibility of many materials. Emboli entrapped within the kidney may be dissolved over a given period of time by the activation of the fibrinolytic system, confounding the test results.

REFERENCES

Chignier, E., Guidollet, J., Heynen, Y., Serres, M., Clendinnen, G., Louisot, P. & Eloy, R. (1983) Biomed. Mat. Res., 17, 623-636.

Chignier, E. & Eloy, R. (1985) J. Biomed. Mat. Res., 19, 115-131.

Fonegra, J., Chignier, E., Clendinnen, G. & Eloy, R. (1982) Surg. Gyn. Obst., 154, 673-680.

Gardais, A., Picard, J. & Tarasse, C. (1969) J. Chromatog., 396, 42-48.

Gott, V.L., Ramos, M.D., Najjar, F.B., Allen, J.L. & Becker, K.E. (1969) in Proc. of the Artificial Heart Programme Conference, Hegyeli, R.J. Ed., V.S. Government Printing Office, Washington, DC.

Kusserow, B. & Larrow, W. (1976) Annual Report PB 252, 720/AS, vol.1, National Technical Information Service, Springfield.

Planche, C.L., Fichelle, J.M., Paul, J., Lethias, C., Eloy, R. & Weiss, M. (1987) J. Biomed. Mater. Res., 21, 509-523.

Radhakrishnamurthy, B., Srinivasan, S. & Berenson, G. (1982) in The glycoconjugates: glycolipids and proteoglycans, vol.4, pp.274-300, Part B", M.I. Horowitz Ed., Academic Press, New York.

Yamagata, T., Saito, H., Habuchi, O. & Suzuki, S. (1968) J. Biol. Chem., 243, 1523-1535.

Weibel, E.R. (1969) Int. Rev. Cytol., 26, 235-302.

Wight, T.N. (1985) Fed. Proc., 44, 381-385.

Chapter VIII

TEST METHODS FOR THE DETECTION OF TOXICITY

Chapter eds.: S. Dawids[1] & R. Eloy[2]

[1]   Institute of Engineering Design, Biomedical Section, Technical University of Denmark, DK - 2800 Lyngby

[2]   INSERM U 37, Chirurgie Vasculaire et Transplantations d'Organes, F - 69500 Bron

INTRODUCTION

Assessment of toxicity of biomaterials is be obtained in several ways. Detection of chemicals leached from the polymer (e.g. sterilizing agents and residual monomers) and low molecular components released from the polymer due to degradation is a primary approach. This chemical detection is well described in official pharmacopeas and will not be covered in this chapter.

The methods covered are in vitro use of unicellular organisms like ciliata and cell tissue cultures. These methods are gradually replacing the use of animals for toxicity testing as described in official pharmacopeas.

Ex vivo and in vivo exposure of polymers to living tissue such as blood or implantation in animals provide more realistic approaches to the performance of the material. Extracorporeal circulation is therefore a necessary tool for investigation of thrombogenicity. Implantation tests are required for long term evaluation of material. This includes investigation of the proximity of cell ingrowth and - for vascular devices -the ability of the surface to provide adhesion of endothelial, and how the material influences the long term overgrowth and patency of vascular implants. Clinically it is well known that artificial vessels below a few mm almost certainly will occlude within very short time and implantation tests are the only methods at present which provide a realistic evaluation of perfonnance.

This chapter contains thus three groups of tests:

      1.   In vitro toxicity test for biomaterials.

S. Dawids (ed.), Test Procedures for the Blood Compatibility of Biomaterials, 535–536.
© 1993 *Kluwer Academic Publishers. Printed in the Netherlands.*

> 2.   In vitro toxicity tests using biological methods.
> 3.   In vitro cytocompatibility tests.

1.   <u>In vitro toxicity test for biomaterials</u>. This group of tests only include Ciliata test where the unicellular organism, tetrahymena pyriformis is used as a simple and sensitive indicator to detect cytotoxic or cytostatic properties in the hygienic, toxicological, medical, pharmacological and environmental areas. In most aspects the ciliata organism is comparable to mammalian cellular tissue cultures.

2.   <u>In vitro toxicity tests using biological methods</u>. These tests encompass the use of cell cultures with color indicators. The Trypan Blue Exclusion test relies on the fact that healthy cells can exclude trypan blue while damaged cells take up the dye. The Neutral Red Release test is based on the intracellular staining of viable cells and release of the dye when damage of the lysosome membranes occurs. The included Cell Growth test has a close similarity to the Ciliata test and is based on temporal detection of cell growth. Finally the DNA Synthesis test is based on temporal detection of cellular uptake and incorporation of ^{3}H-thymidine in DNA.

3.   <u>In vitro cytocompatibility tests.</u> In contrast to cytotoxicity tests which are negative criteria (cellular alterations, cell death, hampered growth etc.), tests for cytocompatibility measure whether both structure and function of tissue in direct contact with material remain unchanged. A test for prostacyclin production from endothelial cells is described. A test for detecting wettability and cellular adhesion is included with direct application to vascular prosthesis. A method for detection of cellular repair i.e. the ability to overgrow mechanically damaged tissue on biomaterials is given. In this respect measurement of cellular growth spreading and adhesion is included as a separate test method.

In vivo tests are dealt with in the chapter on *"Test methods for the detection of biodegradation"* and in *"General Introduction"*.

In Vitro Toxicity Test for Biomaterials

Author: W. Lemm

Dept. of Experimental Surgery, Rudolf-Virchow-Clinic, Location Charlottenburg,
D - 1000 Berlin 19

INTRODUCTION

Medical devices made of synthetic materials either implanted as permanent
prostheses or only contacting the body or body fluids for a short period are not
allowed to affect the health of the patient. Therefore toxicological tests are
absolutely necessary for such devices to prevent a cytotoxic, histotoxic, mutagenic
or allergic like pyrogenic response of the organism.

In general the polymer itself can be considered as non-toxic provided that it is not
sensitive to degradation releasing monomers or oligomers. In most cases plastic
materials are compositions containing large quantities of additives like plasticizers,
stabilizers, catalysts and traces of other compounds. These low-molecular additives
are available for the organism by leaching and could be a risk for the patient's
health.

Poly-vinylchloride (PVC) e.g is the most often used plastic material in medicine.
Up to 40% plasticizer is added to the PVC-powder to obtain a flexible material.
Stabilizers are needed to prevent a depolymerisation of the PVC when exposed to
the high extrusion temperatures.

Polyurethane elastomers do not require plasticizers but stabilizers in an amount of
2-3% to prevent a depolymerisation and side-reactions when extruded at about
300°C. Pellethane 2363-80 AE, a medical-grade poly-etherurethane, contains 1.2-
ethylene-di(fatty-acid)-diamide as stabilizer and extrusion help which might be
leached out.

Plastic devices for therapeutic application have to be sterilized. Under the
conditions of heat and steam sterilization nearly all polymeric materials will be
damaged loosing either their properties or releasing harmful fragments. Therefore

S. Dawids (ed.), Test Procedures for the Blood Compatibility of Biomaterials, 537–544.
© 1993 Kluwer Academic Publishers. Printed in the Netherlands.

ethylene-oxide is widely adopted for the sterilization of plastic devices. Such materials adsorb the toxic ethylene-oxide gas to a large extent so that a period of evaporation must follow the sterilization process.

The complete evaporation of ethylene-oxide and derivatives must be assured by toxicological tests.

Animal experiments are more and more replaced by in-vitro toxicity tests with cell and tissue cultures isolated either from men or mammals. But their experimental effort is considerable, the exact quantification of cytotoxic effects is difficult.

Name of method:

The ciliata test.

Aim of method:

The following in vitro method detects the acute toxicity of polymers, monomers, extracts, additives and the release of toxic compounds by degradation.

Biochemical background of the method:

In the last years bacteria free single cell cultures of ciliata (species: tetrahymena pyriformis) are more and more used for the detection of cytotoxic or cytostatic properties of chemical compounds and materials in the hygienic, toxicological, medical, pharmaceutical field and environmental protection. The ciliata are sensitive to histotoxic noxae, in most cases comparable to that of mammalian cellular tissue cultures.

The ciliata are suspended in their nutrition solution (*Fig. 1*). Under fixed environmental conditions the population rate is reproducible. The increasing optical density of the solution (turbidity measurement) is detected in a photometer at a wavelength of 570×10^{-9} m (*Fig. 2*). The growth rate of cell culture in contact with a material supposed to be toxic will differ and is compared to a contaminated reference culture.

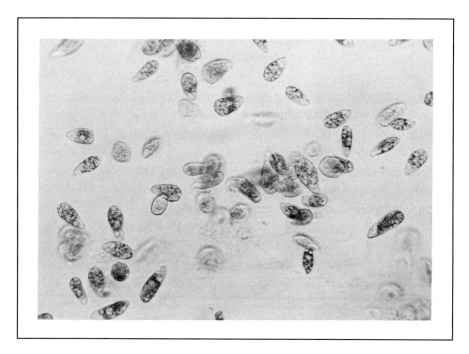

Fig. 1

Population of ciliata in the nutrition solution.

Scope of method:

The ciliata test may replace animal experiments. The ciliata test stands out for an easy performance. The well reproducible results are available within a comparatively short time.

DESCRIPTION OF METHOD

Equipment.

Photometer (wavelength 570 nm).

Ciliata (tetrahymena pyriformis) available from Carolina Comp., South Carolina or from the author.

Sterile nutrition solution: Proteose Pepton 5g, Trypton 5g (both available from Difco Comp. Detroit, Michigan), K_2HPO_4 0.2g, Distilled water 1000 ml, $NaHCO_3$-solution (8.4%) to adjust a p_H-value of 7.2.

Facilities for sterile manipulation (laminar flow bank).

10 ml glass test vials.

Test cuvettes for the photometer.

Pipettes in different volumes.

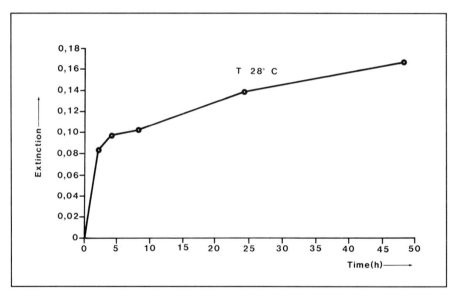

Fig. 2

Growth rate of ciliata at 28°C.

Outline of method

Maintenance of "stand-by" cultures of tetrahymena pyriformis:
The preparations and testing must be carried out under strictly sterile conditions.

30 ml of the nutrition medium is placed in a cell culture flask. Use disposable syringes, pipettes and tips. 0.5 ml of the bacteria free ciliata cell culture is added. Avoid any contamination.

It is recommended to run two separate identical cultures to protect at least one in case of contamination of other microorganisms.

The cultures are stored at 25°C. Every week this procedure should be repeated. These cultures are considered as "stand-by" cell suspensions for use in the toxicity test.

Description of procedure.

0.5 g of the test material (2.0 ml of an aqueous extract) in 10.0 ml of the nutrition medium is sterilized at 110°C for 30 min. and recooled to room temperature. Use glass vials! Any contamination must be strictly avoided.

0.5 ml of the four day old "stand-by" cell culture is added and the glass tubes are safely sealed.

Reference cultures (without test material) are prepared.

Now the test and the reference tubes are stored for four days at 25°C or for two days at 28°C.

Each material (or extract) must be tested in two series with two separate ciliatapopulations. It is recommended to prepare 10 test cycles of each material or extract with two cultures in order to obtain a reasonable statistical evaluation of the results. Extreme deviations of the mean value should not be included in the final calculation.

A cuvette containing only nutrition solution but without cells is inserted in the reference beam of the photometer.

The number of cells developed after four (two) days at 25°C (28°C) is detected photometrically at 570 nm. The extinction is proportional to the number of cells. Test cultures are compared to the reference cultures:

$$\text{Rel. population rate: } E_t/E_r \times 100 \ (\%)$$

E_t = Extinction of test culture
E_r = Extinction of reference culture

542

Empirically a relative population rate below 90% indicates a toxic material or contaminants (*Fig. 3*).

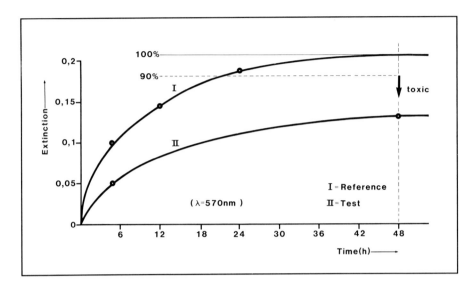

Fig. 3

Proliferation rates of reference cultures and test cultures.

Table I

Toxicity of various biomaterials according to the conditions of the ciliatta-test.

Material	Relative Cell Population	
	in I	in II
Pellethane 2363-80 AE	96 (±5)%	94 (±5)%
Pellethane 2363-80 AE (2 days after EtO sterilization)	44 (±7)%	38 (±6)%
PVC-no-DOP	98 (±9)%	92 (±6)%
PVC-no-DOP (3 days after EtO sterilization)	71 (±11)%	54 (±7)%
PVC-DOP (techn. grade)	47 (±4)%	54 (±6)%
PVC-DOP (sterilized by gamma-radiation)	28 (±6)%	37 (±5)%

RESULTS

A selection of a large number of tested materials is presented in *Table I*.

These results demonstrate very clearly the effect of ethylene-oxide down to residues of 5 ppm on the toxicity of a material.

UV-analysis of an extract of the technical PVC-DOP confirmed a large amount of impurity with mono-(2-ethylhexyl)-phthalat beside the basically non-toxic di-(2-ethylhexyl)-phthalat.

PVC which is not protected against the sterilization conditions with gamma-radiation (at least 2.5 Mrad) by additional quantities of stabilizer releases hydrochloric acid resulting in a toxic response of the single cells.

DISCUSSION

Interpretation of results.

Ciliata cells are aerophillic; therefore living individuals tend to accumulate towards the surface, dead cells precipitate. False results and misinterpretations can be avoided by a careful separation of living and dead cells before the population rate is determined in the photometer.

Relative population rates higher than 100% are observed in such cases where the ciliates accept parts of the test material or extract as an additional nutrition.

Ciliata are also extremely sensitive to traces of detergents. Washed glass vials must be carefully rinsed with distilled water.

Limitations of method.

Only aqueous extracts or dispersions of solid particles can be tested. Extracts obtained by organic solvent extraction might be introduced if the solvent is completely evaporated, and the extract can somehow be incorporated into water.

CONCLUSION

The ciliata test is characterized by a simple handling while safely detecting the presence of small quantities of potentially dangerous compounds. The reasons for its growing importance and application are that it is easy to perform in any laboratory, and the results are well reproducible and available within a short time (48 hours).

Recently bioluminescent-ATP-assays are introduced to the investigation of toxicity of materials with tetrahymena pyriformis to improve the sensitivity of the test.

REFERENCES

Bommer, J., Barth, H.P., Wilhelms, O.H., Schindele, H. & Ritz, E. (1985) The Lancet, 1382-1385.

Gräf, W. (1985) GIT Fachz. Lab. 6, 601-614.

Kidder, G.W. & Dewey, V.C. (1951) in "The Biochemistry and Physiology of Protozoa" (Woff, L. ed.), vol.1, pp. 323-400, Academic Press, New York.

In Vitro Toxicity Tests Using Biological Methods

Authors: M.F. Sigot-Luizard & R. Warocquier-Clerout
Laboratoire de Biologie Cellulaire Experimentale. Univ. de Compiègne
F - 60206 Compiègne

INTRODUCTION

To date most of the in vitro culture tests utilized to assess the cytotoxicity of a material are mainly based upon the observations of cellular alterations. The reliability of these methods has been tested on many classes of biomaterials especially for those defined in the pharmacopeia and other normalization bodies as acceptable polymers.

Progress in the techniques of in vitro culturing of human and animal cells or of differentiated tissues together with advances in molecular biology has favoured an expansion of in vitro cell culture techniques. This has also been promoted by the increasing legal and ethical difficulties in performing animal experimentation.

On the other hand, performance of quantitative analysis of specific cytocompatibility of a material is no longer restricted to the detection of cellular alterations, but is extended to the phenomena related to the material/tissue interface aiming at the future utilization of the particular biomaterial. In vitro culture measurements are to day standardized to an extent that any change in the environmental parameters such as culture substrate, culture medium, cell strains, addition of growth factor or drugs will cause measurable variations in the growth kinetics and cellular metabolism. Thus the reproducibility of these techniques rely on rigorously well-defined conditions in performing the tests.

Two approaches have been recommended in the general evaluation of cell viability:

1. the agar diffusion test (*Guess et al., 1965*) which allows detection of diffusible and leachable toxic products from the biomaterial or from extracts diffusing through the agar,

S. Dawids (ed.), Test Procedures for the Blood Compatibility of Biomaterials, 545–567.
© 1993 *Kluwer Academic Publishers. Printed in the Netherlands.*

2. direct contact method (*Rosenbluth et al., 1965*) where the toxicity is assessed in direct contact with monolayer cell culture.

These methods have been described in detail elsewhere (*ASTM F 813-83, ASTM F 895-84*). A survey with comparative studies (*Johnson et al., 1983*) describes the sensitivity of the methods and shows a good correlation with in vivo experimentation.

A group of tests using cell cultures in the evaluation of biomaterials is gathered in this contribution which will deal with classical techniques to analyze the **cytotoxicity** of all classes of materials, most of them considered useful for a general screening for cytotoxicity. Techniques on the aspect of **cytocompatibility** which allows detection of morphological structures and quantification of specific functions involved in materials in contact with blood is described in the following contribution.

The tests in the present context are:

Colour indicator tests on cell cultures:
Trypan Blue Exclusion test, relying on the fact that healthy cells can exclude trypan blue, while damaged cell membranes will lead to uptake of the dye.
Neutral Red Release test, based on staining of viable cells and release of dye when the lysosome membranes are damaged.
Cell growth test, based on temporal detection of cell growth.
DNA synthesis test, based on temporal measurement of cellular uptake and incorporation of ^{3}H-thymidine in DNA.

Equipment used for these tests encompasses basically the following components.

Basic equipment:
Sterilization equipment (autoclave, poupinel, filtration membranes)
Laminar flow bench
Carbon dioxide-air incubator (37°C)
Deep-freezer
Thermostated water-bath
Standard centrifuge (applying > 1500 g)
Disposable sterile centrifuge tubes
Inverted optical microscope with magnifications of 40x, 100x and 200x
Standard haemocytometer, double chamber (Thoma, Malassez)

Low speed magnetic stirrer (1 rev./sec)

Multiplate spectrophotometer (absorbance reading at 540 μm)

Sterile 35 mm tissue culture grade Petri dishes

Disposable multiwell plates (96 wells)

75 cm^2 flasks for culturing (Dow Corning)

Hanks solution without Ca and Mg

Hanks solution added 0.1% trypsin (2 x cristallized)

Trypsin® solution 0.1% in phosphate buffered saline (PBS) without Mg and Ca

For most cell cultures:

> 1 vol. of 0.25% Trypsin (Sigma, France) is mixed with 1 vol. of 0.20% EDTA (Sigma, France). The mixture is prepared just prior to use.

For HUVEC cultures:

> 1 vol. of 0.25% Trypsin is mixed with 3 1/2 vol. of PBS and 1/2 vol. of 0.20% EDTA (to obtain Trypsin 0.05% and EDTA 0.02%)

Minimum essential medium (MEM) sterilized by filtration and added

> 5% calf serum
> 200 IU/ml penicillin
> 100 μg/ml streptomycin.

Minimum essential medium (MEM) without glutamine and NaHCO$_3$, added

> 0.292 g/l glutamine
> 2.20 g/l NaHCO$_3$
> 5% fetal calf serum
> no antibiotics

Each separate test may require supplementary items which are given under the particular test.

REFERENCES

ASTM F 813-83. Direct contact cell culture evaluation of materials for medical devices.

ASTM F 895-84. Agar diffusion cell culture screening for cytotoxicity.

Guess, W.L. et al. (1965) "Agar diffusion method for toxicity screening of plastics on cultured cell monolayers". J. Pharm. Sci., 54, 1545-1547.

Johnson, H.J. et al. (1983) "Biocompatibility test procedures for materials evaluation in vitro. I. Comparative test system sensitivity". J. Biomed. Mater. Res., 17, 571-586.

Rosenbluth, S.A. et al. (1965) "Tissue culture method for screening toxicity of plastic materials to be used in medical practice". J. Pharm. Sci., 54, 156-159.

Name of method:

Colour indicator tests on cell cultures.

Aim of methods:

To detect low level cytotoxic effects of polymers or extracts of polymers in cell cultures using dyes as indicators.

Biophysical background for the methods:

The cytotoxicity of a xenobiotic causes cellular alterations and may even inhibit essential cell functions. The classical criteria for these morphological changes (*Holmberg, 1960*) are characterized by cell viability, rate of cell growth and changes in metabolism. Cytotoxicity tests are usually performed on monolayer cell cultures grown onto a biomaterial or in a culture medium containing extracts leached from the test material. Continuous cell lines immortalized by chemical irradiation or by viral transformation are used as proposed by the pharmacopoeia and normalization bodies (cell types: L-929, MRC 5, etc.). However, finite cell lines such as endothelial cells from humans and animals are now increasingly used to assess both cytotoxicity and cytocompatibility of materials in contact with blood.

The analysis of cell viability takes advantages of modification or loss of the cell membrane integrity or of damage of structures within the cells which modulate their capacity to

1. exclude vital dyes from entering the cell (Trypan Blue Exclusion, TBE, test),
2. incorporate vital dyes (Neutral Red Release, NRR, test),

Trypan blue exclusion (TBE) method.
The possible toxic effect on cells of a polymer or its extract can be detected on exposed cells by their capacity to exclude trypan blue. The dye does not enter healthy cells, because the plasma membrane is normally impermeable to this molecule. Loss of the membrane integrity changes the ability to resist uptake. This technique (*Guess et al., 1965*) is now accepted as a general screening of cytotoxicity by Ph.Eur. (11th Edition, 1989). The advantage of this method is that it can be carried out in a short time and that computerized equipment makes the use of dyes relatively easy and provides reliable measurements. A useful technique (*Combrier, 1988*) forms the basis of this applications.

Neutral red release (NRR) method.

This method (*Borenfreund & Puerner, 1984*) combines the use of a visual morphological cytotoxicity assay with the quantitative neutral red spectro-photometric test for the assessment of potential cytotoxicity of extracts from the material. It possesses the same advantages as the TBE method.

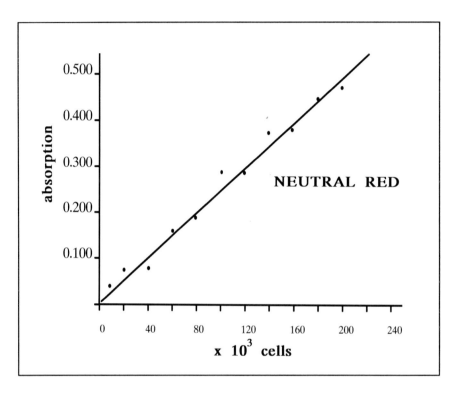

Fig. 1

Optical density of neutral red as a function of cell number.

Neutral red exclusively stains viable cells. The dye transgresses the lipidic cell membrane and binds to anionic sites of the lysosomal matrix by electrostatic bonds. Any alteration of the lysosomal membranes will lead to a decrease and loss of the fixation of the dye. This technique is based on a correlation between cell number and optical density of neutral red as shown in *Fig. 1*. It is used as a general screening of the cytotoxicity of extracts from materials. In Australia (AS 2696, 1984) it is considered as a standard method and has been adopted by AFNOR as part of a cytotoxicity test.

Scope of methods:

The tests are applicable for general in vitro screening purposes of polymers or extracts of polymers. They are recognised as general screening tools or as part of extended cytotoxicity testing. They are generally performed together.

DETAILED DESCRIPTION OF METHODS

Equipment

Basic equipment for cell culture handling (as given in **INTRODUCTION**) supplemented with:
Trypan blue solution 0.2% in isoton sodium chloride (0.9%)
Neutral red (stock solution), colour index no 50040, 0.4% in water. Solution should
      be kept in dark. Working staining solution is prepared by 1/80 dilution
Acid solution: ethanol 50%, glacial acetic acid 1%
Fixation solution: formaldehyde 40% in water, $CaCl_2$ 10%

Outline of the TBE method.

To carry out the test requires standard tissue culture facilities. Cell cultures used in this assay are of the ATCC CCL1 NCTC clone, 929 strain (clone of strain L, mouse connective tissue) designated L 929. This particular cell line is well established, well characterized and readily available and provides reproducible results.

The solid test material is precut to give uniform square fragments with a 3-4 mm side length in a sufficient number to cover one tenth of the surface in each Petri dish (several fragments may be associated). If extracts from test materials is to be used, one or several fragments are immersed in culture medium and incubated at 37°C for 24 hours in a humidified atmosphere containing 5% CO_2. The mixture should contain 0.2 g material/ml culture medium. After extraction the supernatant is pipetted off. In a test procedure both positive and negative control materials must be included. All samples must be sterile. Extracts from material or solid samples are placed onto the cellular monolayer. Possible toxic effects are assessed after 24 hours incubation. Trypan blue stained dead cells are counted (in relation to non-stained viable cells) in a haemocytometer.

Outline of the NRR method.

Preparation is in principle equal to that of the trypan blue exclusion method. The mice fibroblasts (L-929) are cultured routinely with passages twice a week, seeded in 75 cm^2 flasks at a density of 2 x 10^4 cells/cm^2 in Minimum Essential Medium supplemented with 5% fetal calf serum and ,if required, antibiotics.

Extracts of materials to be tested are treated as follows (AFNOR): Fragmented samples of the material are incubated in an extraction vehicle under sterile conditions using glutamine deprived culture medium with or without 5% calf serum for 120 hours at 37°C. Extracts of positive control materials should be processed in parallel. The ratio: tested sample surface/extaction vehicle volume must be 5 cm^2/ml. After the incubation serial dilutions of extracts are performed to give the following final concentrations: 10%, 50% and 100% extract in medium. After 24 hours of seeding the cultures are treated with various concentrations of extract. They are then stained with neutral red solution for 3 hours. The cells are centrifuged after 3 hours, and the neutral red solution is discarded. The optical density is then read.

Description of the TBE procedure.

Six Petri dishes are seeded with 2.5 x 10^5 fibroblastic cells (L-929) diluted in 2 ml of culture medium and incubated for 24 hours at 37°C in 5% CO_2 enriched humidified atmosphere. The culture medium from 3 of the Petri dishes is discarded and 1 ml fresh medium is added to each. One tenth of the surface of these 3 dishes is covered with the fragments of the material to be tested. In the remaining 3 dishes 1 ml of the medium is replaced by 1 ml of the extract. All the dishes are then incubated for another 24 hours.

After the incubation, extracts and samples are removed. One ml of trypsin solution is added to the cultures to induce release of the cells from the surface. Then 1 ml of trypan blue solution is added to stain the cells for 2 min after carefully pipetting to create a cell suspension. One drop of this suspension is transferred for counting in a standard double chamber haemocytometer and mounted on a microscope. Unstained and trypan blue stained cells are then counted according to normal procedure (see **Equipment**). Cell numbers thus obtained are multiplied by 10^3 to give the total number of cells per dish. The respective percentage of viable and dead cells should be calculated. Duplicate countings are performed for each Petri dish. The same procedure should be done with the positive and negative control materials. Results are expressed as percentages of dead cells relative to the total number of cells as described in the NRR method below.

Description of the NRR procedure.

Using a microwell plate 9 x 10^3 cells are seeded per 200 μl culture medium in each well. The plate is incubated at 37°C in humidified 5% CO_2 enriched atmosphere until confluence. Normal cell morphology is ensured under microscope before discarding the culture medium. When confluence is achieved, the culture medium is removed and replaced by 200 μl of extract diluted in the above mentioned concentrations (10%, 50%, 100%). Five wells are required for each dilution of material extract as well as 5 for positive control material, 5 for extraction vehicle (blank) and 5 for culture medium. After 24 hours of incubation the extract medium is discarded and the cell monolayer is carefully washed with Hanks solution. To each well is added 200 μl diluted neutral red solution and incubated further for 3 hours at 37°C. The cultures are then observed under a microscope to ensure cell morphology. The dye solution is carefully discarded, and the cell monolayers are treated with 200 μl calcium formaldehyde solution for 1-2 min. This solution is then discarded, and the pH sensitive staining is developed by adding 200 μl acetate-ethanol solution. The microplates are thoroughly shaken for 15-20 min, and the optical density of the wells are read in a multiplicate spectrophotometer. The degree of viability is expressed by the optical density relative to the control.

The fraction of viable cells is calculated as follows:

$$\% \text{ viability} = \frac{\text{OD in control well}}{\text{OD in extract well}} \times 100$$

RESULTS

The TBE and NRR test methods are generaly applied simultaneously to assess the cytotoxicity of leaching components from polymers.

The tests have been carried out on 4 standard Latex catheters using the following parameters: incubation of samples for 25 hours at 37°C in MEM supplemented with 2.5% calf serum (using 1 g catheter per 10 ml medium). Subsequent dilutions (10, 20, 30, 40, 50 and 60%) were made to achieve serial extract lines. In *Table I* the results are shown.

Table I

Comparative results provided by two staining methods of assessment of cellular viability expressed as a percentage of viable cells. (By permission of Combrier, E., 1988).

	A		B		D		E	
	T.B.	N.R.	T.B.	N.R.	T.B.	N.R.	T.B.	N.R.
CONTROLS	98 ± 00	97 ± 01	98 ± 00	100 ± 00	99 ± 00	100 ± 00	99 ± 00	100 ± 00
10%	58 ± 00	5 ± 10	98 ± 01	103 ± 02	98 ± 00	104 ± 03	98 ± 00	90 ± 05
20%	0 ± 00	4 ± 00	99 ± 00	95 ± 07	98 ± 01	96 ± 01	99 ± 00	70 ± 12
30%	0 ± 00	3 ± 00	98 ± 00	92 ± 04	98 ± 00	95 ± 01	98 ± 00	46 ± 07
40%	0 ± 00	4 ± 00	96 ± 01	78 ± 04	98 ± 00	92 ± 01	98 ± 01	38 ± 02
50%	0 ± 00	4 ± 00	64 ± 10	36 ± 13	98 ± 01	92 ± 02	98 ± 01	24 ± 02
60%	0 ± 00	4 ± 00	0 ± 00	8 ± 02	98 ± 00	87 ± 05	97 ± 01	24 ± 01

T.B.: Trypan Blue Staining   N.R.: Neutral Red Staining

DISCUSSION

In general the dose effect can be observed as demonstrated in the results above. One can see in product A that the TBE method is a less sensitive indicator of cell damage. In 10% concentration TBE shows that 50% of the cells are viable whereas only 5% are viable in the NRR method. The dose effect is also evident. Product B induces an increase in cytotoxicity with the rising concentration of extract. In product E both effects combine. The TBE method does not reveal any alteration of the viability whereas the NRR indicates the cytotoxic effect at 20% concentration. The NRR method thus appears to be more sensitive compared to the TBE method. Since neutral red specifically binds to the anionic sites of the lysosomal membrane, any alteration of this membrane will reflect itself in a decrease of staining although the cells still exhibit morphology of healthy cells.

Reduced lysosomal staining may be considered an early indicator of cell damage and a sensitive global measure of cytotoxicity. The linear relationship between the number of live cells and the optical density reading after staining may provide a sensitive early link to a decrease in viability and growth inhibition. However, the global aspect of the NRR microtitration does neither directly allow assessment of the degree of lethality nor the level of cellular alterations and their sites. The presence of cell debris - which is more or less distinct from dead cells - may morphologically give rise to an overestimation of the mortality in microscopic estimation of the cells.

Related testing methods.

Additional use of fluorogen substrates such as fluorescein diacetate may help to avoid these drawbacks. The fluorescein diacetate is a non-fluoroscent component which easily enters live and dead cells through the plasma membrane. Acetate is then cleaved by the esterases inside the cells and the highly fluoroscent fluoroscein is released. This molecule cannot diffuse through the membrane of intact cells but leaches from damaged cells. Therefore live cells appear clearly labelled.

Supravital dyes do not only require membrane integrity, but an enzymatic hydrolysis must occur to liberate the fluorochrom. It has been demonstrated that esterase activity plays an important role in detoxification processes. Fluoroscent staining of specific cell functions is now a widely used technique in other aspects. By coupling fluoroscent and exclusion dyes like erythrosine B, it is possible to differentiate between live and dead cells and to count live and dead cells with a fair degree of accuracy. Quantitative results may be obtained fairly easily from microscope examination. Automatic cytofluorometric measures may be carried out quicker and provide greater reliability by suppressing the subjective error of the experimenter (*Combrier et al., 1988*), but does on the other hand require advanced equipment.

^{51}Cr-labelled release method is widely used for determining cell viability through release of Cr (*Spangberg, 1973*). The method uses Na_2 $^{51}CrO_4$. In its reduced form $^{51}Cr^{2+}$ enters the cell and is oxidized into $^{51}Cr^{3+}$ and will bind covalently to basic aminoacids of the intracellular proteins. When the cellular membrane is damaged, the cell will lyse and $^{51}Cr^{3+}$ diffuses out with the intracellular protein into the culture medium. The amount of labelled Cr in the culture medium is considered proportional to the degree of cell lysis.

This technique is used as a general screening procedure for cytotoxicity, recommended by the FDA (Tech. Rep. no. 9, 1980) and proposed with modifications by AFNOR. However, this test requires equipment and facilities to handle low radioactive substances. (Editors note: The authors have not included further information on this test).

REFERENCES

Borenfreund, E. & Puerner, J.A. (1984) "A simple quantitative procedure using cultures for cytotoxicity assays (HTD/NR-90)". J. Tissue Cult. Method., 9, 7-9.

Combrier, E. (1988) "Etude comparative des méthodes d'évaluation de la cytoxicité in vitro". Thèse du Laboratoire de l'Ecole Practique des Hautes Etudes.

Combrier et al. (1988) "Flow cytometric assessment of cell viability: A multifaceted analysis". Cytotechnology

Holmberg, B. (1960) "On the permeability to lissamine green and other dyes in the course of cell injury and cell death". Exp.Cell.Res., 22, 406-414.

Spangberg, L. (1973) "Kinetic and quantitative evaluation of material cytotoxicity in vitro". Oral Surg., 35, 389-401.

Name of method:

Cell growth test method.

Aim of method:

In vitro routine toxicity test using cell growth assessment in cell cultures.

Biophysical background for the method:

This test is generally coupled with measurement of the total cell protein or measurement of DNA synthesis activity. Continuous cell lines such as L-929 are routinely applied for tests of viability. On the other hand differentiated cell lines seem to be preferred if measurement of protein and DNA synthesis are desired. In the presence of biomaterials intended for implantation which will be in direct contact with different tissues, it seems advantageous to chose cells closely related to specific tissues or organ of the implantation site. Thus a culture of the endothelial cells is advisable to use for blood contacting materials. Such cultures are routinely obtained from samples of human umbilical cord veins (*Jaffe, 1973*) with refinements (*Guillot, 1988*).

Scope of method:

The test can be used for general screening purposes although the described approach makes it more an experimental method.

DETAILED DESCRIPTION OF METHOD

Equipment.

Basic equipment for cell culture handling (as given in **INTRODUCTION**) supplemented with:
Cell counter (Coultronics, USA)
Culture flasks 25 cm^2 (Falco n)
4-well plates (Nuclon®)

Primary cell culture medium: Standard medium 199 to which is added 100 IU/ml penicillin and 100 μg/ml streptomycin, 200 mMol L-glutamin and 20% fetal calf serum (Boeringher)

Fetal calf serum 20% (Boeringher)

Endothelial cell growth supplement (Collaborative Research, USA)

Outline of method.

The measurement of cell proliferation enables one to assess the cytotoxicity of a material or its extract by measuring the growth rate of the cells which have been seeded onto the material or grown in culture medium supplemented with extract from the material to be tested. It has been demonstrated that cell growth may be inhibited although the cell viability is preserved. Thus this technique can be used as part of a general screening for cytotoxicity.

Description of procedure.

The material to be tested is precut to provide a circular disc (15 mm diameter) which covers the bottom of the culture well. The culture well-plates are also used with control material which is treated in a simular fashion. If extract is used this is prepared by incubating 5 cm^2 of the surface of the material in 1 ml of Minimum Essential Medium without serum for 2 hours at 37°C.

Human umbilical cords are available from maternal wards where normal vaginal delivery is performed. The cords are isolated and stored in cold sterile buffer for les than 24 hours before harvesting. Human umbilical vein endothelial cells (HUVEC) are routinely extracted (*Jaffe, 1973*). The primary cultures are grown in primary cell culture medium. Secondary cultures are grown in the same medium supplemented with 50 μg/ml endothelial cell growth supplement and 50 μg/ml heparin (Sigma). The material to be analysed is deposited in the bottom of a 4-well plate Nunclon® and culture medium is added before seeding with HUVEC. Uncovered well-plates are used as control surface. Cell proliferation is measured by cell counting at daily intervals until a plateau is reached. The cell counting is carried out as follows:

Primary preconfluent cultures are briefly treated by 3 ml/25 cm^2 of trypsin-EDTA. The solution is discarded just when the cells begin to detach. The cells are dispersed by pipetting fresh medium over the monolayer to provide a single cell suspension. Cells are counted in a haemocytometer and diluted to a final concentration of approx. 40,000 cells/ml. Each multiwell-plate containing the material and control well-plate are inoculated with 1 ml of cell suspension.

To test extracts of material, secondary cultures are grown in plastic multiwell-plates as described above and incubated for 24 hours at 37°C. After the incubation period the culture medium is discarded and 1 ml of extract is added to each well. Control cultures receive 1 ml of fresh medium.

Culture plates are incubated at 37°C in a humidified 5% CO_2 enriched atmosphere. At regular time intervals up to 12 days triplicate cultures are trypsinized for cell counting. Growth curves are drawn by plotting the cell numbers versus time. More objective comparison is provided by calculating the population doubling number (PDN) according to the following equation:

$$PDN = \log_2 \left({}^{Nf}/_{Ni} \right)$$

where Nf is the final number of cells in confluent control cultures or in the corresponding tested culture. Ni is the initial number of cells seeded x attachment efficiency (i.e. viable cells counted after 24 hours of seeding).

RESULTS

An example of in vitro model material for cytocompatibility testing is a polymer used for coating Dacron® vascular prostheses in order to reduce permeability and to enhance endothelialization (*Warocquier-Clerout et al., 1986, Biomaterials*). This material consists of a mixture of albumin and gelatin which is crosslinked by carbodiimide which is directly polymerized over the bottom of each well before seeding. In *Fig. 1* the growth of HUVEC cells are given, seeded on the control plastic and on the protein matrix respectively. Control cell cultures reach confluence in 7 days when cell density approximates 150.000 cells per well whereas HUVEC grows more slowly onto the matrix indicating that they are still exponentially growing by day 7. PD numbers were 2.17 and 1.85 for control and test culture respectively (*Warocquier-Clerout et al., 1986, ESAO Proceedings*).

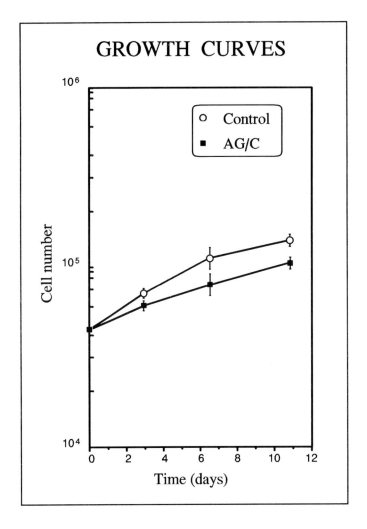

GROWTH CURVES

Fig. 1

Comparative growth curves of HUVEC seeded at an initial density of 2×10^4 cells/cm^2 either on plastic dishes (O) or on dishes coated with carbodiimide crosslinked A + G (■). Values are the means ± SD; n = 4, p < 0.010, Student T-test.

DISCUSSION

Cell growth can only be evaluated in relation to control cultures. A slight cell damage will delay cell growth, but may not influence the viability of the cells. Cell growth test should be used in conjunction with other tests mentioned in this contribution.

REFERENCES

Guillot, R. (1988) "Culture de cellules endothéliales. Technique en culture de cellules animales". Ed. INSERM, 257-289.

Jaffe et al. (1973) "Culture of human endothelial cells. Identification by morphologic and immunologic criteria". J. Clin. Invest., 52, 2745-2756.

Warocquier-Clerout, R. et al. (1986) "F.p.l.c. analysis of the leakage products of proteins crosslinked on Dacron® vascular prostheses". Biomaterials, 8, 118-123.

Warocquier-Clerout, R. et al (1986) "Comparative cytocompatibility of crosslinked collagen and albumin with glutaraldehyde and carbodiimide towards human umbilical vein endothelial cells". Proceedings XIII annual meeting, ESAO, 106-108.

Name of method:

DNA synthesis test method.

Aim of method:

In vitro toxicity test to measure the amount of DNA synthesis in cell culture.

Biophysical background for the method:

Cell proliferation is generally coupled closely with a proportionate increase of DNA in the cell culture. Measurement of DNA synthesis activity is therefore a good indicator of cell growth. Although continuous cell lines such as L-929 are generally used for tests measuring cell viability, differentiated cell lines seem to be preferentially selected for measurement of DNA synthesis. Measurement of DNA synthesis is carried out by incorporation of radiolabelled metabolite precursors. They provide the possibility of investigating the intimate process of synthesis of macromolecules. The present method can improve the quantitative measurement of cytotoxicity. It is carried out by incubating a cell culture with ^{3}H-thymidine (^{3}H-TdR) precursor which is built into the DNA according to the rate of growth (*Gimbrone et al., 1974*). For analysis of the incorporated ^{3}H-TdR, the DNA is made insoluble with trichloroacetic acid and the radioactivity of acid insoluble DNA is measured.

Scope of method:

The method can be used in vitro as quantitative screening for growth inhibition in cell cultures exposed to polymers or extracts of polymers.

DETAILED DESCRIPTION OF METHOD

Equipment.

Basic equipment for cell culture handling (as given in **INTRODUCTION**) supplemented with:
4-well plates (Nunclon®, Nunc, DK)
Liquid scintillation spectrometer with scintillation vials

Vacuum filtration device with glass microfiber filters (2.5 cm diameter)
^{3}H-Thymidine precursor (Commissariat à l'Énergie Atomique, CEA, France)
Trichloroacetic acid 10% (used as cold solution)
Ethanol 90%
Organic scintillation liquid (OCS, Amersham, UK)
The laboratory should possess facilities for handling of low radioactivity products.

Outline of method.

The material to be tested is precut into circular discs with a diameter of 15 mm.
Extracts of the polymer are prepared from the material with approx. 5 cm^2
surface/1 ml by incubation for 2 hours at 37°C in Minimum Essential Medium
with 5% calf serum.

4-well plates (Nunclon$^®$) are fitted with circular discs of test material as well as
with negative and positive control material prior to seeding with 4 x 10^4 cells per
well obtained from a primary human unbilical vein endothelial cell (HUVEC)
culture.

At time intervals up to confluence one well is labelled with 0.5 μCi of ^{3}H-TdR. The
The ^{3}H-TdR incorporation is blocked after 16 hours by adding an excess of cold
thymidine to provide a final concentration of 10^{-2} M. The cells are then briefly
trypinized to induce release and the cells are dispersed by pipetting fresh medium
over the monolayer to obtain a single cell suspension. The cell suspension is
transferred to a vial and an equal volume of cold 10% TCA is added. The treated
cell mixture is then allowed to rest 20 min. at 4°C. The suspension is then filtered
on glass fiber filter, which will retain the acid precipitable cell fraction. This is
successively rinsed with cold 5% TCA and 90% alcohol solution. This cell fraction
is then transferred to scintillation measurement.

Description of procedure.

The HUVEC cultures are grown together with positive and negative controls on
4-well plates. The cell cultures are labelled with ^{3}H-TdR for 16 hours on day 2, day
6 and day 10. At the same time the exponential preconfluent and confluent phases
of the cultures are measured on the remaining well culture not being labelled. This
is done by cell counting in a haemocytometer. The cells which incorporate
radioactivity are measured as cpm (counts per minute) per 1,000 cells. Specific

564

procedures depend on the equipment and are given in the manuals for the particular equipment.

Safety aspects.

Radioactive labelling experiments require a safety laboratory registred for low risk radioactivity.

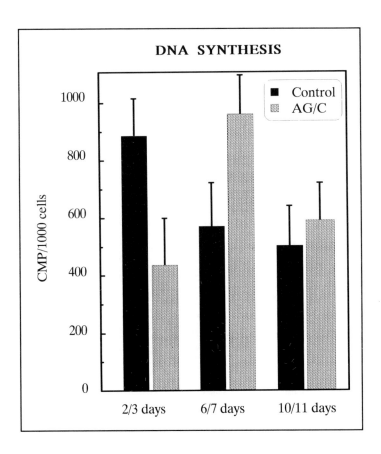

Fig. 1

[3]H-Thymidine incorporation into TCA-insoluble activity of HUVEC at successive stages of culture growth. Medium was exchanged 6 hours before labelling with 0.1 μCi/ml ^{3}H-Thymidine for 17 hours. Each bar represents the average incorporation ± SD of triplicate cultures. HUVEC were grown either on plastic dishes (■) or on dishes coated with carbodiimide crosslinked A + G (▒).

RESULTS

In *Fig. 1* results are given of a test concerning negative control compared to a polymer coated with crosslinked albumin gelatine. From the control results it can be seen that the maximum of radiolabelling corresponding to a maximum of DNA labelling gradually decreases from day 2 to day 10. Inhibitory influence as is shown in the figure results in a delay of DNA synthesis but without any further inhibition. A similar maximum value of specific radioactivity is reached by day 6 in contrast to day 2 on the control (*Lee, 1988*). At the 10th day the thymidine incorporation showed no significant difference.

DISCUSSION

Several factors seem to influence the uptake of thymidin. The cell substrate relationship has been shown to be affected by the protein composition of the substrate, its charge and its wettability which, in turn, influences the efficiency of cell attachment to various substrates. These factors may fluctuate, and the cell growth kinetics vary in response to metabolic regulatory processes which demonstrate that cells adapt to altered physiological conditions. Comparative studies of growth curves of cells grown on test material and control material respectively will enable an analysis of the cytoxicity assessed in a long term experiment which may overlap several cell cycles, thus giving the cells a chance to survive toxic injuries. Whereas a short term toxicity which is reversible may be reflected by lower numbers of cells adhering to the material, a more quantitative response may be provided by measuring the incorporation of tritrated thymidine into DNA which is being synthesized. The method allows not only a quantification of the DNA synthesis, but will also detect a possible delay in the S-phase of the cell cycle using comparatively labelled cell cultures. - The draw-back of the method is that it requires more than one week to carry out.

Related testing methods.

Rapid photometric and fluorometric methods have been proposed as alternative measures methods to detect the proliferation of adherent cells in 96-well plates. Cell monolayers are in these methods stained with May-Grünwald Giemsa or neutral red (*Klein-Soyer et al., 1987*). The linear correlation between the density of stained cells and their absorbance are measured in a spectrophotometer which allows reasonably good quantification. Similar results can be obtained using a DNA specific fluoroscent dye from Hoechst: 33258 or DAP I (*McCaffrey et al., 1988*). These methods do not require equipment for radioactive measurements, but do - on the other hand - require equipment for fluorimetry and spectrometry. The fluorimetric or spectrometric measures can mainly be used for applications where extracts are to be analyzed for the toxicity of products. They are not suitable for analyzing surface cytocompatibility.

Another method is to measure cell proteins (*Lowry et al., 1951; Bradford, 1976*) coupled with measurement of cell proliferation as indicant of cytotoxicity. This assessment of cytotoxicity of biomaterials is adapted by AFNOR where it is described in detail. (Editors note: The authors have not provided details on this standard test).

It must be noted that all the techniques to assess the cytotoxicity of a material or an extract must be considered more or less complementary. A detailed analysis of the possible cellular alterations induced by the material or its leaching products can only be considered complete if all aspects are considered. Statistical analysis of the results may increase the prediction of the tests and reduce the subjective interpretation due to the operator. Furthermore, the cellular staining with metabolic supravital dyes offers a good opportunity to reveal mechanisms and preferential toxicity sites which may be quantified in a multiparametric analysis using flow cytometry. The usefulness of such methods may explain why staining techniques are often preferred to isotope methods such as ^{51}Cr release in the detection of the integrity of the cell membrane (*Slezak et al., 1988*).

REFERENCES

Bradford, M. (1976) "A rapid and sensitive method for the quantification of microgram quantities of protein using the principle of protein-dye binding". Anal. Biochem., 72, 248-254.

Gimbrone, M.A. et al.(1974) "Human vascular endothelial cells in culture". J. of Cell Biol., 60, 673-684.

Klein-Soyer, C. et al. (1987) "Rapid and sensitive photometric quantification of the profifertion of adherent vascular endothelial and smooth muscœe cells in 96-multiwell plates after May-Grünwald Giemsa staining". Nouv. Rev. Fr. Hematol., 29, 311-315.

Lee, Y.S. (1988) "Cytocompatibilité comparée de differents supports protéiques réticulés. Application aux prothèses vasculaires". Thèse Université de Compiègne.

Lowry, O.N. et al. (1951) "Protein measurement with the folin phenol reagent". J. Biol. Chem., 193, 265-275.

McCaffrey, T.A. et al. (1988) "A rapid fluorometric DNA asssay for the measurement of cell density and proliferation in vitro. In Vitro, 24, 247-252.

In Vitro Cytocompatibility Tests

Authors: M.F. Sigot-Luizard & R. Warocquier-Clerout
Laboratoire de Biologie Cellulaire Experimentale. Univ. de Compiègne
F - 60206 Compiègne

INTRODUCTION

While cytotoxicity involves negative criteria such as cellular alterations, cell death, hampered growth etc. then cytocompatibility is more vague but suggests positive criteria. A material will be considered as cytocompatible if both structure and functions of the tissue in direct contact with it remain unchanged. The maintenance of these functions depends directly upon the quality of the material surface. Thus a material may induce reorganisation of a tissue without exhibiting cellular damage, but the previous (original) structure and functions will be lost. Therefore research and development of biomaterials may provide surface properties of the material which satisfy these requirements. Most research on blood contacting material is concerned with this aspect. As non-thrombogenicity is a main property of the vascular endothel recent studies aim at finding materials which enhance neo-endothelialization. Cell culture experiments may be carried out with endothelial cells and with tissues after a control of the structural and functional state. Tests of these properties are described in the following.

Morphology is best studied by staining the cells. The classical method of staining with silver nitrate is used to outline the endothelial cell boundaries. The silver is deposited at the boundaries visualizing a well characterized cobblestone morphology of the normal endothelium (*Garbarsch et al., 1970*). This method is simple and rapid. It is routinely used for identification of endothelial cells in culture. - Transmission electron microscopy allows the visualization of intracellular organelles. The so-called Weibel-Palade bodies are specific to the human endothelial cells (*Weibel & Palade, 1964*).

The tests in this context are:

S. Dawids (ed.), Test Procedures for the Blood Compatibility of Biomaterials, 569–594.
© 1993 *Kluwer Academic Publishers. Printed in the Netherlands.*

In vitro prostacyclin test, based on release of prostacyclin measured as the stable metabolite 6-Keto-PGF$_{1\alpha}$.

Wettability and cellular adhesion test, based on measurement of adhesion of endothelial cell cultures on protein coated surfaces.

Cellular repair test, based on the ability of the cells to repair a well defined circular damage of a monolayer of endothelial cells.

Cell growth and adhesion test, based on measurement of cell multiplication and spreading from a vascular explant.

Equipment for these tests encompasses basically the following components:

Basic equipment:
Sterilization equipment (autoclave, poupinel, filtration membranes)
Laminar flow bench
Carbon dioxide-air incubator (37°C)
Deep-freezer
Thermostated water-bath
Standard centrifuge (applying > 1500 g)
Disposable sterile centrifuge tubes
Inverted optical microscope with magnifications of 40x, 100x and 200x
Standard haemocytometer, double chamber (Thoma, Malassez)
Low speed magnetic stirrer (1 rev./sec)
Multiplate spectrophotometer (absorbance reading at 540 μm)
Sterile 35 mm tissue culture grade Petri dishes
Neutral red (stock solution), colour index no 50040, 0.4% in water. Solution should
    be kept in dark. Working staining solution is prepared by 1/80 dilution
Acid solution: ethanol 50%, glacial acetic acid 1%
Fixation solution: formaldehyde 40% in water, CaCl$_2$ 10%
Disposable multiwell plates (96 wells)
75 cm^2 flasks for culturing (Dow Corning)
Hanks solution without Ca and Mg
Hanks solution added 0.1% trypsin (2 x cristallized)
Trypsin® solution 0.1% in phosphate buffered saline (PBS) without Mg and Ca.
For most cell cultures:
    1 vol. of 0.25% Trypsin (Sigma, France) is mixed with 1 vol. of 0.20% EDTA
    (Sigma, France). The mixture is prepared just prior to use.
For HUVEC cultures:
    1 vol. of 0.25% Trypsin is mixed with 3 1/2 vol. of PBS and 1/2 vol. of 0.20%
    EDTA (to obtain Trypsin 0.05% and EDTA 0.02%)

Minimum essential medium (MEM) sterilized by filtration and added

5% calf serum

200 IU/ml penicillin

100 μg/ml streptomycin.

Minimum essential medium (MEM) without glutamine and $NaHCO_3$, added

0.292 g/l glutamine

2.20 g/l $NaHCO_3$

5% fetal calf serum

no antibiotics

Each separate test may require supplementary items which are given under the particular test.

REFERENCES

Garbarsch, C. et al. (1970) "Scanning electron microscopy of aortic endothelial cell boundaries after staining with silver nitrate". Angiologica, 7, 365-373.

Weibel, E.R. & Palade, G.E. (1964) "New cytoplasmic component in arterial endothelia". J. Cell. Biol., 23, 101.

Name of method:

In vitro prostacyclin test.

Aim of method:

To evaluate the integrity of endothelial cells in cell cultures by release of prostacyclin.

Biochemical background for the method:

Specific functions of the endothelial cells can be used to estimate the integrity of the cultured endothelial cells. One of these is the synthesis and release of prostacyclin PGI_2.

Prostacyclin PGI_2 is liberated in small amounts from healthy endothelial cells in cultures. PGI_2 acts in vivo as a potent vasodilator and as an inhibitor of the platelet activation in respect to aggregation and attachment to endothelial surfaces (*Weksler et al., 1977*). Endothelial cell cultures secrete a basal amount of PGI_2 depending on cell density and incubation time. In contrast to this, resting cells in vivo neither synthesize nor store PGI_2. A stimulus seems necessary to cause PGI_2 production. The kind of stimulus includes exposure to substances such as thrombin, angiotensin II, bradykinin, histamin, Ca-ionophore A 23187 and arachidonic acid. Mechanical agitation and change of culture medium leading to cell membrane injury causes release of arachidonic acid from membrane phospholipids (*de Caterina et al., 1985; Evans et al., 1984; Schror, 1985*). The injury of the cell membrane leads to a change in the composition of the cellular substrate which, in turn, is responsible for specific membrane alterations. Thus the analysis of PGI_2 production by endothelial cells grown on a test material and on a control substrate may provide, together with the cellular responses on an identical external stimulus, information relative to the cytocompatibility of the particular material. Extract of the material can also be tested as a potential stimulus. This method is proposed as an experimental study of the cytocompatibility.

The amount of PGI_2 released into the medium or synthesized in cell cytoplasma can be estimated routinely by radioimmunoassay of the stable metabolite 6-Keto-$PGF_{1\alpha}$.

Scope of method:

The method can be used as an experimental test. However, the commercial kits make it possible to perform reliable screening. It can only be used in comparative set-up, because the production and release of PGI_2 varies in each cell culture and is highly dependent on culture conditions.

DETAILED DESCRIPTION OF METHOD

Equipment.

Basic equipment for cell culture handling (as given in **INTRODUCTION**) supplemented with:
6-Keto-$PGF_{1\alpha}$ kit (Amersham, UK; New England Nuclear, USA). (Detailed guidelines are enclosed).

Outline of method.

The ready to use 6-Keto-$PGF_{1\alpha}$ assay system contains detailed description together with reagents. Briefly, the principle of the assay is the following: It is based on competition between unlabelled 6-Keto-$PGF_{1\alpha}$ and a fixed quantity of the tritium labelled compound for binding to a protein which has a high specificity and affinity for 6-Keto-$PGF_{1\alpha}$. With known fixed amounts of antibody and radioactive ligand, the amount of radioactive ligand bound by the antibody will be inversely proportional to the concentration of added radioactive ligand. The measurement of the protein bound radioactivity enables the amount of unlabelled 6-Keto-$PGF_{1\alpha}$ in the sample to be estimated. Separation of the protein bound 6-Keto-$PGF_{1\alpha}$ from the unbound ligand is achieved by adsorption of the free 6-Keto-$PGF_{1\alpha}$ on to dextran coated charcoal followed by centrifugation. Measurement of the protein bound radioactivity enables calculation of the amount of unlabelled 6-Keto-$PGF_{1\alpha}$ in the sample. Measurement of the radioactivity in the supernatant quantifies the amount of radioactive ligand bound by the antibody. The concentration of unlabelled 6-Keto-$PGF_{1\alpha}$ in the sample is then estimated from a linear standard curve.

Description of procedure.

The exact procedure is provided in the guidelines of the kit. The culture medium is daily changed and fresh medium is added with or without 20 μmol arachidonic acid which stimulates PGI_2 production.

RESULTS

The results in *Table I* gives a summmary of the amount of PGI_2 released in culture medium by human unbilical vein endothelial cells (HUVEC) grown on non-coated (control material) and protein coated 4-well plates Nunclon®. From aliquots 100 μl Radio Immuno Assays (RIA) are performed. These are taken from cell supernatants after 30 min. incubation.

Table I

6-Keto-Prostaglandin F_2 ng/1000 cells.

Incubation Time (days)	1	2	3	4	5
Control culture	8.3 ± 0.3	9.4 ± 0.6	18.1 ± 0.8	12.1 ± 0.8	11.2 ± 3.2
Stimulated control culture	50.7 ± 2.02	35.6 ± 1.6	34.2 ± 1.1	30.1 ± 19.1	11.1 ± 4.8
Stimulated coated culture	57.3 ± 17.4	95.5 ± 33.1	64.4 ± 11.2	37.7 ± 4.9	64.5 ± 16.6

Comparative PGI_2 production by HUVEC grown on plastic plate and protein coated plate. PGI_2 production was stimulated by adding 20 mol arachidonic acid to fresh medium 30 min. before taking an aliquot for Radio Immuno Assay. Basal production was measured in control culture. Values are means of 3 determinations ± SE.

The basal PGI_2 production on negative control polymer is 8.3 ± 0.3 pg/1000 cells at day 1 of the control culture without stimulus. In contrast to this the amount of PGI_2 increases 6-fold in the presence of arachidonic acid. The PGI_2 which is produced in response to stimulation reaches a maximum during the exponential growth phase. It then decreases gradually. Cells grown on carbodiimide crosslinked albumin-gelatin coated plates will release greater amounts of PGI_2 after of stimulation at day 2 and 3 compared to control cultures while the kinetics of cell growth are slightly lower.

DISCUSSION

Results of the release of PGI_2 production suggest that the cross linked proteins coating surface may affect cell membrane which, in turn, stimulates the PGI_2 release. This induced process which depends on RNA and protein synthesis (*Nawroth et al., 1984*) is believed to protect the cells from a harmful influence of the substratum (*Johnson et al., 1985*). A basal production of 31.2 ± 7.2 pg/1000 cells is expected after 24 hours incubation of HUVEC in medium containing 20% human serum.

Several publications have reported quantitative results of PGI_2 production (*de Caterina et al., 1985*). The amount of PGI_2 seems to vary highly from one cell culture to another and is thus greatly dependent on the culture conditions and passage. This method of assessment of the cytocompatibility of a material should therefore be performed as a comparative method, and for this reason requires rigourously controlled culture conditions.

Related testing methods.

Release of von Willebrand factor (vWF) can likewise be investigated on confluent endothelial cells after 24 hours of incubation in serum free medium containing 1% human albumin and 5 U/ml human thrombin. A number of commercial kits can now be used to measure the rate of release of vWF either using radioimmunoassay (Immunotech) or using ELISA method (Asserachrom-Stago). Primary cultures of endothelial cells originating from human umbilical vein synthesize approx. 10 - 15 $mU/10^5$ cells of vWF in 24 hours (*Klein-Soyer & Cazanave, 1986*). It is used routinely in the identification of endothelial cells.

REFERENCES

de Caterina, R. & et al. (1985) "Nitrates and endothelial prostacyclin production: studies in vitro." Circulation, 71, 176-182.

Evans, C.E. et al. (1984) "Prostacyclin production by confluent and non-confluent human endothelial cells in culture". Prastaglandins Leukotrienes and Medicine, 14, 255-266.

Johnson, A.R. et al. (1985) "Arachidonic acid metabolites and endothelial injury: studies with cultures of human endothelial cells". Federation Proc., 44, 19-24.

Klein-Soyer, C. & Cazenave, J.-P. (1986) Physiopathologie de l'hémostase et de la thrombose in Sultan, A.M. FISCHER; Prgrès en hématologie. Doin ed., Paris, 83-93.

Nawroth, P.P. et al. (1984) "Prostacyclin production by pertubed bovine aortic endothelial cells in culture". Blood, 64, 801-806.

Schror, K. (1985) "Prostaglanding, other eicosanoids and endothelial cells". Basic Res. Cardiol., 80, 502-514.

Weksler, B.B. et al. (1977) "Synthesis of prostaglandin I_2 (prostacyclin) by cultured human and bovine endothelial cells". Proc. Natl.Acad. Sci., USA, 74, 3922-3926.

Name of method:

Wettability and cellular adhesion test.

Aim of method:

The method can be used as experimental screening.

Biochemical background for the method.

This procedure assesses the effect of surface polymer properties covering wettability and protein adsorption on in vitro adhesion and proliferation of cultured human umbilical vein endothelial cells (HUVEC). Several published results (*Weiss, 1960; Baier, 1968; Grinell, 1978*) have demonstrated the importance of polymer surface wettability for cell adhesion. It has been shown (*van Wachem, 1987*) that protein from serum contains substances which might affect proliferation and adhesion of HUVEC. This correlation between wettability, protein adsorption and cell behaviour is utilized.

Scope of the method:

The procedure, which was originally developed by van Wachem et al., can be used as an experimental technique to screen the polymer surface for cytocompatibility.

DETAILED DESCRIPTION OF METHOD

Equipment.

Basic equipment for cell culture handling (as given in **INTRODUCTION**) supplemented with:
Tissue culture polystyrene (TCPS) flasks, Corning, N.Y.
Modified "Bionique" chamber, Corning, N.Y.
Partially purified human plasma fibronectin (FN^c) (according to *Ruoslahti et al., 1978*) for precoating
Culture medium consisting of medium 199 (40%), medium RPMI 1640 (40%), human serum 20%

Outline of method.

Human endothelial cells are harvested from umbilical cord veins according to the method of Jaffe (*Jaffe et al., 1973*). The cells are used after second or third passage when cultures reach confluence.

Polymers are precoated with protein solutions by exposing the surface for two hours after which the protein solutions are removed.

The wettability of the polymer surface is measured as described by Baszkin (see chapter on *Test methods for surface analysis: Contact angles and surface free energies of solids*).

HUVEC are seeded on the precoated materials and cultured up to confluence. Cell proliferation and adhesion are expressed by the number of cells per cm^2 and related to the percentage of attached cells on the reference material (tissue culture polystyrene) coated with partially purified human plasma fibronectin (FN^c).

Description of procedure.

For adhesion experiments the cells are seeded at a density of 4×10^4 cells/cm^2. They are incubated at 37°C in a humidified 5% CO_2 enriched atmosphere. Cell morphology can be controlled under microscope. For adhesion determination, the culture medium is replaced thereby removing non-seeded cells. The cells are then released using trypsin solution until the cells begin to release. They are then transferred to a haemocytometer and counted as earlier described. Detection of cell proliferation is carried out in a similar fashion at pre-set days after seeding.

Renewal of culture medium and simultaneous cell count in a haemocytometer is performed at two days intervals.

RESULTS

The different materials commonly used are listed in *Table I*. The simultaneous contact angles (*Andrade et al., 1979*) are calculated according to the captive bubble method (see chapter on *Test methods for surface analysis: Contact angles and surface free energies of solids*). Precoating for two hours at room temperature was performed with: pooled plasma, serum, 2 mg/ml partially purified fibronectin (FN^c), 40 mg/ml albumin (Alb), 30 mg/ml immunoglobulin G (IgG), 50 μg/ml fibrinogen (Fg), 50 μg/ml pure fibronectin (FN), 0.1 mg/ml high density lipoprotein (HDL).

Table I

Typical contact angle measurement on polymeric materials. The values can be used to calculate the free surface energy. The measurements are closely related to cell and tissue adhesion to the surface. (By permission of van Wachem et al., 1984)

Surface	Code	Contact Angle (θ)
Tissue culture polystyrene, Costar	TCPS	34°
Polystyrene	PS	77°
Tissue culture polystyrene terephtalate, Falcon	TCPETP	44°
Polyethyleneterephtalate	PETP	65°
Polymethylmethacrylate	PMMA	61°
Polycarbonate	PC	83°
Fluoro-ethylene-propylene copolymer	FEP	102°
Poly-L-lactic acid	PLLA	71°
Polyurethane, Biomer	PUR	37°
Cellulose, Cuprophane	CE	22°
Cellulose-2.5-acetate	CA 2.5	30°
Cellulose-3-acetate	CA 3	52°
Hard glass, Corning	GLASS	37°

The HUVEC behave quite differently dependent on the material as shown in *Fig. 1A (a & b)*. This difference might be correlated with the wettability as shown in *Fig. 1B (c)* where the contact angle of the polymer is plotted as a function of the adhesion of HUVEC. Moderate wettable surfaces generally exhibit a good cell adhesion whereas both more hydrophobic and hydrophilic surfaces exhibit poor adhesion of the HUVEC. These results can be modified by adsorption of proteins from serum containing culture medium as shown in *Fig. 1B (d)*. Proteins such as fibronectin promotes cell adhesion whereas HDL, albumin and IgG inhibit the adhesive property of HUVEC in vitro.

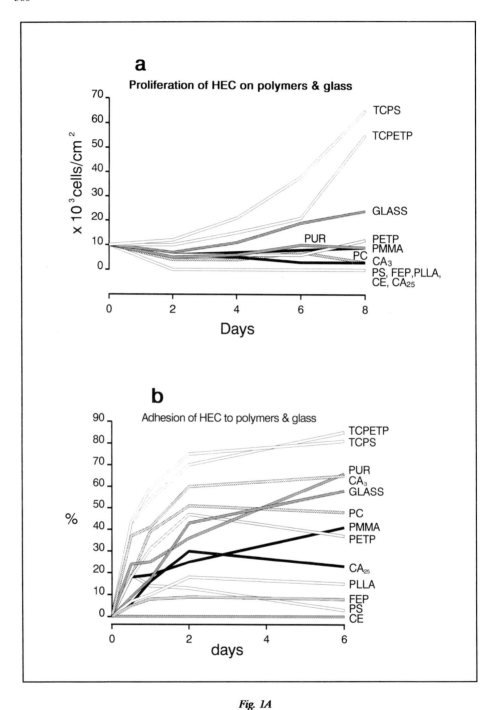

Fig. 1A

*HEC behaviour: proliferation (a) and adhesion (b). (By permission of van Wachem et al.,
1982)*

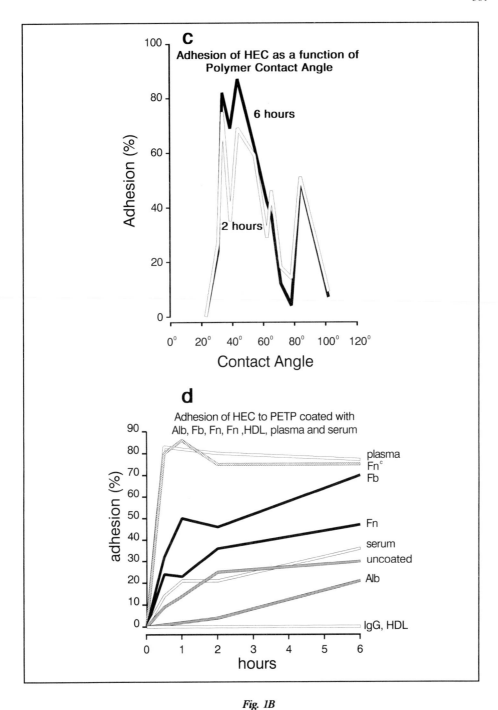

Fig. 1B

HEC behaviour related to wettability (c) and protein adsorption (d). (By permission of van Wachem et al., 1982).

582

DISCUSSION

Precoating with proteins appears to have a profound effect on adhesion of cells. The importance of the early phenomena of adhesion is evident when the material is in contact with the living tissue. The native surface properties of the polymer are modified by rapid protein adsorption and consequently its in vitro cytocompatibility. Detailed analyses in vitro of this phenomenon are given in the literature for materials such as fibronectin, high density lipoproteins and immunoglobulin G (*van Wachem et al., 1984*). This test needs further experience before it can be used as predictor to tissue compatibility.

REFERENCES

Andrade, J.D. et al (1979) "Surface characterization of poly(hydroxyethyl metarylate) and related polymers. 1. Contact angle method in water". J. of Polymer Sc: Polymer Symposium, 66, 312.

Baier, R.E. (1968) "Adhesion: Mechanisms that assist or impede it". Science, 162, 1360-1368.

Grinnell, F. (1978) "Cellular adhesiveness and extracellular substrates". Int. Rev. Cytol., 53, 65-144.

Jaffe, E.A. et al. (1973) "Culture of human endothelial cells. Identification by morphologic and immunologic criteria". J. Clin. Invest., 52, 2745-2756.

Ruoslahti, E., Vuento, M. & Engwal, E. (1978) "INteraction of fibronectin with antibodies and collagen in radioimmuno assay". Biochem. Biophys. Acta, 534, 210-218.

van Wachem, P.B. et al. (1984) "The interaction of cultured human vascular endothelial cells and biomaterials". J. of the Europ. Soc. for Art. Org., Proceedings XI Annual Meeting, ESAO, 2. supplement 1, 98-102.

van Wachem, P.B. et al. (1987) "The influence of protein adsorption on interactions of cultured human endothelial cells with polymers". J. of Biomed. Mat. Res., 21, 701-718

Weiss, L. (1960) "The adhesion of cells". Int. Rev. Cytol., 9, 187-225

Name of method:

Cellular repair test.

Aim of method:

The method is suitable for screening purposes to estimate the ability of cell cultures to overgrow the tissue after a mechanical damage.

Biochemical background for the method:

The cytocompatibility of the material can be assessed by means of its ability to repair after a mechanical injury (*Klein-Soyer et al., 1986*). This capacity of repair is dependent on tissue/material interface. The endothelium is responsible for the vascular integrity. Thus cellular damage will lead to considerable pertubations of endothelial cell properties which, in turn, will affect coagulation, fibrinolysis and inflammatory reactions.

Scope of method:

This technique can be used as a general screening of the cytocompatibility of a material related to its property in inhibiting cellular repair on its surface.

DETAILED DESCRIPTION OF METHOD

Equipment.

Basic equipment for cell culture handling (as given in **INTRODUCTION**) supplemented with:
Petri dishes 35 mm diameter
Calibrated disks of cellulose polyacetate paper (Sepraphore III Gelman n° 51003, Ann Arbor, Mich. USA) or
Millipore filter HAWP 0.45 μmol (Millipore, S.A., Molsheim, France) sterilized under UV light or obtained sterile
Adhesive grids, Linbro 76-63-01, Flow Laboratories, Puteaux, France)
Partially purified human plasma fibronectin (FN^c) (according to *Ruoslahti et al., 1978*) for precoating

Extracellular matrix
Transglutin® Type I collagen
Special culture medium M199 1 vol RPMI 1640 1 vol HEPES 10 μmol/l, glutamin
2 mmol/l, penicillin 100 U/ml, streptomycin 100 μg/ml

Outline of method:

Human unbilical vein endothelial cells (HUVEC) are obtained according to the
method of *Jaffe* with some modifications (*Klein-Soyer et al., 1984*). The endothelial
cells are grown on different protein coatings until a confluent monolayer has
formed. On a circular area of the HUVEC monolayer a mechanical injury is
exerted. The ability of the cells to repair this damage is assessed by measuring the
percentage of cell regeneration as a function of time.

Description of procedure.

HUVEC are grown to confluence. The culture medium is discarded, and a
calibrated disc is applied in the center and then carefully removed. Cellulose
polyacetate paper of three different diameters (4, 5 and 6 mm) is used. Two ml of
fresh culture medium are added immediately. This technique causes selective cell
detachment, but preserves the endothelium cell membrane (ECM) in the area of
detachment.

The mechanical damage of the HUVEC monolayer is controlled by an indirect
immunoperoxidase technique which demonstrates the absence of vimentin, a
cytoskeletal protein in the base of the cell. Quantitative analysis of the cell
regeneration is carried out as follows: Adhesive grids (the same size as the lesion
with dividing lines into 0.5 mm^2 squares) are mounted at the bottom of the culture
dish at time zero. The culture dishes which are removed every 24 hours are rinsed
3 times with Phosphate Buffered Saline and fixed with absolute methanol for 10
min at 4°C. The HUVEC are then stained with May-Grünwald Giemsa. The
injured area is measured with an inverted microscope by counting the squares
covering the lesion in the standard fashion.

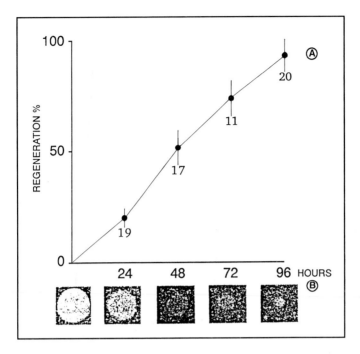

Fig. 1

*A) Time course of the regeneration of the lesion. Values are means ± SEM (vertical bars)
of the number of experiments. B) Macroscopic aspect of the lesion at the various time points
of regeneration (May-Grünwald-Giemsa stain, original size of the lesion at time 0 is 6 mm
diameter). (By permission of Klein-Soyer et al., 1986).*

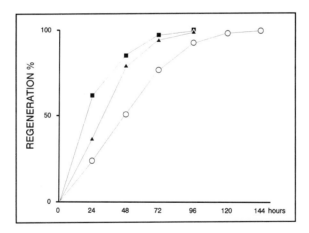

Fig. 2

*Time course of the regeneration as a function of the size of the lesion. Original diameter of
the lesion at time zero:* O *: 6 mm;* ▲ *: 5 mm;* ■ *: 4mm. Each point is the mean of two
experiments. (By permission of Klein-Soyer et al., 1986).*

RESULTS

In *Fig. 1, 2* and *Table I* the regeneration pattern of a lesion is shown. The first 48 hours show a linear decrease of the damaged area. After 72 hours the repair rate is the slower to become completely repaired after 120 hours. The HUVEC should show a morphology and cell density similar to the HUVEC in the surrounding non-damaged area.

Table I

Time of repair of endothelial lesions 6 mm diameter on protein coated discs.

Protein coating	Number of experiments	Regeneration time for 50% coverage of the lesion: T_{50} (h). Mean ± SEM.
Plasma enriched in FN^c	(14)	47.7 ± 0.9
FN^c	(3)	46.2 ± 1.0
ECM	(4)	46.2 ± 1.0
Transglutine	(2)	45.9 ± 0.3
Type I collagen	(2)	48.2 ± 1.6

No significant difference among T_{50} using a one way analysis of variance. FN^c: fibronectin, ECM: extracellular matrix. (By permission of Klein-Soyer et al., 1986).

DISCUSSION

The duration of the repair process depends on the initial size of the lesion and the protein coating of the culture dishes. As the process combines cell migration and cell multiplication, the results are best expressed as the required time to cover 50% of the initial lesion (T_{50}) as a function of the sixe of the lesion at zero time. The described cell model allows a quantitative measurement of the integrated effect of many compounds involved at the blood/material interface (vasoactive substances, vascular protectors, antiplatelet drugs, etc.). It must be performed in such experimental conditions including the mechanical damage of confluent endothelium, presence of ECM after HUVEC detachment etc. that the model is representative for in vivo events. It should be realized that individual cell cultures grow with different patterns which underlines the need for comparative tests.

REFERENCES

Klein-Soyer, C. et al. (1984) "Effect of an extract of human brain containing growth factor activity on the proliferation of human vascular endothelial cells in primary culture". Biol. cell., 52, 9-20.

Klein-Soyer, C. et al. (1986) "A simple in vitro model of mechanical injury of confluent cultured endothelial cells to study quantitatively the repair process". Thrombosis and Haemostasis. F.K. Schattauer Verlag, GmbH (Stuttgart), 56, 232-235.

Ruoslahti, E., Vuento, M. & Engwal, E. (1978) "Interaction of fibronectin with antibodies and collagen in radioimmuno assay". Biochem. Biophys. Acta, 534, 210-218.

Name of method

Cellular growth, spreading and adhesion test.

Aim of method:

To assess in vitro the ability of a polymer to support growth, spreading and adhesion of cells maintaining normal tissue morphology.

Biophysical background for the method:

An organotypic cell culture method is used to assess the compatibility of a material. Three properties are measured at the tissue/material interface: Cell multiplication, cell migration and cell adhesion. The 3 biological properties appear to play a major role in e.g. the endothelialization process. Newly formed endothelial cells must divide and cover the material surface and exhibit normal metabolic activities. First of all, the cells should adhere strongly to the material (e.g. vascular prosthesis wall) to be able to withstand a blood flow force.

Organotypic culture testing uses a tissue, even a whole organ, which has the advantage of preserving interactions between the different cell types necessary for the functions of the tissue. Moreover, the tissue or the organ is cultured at the air/culture medium interface. Such culture conditions closely reproduce the in vivo conditions. It is proposed as a routine test by AFNOR where it is described in detail.

Scope of the method:

This technique (*Sigot-Luizard, 1988*) can be used as a general in vitro screening of the cytocompatibility of a material.

DETAILED DESCRIPTION OF THE METHOD

Equipment.

Basic equipment for cell culture handling (as given in **INTRODUCTION**) supplemented with:

Drafting device for area detection
Planimetric measurement device
Roller shaker
Sterile Petri dishes bacterial grade, 60 mm

Reagents

Culture medium: M 199 with 10% fetal calf serum, 0.02 mol tricin, 1% L-glutamin. To this is added the equal volume of Bacto agar 1% (DIFCO) in gel solution without antibiotics.
Isotonic counting solution (ISOTHON II™ Coultronics)
Neutral Red 2,5% in ISOTHON II solution
Trypsin solution: 1 vol 0.25% trypsin mixed with 1 vol of 0.20% EDTA

Outline of the method:

The endothelium is obtained by dissecting a chick embryo or an adult rat aorta. The material samples should be plain film of 1-3 mm thickness and with an area of 1 mm^2 on an average.

Description of procedure.

The aorta is split longitudinally and cut into pieces measuring 1 mm^2. Each fragment is deposited on the agar nutrient medium and covered with the material which faces the endothelial side (*Fig. 1*). The aorta explants are cultured in contact with the material for 7 days after which the cells have migrated on to the material. Quantification of the cell multiplication and migration is made by calculating the cell density (number of cells per unit area of migration). Cell adhesion is defined by the sensitivity of migrated cells to release upon exposure of trypsin, measured as the percentage of cells released as a function of time.

After 7 days of culture the evaluation is made on the material covered by the explant and the cell layer in the following way:

Area calculation is obtained by placing a drop of neutral red on each sample. Both total area of the culture (ST = explant + cell layer) and the area of the explant (SE) are outlined and measured with a drafting and planimetric device. Then the area of cell migration is calculated as SV = ST - SE.

The cell dissociation is estimated after removing the explants from the culture dish. The material which is covered with cellular tissue is treated with trypsin-EDTA solution diluted 1:10 with saline in periods from 5 min. to 1 hour at 37°C under gentle shaking. The cells which detach during the treatment are detected in a coulter counter. Aliquots are taken at the periods: 5, 10, 20, 30 and 60 min. The initial volumes removed for testing are replentished by addition of fresh trypsin-EDTA solution. Non detached cells at the end of the enzymatic treatment are further treated with undiluted trypsin-EDTA solution for 15 min. to determine the total amounts of cells present.

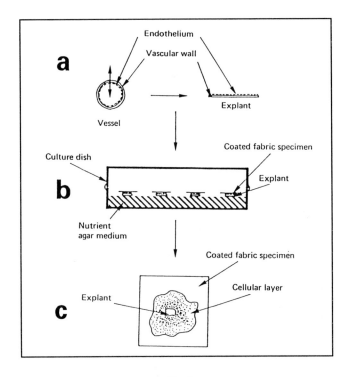

Fig. 1

*Culture technique. **a**) Preparation of the explant. **b**) Seeding of cells. **c**) Schema of the sample (coated fabric specimen + explant + migrating cell layer) after 7 days of culture.*

RESULTS

Successive results of cell numeration provide both total cell number per explant and the cumulative percentage of cells released after 5, 10, 20, 30 and 60 min. of

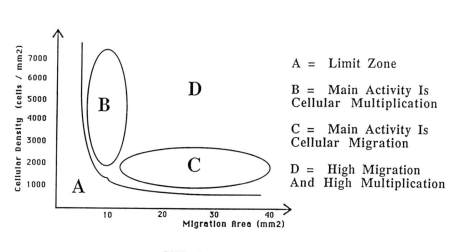

DIAG. I

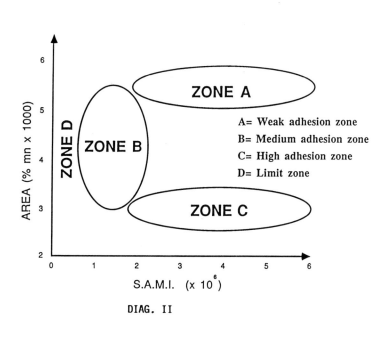

DIAG. II

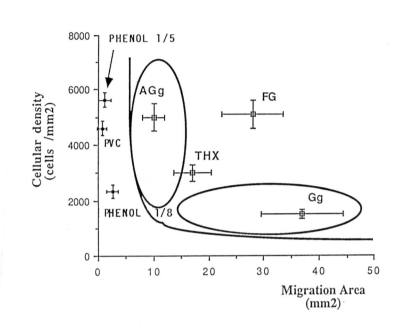

DIAG. III

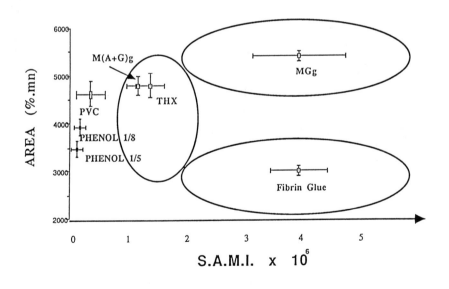

trypsinization time. Percentages of released cells are plotted versus trypsinization times. The area A below the graph and along the X-axis is calculated by the Simpson integration method, and a static adhesion modulated index (SAMI) is defined as the product of area A and the total cell number which provides an important parameter of the adhesion process (*Duval et al., 1988*). Cell migration and multiplication behaviour are expressed by cell density as a function of migration area as shown in *Diagram I*. Cell adhesion behaviour is expressed by the area A versus SAMI as shown in *Diagram II*.

Polyester (Dacron®) coated with one of the following proteins (and tested comparatively to polymer control surfaces): fraction V from Bovine albumin (Sigma) named Ag, pig skin gelatin (Rousselot) named Gg, mixed gelatin and albumin named AGg are all cross-linked with glutaraldehyde (*Duval et al., 1988*). Fibrin glue named Fg (Tissucol® Immuno) coating is included in the test. The negative control is THX treated for cell culture (Thermanox Lux Corp.). The positive control is a polyvinylchloride (Travenol, Nivelles, Belgium) and phenol solutions at different concentrations that are added to the culture medium (Ph 1/5, Ph 1/8). When the aorta fragments are cultured in contact with THX, the results are as shown in *Diagram III* and *IV*. The endothelial cells behave quite differently depending on the substrate used.

DISCUSSION

The two diagrams provide a more accurate and sensitive interpretation compared to measurements such as cell density for cell growth and migration or static adhesion modulated index (SAMI) for cell adhesion. Furthermore, cell density in relation to the migration area allows an estimate of both properties growth and migration to be separated.

Although AGg and Fg show identical cell densities they appear in different zones of the diagram. Similarly the results expressed by A as a function of SAMI allow elimination of errors due to identical products obtained from high cell number and low area value A or reversely. Fg and Gg both show two high SAMI values corresponding to two different trypsin sensitivities. The Gg coating response reveals a good correlation between the two diagrams, since it enhances high cell migration which diminishes cell adhesion as well as trypsin sensitivity measurements. The cell number values classify this protein (Gg) coating in the weak adhesion zone. These

594

two diagrams allow a classification of materials from their function of their biological requirements. Moreover, including the phenol solution positive control allow a definition and graduation of the toxicity of the material. As optional a microcomputer system and a program package may allow one to obtain statistical results of all the measurements and to draw the diagrams accurately.

These methods using endothelial cells or tissue cultures make it possible to investigate some important parameters involved at the material tissue interface which could be masked in normal in vitro cell studies. A better knowledge of the mechanisms involved in these interface phenomena may allow the design of materials with a surface endowed with the biological properties required to satisfy the conditions of cytocompatibility. Such in vitro methods do, however, neither take into account the influence of blood flow force on the morphology nor the functions of endothelial cells nor the secretion of products of other blood cells etc. Such factors may modify the in vivo response of the endothelial cells. The above tests do nevertheless appear to be quite reliable and sensitive for an initial screening of the cytocompatibility of materials in contact with blood.

REFERENCES

Duval, J.L. et al. (1988) "Comparative assessment of cell/substratum static adhesion using an in vitro organ culture method and computerized analysis system". Biomaterials, 9, 155-161.

Sigot-Luizard, M.F. (1988) "Mise au point d'un modèle d'evaluation quantitative de la cytocompatibilité des biomatériaux en culture organotypique in vitro". Les colloques de l'INSERM.

Chapter IX

GLP-GMP RULES FOR TESTING OF POLYMERS

Chapter ed.: S. Dawids
Institute of Engineering Design, Biomedical Section, Techn. Univ. of Denmark, DK - 2800 Lyngby

INTRODUCTION

In laboratories and in production plants the quality control is a critical factor. The need for cleanliness, uniform, reproducible procedures and the quality control has led to the development of two sets of rules in almost all industries: "Rules for Good Laboratory Practice" (GLP) and "Rules for Good Manufacturing Practice" (GMP). The authorities have a considerable interest in such rules or guidelines as they can support insight in the quality assurance of the products. The creation of standardised national guidelines have been attempted. However, once such guidelines have been set up in various countries, they have also served as a political tool to support the country's national bioengineering industry. This has resulted in several sets of guidelines or rules of which the most important ones come from the United States, Japan and Europe.

In respect to GLP there are wide differences in legislation. But in general the rules require that the manufacturer performs laboratory studies and submits the results of these studies to a governmental authority for assessment of the potential hazards to human health and to the environment. In this respect the growing awareness of hazards has had the consequence that several OECD member countries have established uniform criteria for the performance of studies. Some countries have set up ethical commities, rules for informed consent etc. In many respect this is an improvement as it helps to keep the non-professional and non-serious projects from being performed. However, it does add dramatically to the overall expense of such studies.

To avoid different schemes of implementation that could impede international trade of e.g. chemical products an international harmonization of test procedures

S. Dawids (ed.), Test Procedures for the Blood Compatibility of Biomaterials, 595–598.
© 1993 *Kluwer Academic Publishers. Printed in the Netherlands.*

and of Good Laboratory Practice is increasingly acknowledged as a basis of mutual acceptance.

In **USA** standards for GLP have been enforced in 1978 with later amendments. These rules closely describe the guidelines for non clinical laboratory studies and the rules for inspection and for disqualification.

In **Switzerland** a guide exists for GLP and for non clinical laboratory studies (the Intercantonal Agreement concerning the control of drugs: "IKS Regulativ").

In the **United Kingdom** GLP rules have been accepted in 1982 and are closely related to the principles of GLP in the earlier OECD guidelines. A GLP-unit (concerning toxicology and environmental protection) has been set up under the auspices of DHSS medical division.

In **JAPAN** GLP standards for safety studies on drugs were enforced in 1983. The text with amendments has been published in the form of a notice from the director general of the Pharmaceutical Affairs Bureau.

An international group from main countries worked out a document in 1980: "Principles for Good Laboratory Practice". This paper included guidelines for common managerial and scientific practices together with the experience from various national and international sources. However, it was only meant as guidelines and had no legal value. These principles are generally quoted in publications on the topic.

The purpose of GLP is to ensure that quality of test data are comparable and can form the practical basis for mutual acceptance of data.

In essence the GMP guidelines which have emerged through the last 3 decades aim at ensuring a safe product of uniform (high) quality and - in case of failures or errors - to provide a possibility to go back through each individual step of the manufacturing to the raw materials in order to identify the site at which the breach of quality occured. The same applies for GLP. These rules will (if followed) ensure:

1)   that reliable (one could say "honest") data and an accepted level of quality are achieved in performed studies in the laboratory or in clinical trials,

2)   that general ethical limitations and obligations are followed.

On the other hand GMP aims mainly at the quality assurance aspect in relation to purity of the raw materials, cleanliness of production and quality of test procedures. Much discussion has been wasted on the GMP rules because political aspects are involved. However, official requirements have been issued in many industrial countries. In Europe most of the GMP publications relate closely to the guidelines set up by *EUCOMED* (European Confederation of Medical Suppliers Association). These guidelines are based on earlier proposals issued by e.g. OECD, 1981 and WHO. In many respects the regulations adhere to the "Standards of Good Manufacturing Practice for Pharmaceutical Goods" (1972) which describe in detail guidelines for production facilities, conduct of personnel and control procedures of each step of production.

An important purpose of the nationally implemented guidelines relates to the inspection according to the *Pharmaceutical Inspection Convention (PIC)*. According to this convention an approval by authorized inspectors will be recognized in all the countries which have signed the convention.

The following guidelines are - if they are fully respected - sufficient to obtain full acknowledgement by the authorities for the laboratory data, for the production facilities and for the product. It is important to note that many producers go even further in their striving for improved uniform quality of the products, not only to meet legal measures but for the sake of true professional pride in developing good and safe products.

REFERENCES

EUCOMED: Guide to Good Manufacturing Practice for Sterile Medical Devices. 84/9 (1984). European Confederation of Medical Suppliers Association. 551 Finchley Rd, Hampstead, London NW3 7BJ, UK.

Guide to Good Manufacturing Practice for Sterile Medical Devices and Surgical Products (1981), (ISBN 0 11 3200770 0), Her Majesty's Stationary Office.

Pharmaceutical Inspection Convention (PIC): Convention for the mutual recognition of inspections in respect of the manufacture of pharmaceutical products (and explanatory notes (Geneva, 1970).

British Standards Institution, London 1991.
    Sterilization of medical devices. - Validation and routine control of ethylene oxide sterilization.
        EN AB001 - REQUIREMENTS
        EN AB002 - GUIDANCE
British Standards Institution, London 1991.
    Sterilization of medical devices. - Validation and routine control of sterilization by irradiation.
        EN AB003 - REQUIREMENTS
        EN AB004 - GUIDANCE
British Standards Institution, London 1991.
    Sterilization of medical devices. - Validation and routine control of steam sterilization.
        EN AB005 - REQUIREMENTS
        EN AB006 - GUIDANCE

British Standards Institution, London 1991.
    Quality systems

EN 29001	*Model for quality assurance in design/development, production installation and servicing (including EN 46001 - Specific quality systems requirements for medical devices)*
EN 29001	*Model for qualitiy assurance in production and installation (including EN 46002 - Specific quality systems requirements for medical devices)*
EN 29004	*Quality management and quality systems - GUIDANCE*

Rules for Good Laboratory Practice (GLP)
General Considerations

Author: S. Dawids

Institute of Engineering Design, Biomedical Section, Techn. Univ. of Denmark, DK - 2800 Denmark

INTRODUCTION

The principles of Good Laboratory Practice were initially mainly applied to the testing of chemicals and chemical products to obtain unbiased data on their properties and/or their safety with respect to human health and to the environment. Biological studies are normally evaluated by GLP, and data from such studies should thus be developed for the purpose of meeting regulatory requirements.

Modifications of the different national sets of rules aim at providing a reliable basis for evaluation concerning a national GLP inspection and for the rules for toxicity studies in general. This means that the described procedures will provide data which enable the authorities to have adequate insight of the quality of control and of monitoring within the laboratory in question. In brief the guidelines describe "correct behaviour" which enables the responsible(s) of the laboratory to obtain pertinent data of the production.

Basic concept and principles of Good Laboratory Practice.

The principles of GLP represents the sum of the activities necessary to ensure that testing procedures procedures of chemicals and chemical products obtain unambiguous data on properties and on the safety during handling with respect to human health and to the environment. The basic principles include:

- a system for ensuring safety procedures,
- allocation of responsabilities for safety with respect to human health and to the environment,
- well kept premises equipment and well defined raw materials,
- properly informed personnel,

S. Dawids (ed.), Test Procedures for the Blood Compatibility of Biomaterials, 599–614.
© 1993 *Kluwer Academic Publishers. Printed in the Netherlands.*

- documented procedures for the measures taken to secure ensuring safety procedures,
- appropriate records to monitor the results of procedures.

GLOSSARY OF TERMS

Below are given definitions which are used in the GLP rules.

Batch	A specific quantity or lot of a test or reference substance produced during a defined cycle of manufacture in such a way that it could be considered to be of a uniform character and should be designated as such.
Good Laboratory Practice (GLP)	The rules describe the organizational process and the conditions under which laboratory studies should be planned, performed, monitored, recorded and documented.
OECD Test Guideline	A test guideline according to the recommended OECD procedures for use in member countries.
Quality Assurance Programme	An internal control system designed to ensure that the study is in accordance with GLP.
Raw Data	All original laboratory records, documentation or verified copies thereof which are the result of the original observations and activities in the study.
Reference Substance (Control Substance)	A well defined chemical substance or any mixture other than the test substance used to provide a basis for reference with the test substance.
Sample	Any quantity of the test or reference substance.
Specimen	Any material derived from a test system for examination, analysis or storage.

Sponsor	The legal person(s) or entity who commissions and/or supports the studies.
Standard Operating Procedure (SOP)	The written procedures describing the particular routine of laboratory tests or activities which normally are not specified in detail in the general study plans or test guidelines.
Study	An experiment or set of experiments in which a product is examined to obtain comprehensive data on its properties, performance and/or its safety with respect to human health and environment.
Study Director	The person responsible of the overall conduct of the study.
Study Plan	The document which defines the entire scope of the study.
Test Facility	Persons, premises and operational unit(s) which are necessary for conducting the study appropriately.
Test Substance	A chemical substance or a mixture which is under investigation.
Test System	Any system or combination of systems consisting of animal(s), plant(s), microbial as well as other cellular, subcellular, chemical or physical component(s) used in the study.
Vehicle (Carrier)	Any agent which serves as a carrier used to mix, disperse or solubilise the test or reference substance to facilitate the administration to the test system.

TEST FACILITY ORGANISATION AND PERSONNEL

Responsabilities of the management.

The management of the test facility should ensure that the principles of GLP are followed.

The responsabilities include as a minimum that

- qualified personnel, facilities, equipment and material are available,
- a record is maintained of the qualifications, training, experience and job description for each professional and technical individual,
- the personnel clearly understands the functions they are to perform and, where necessary, that sufficient training for these functions are provided,
- health and safety precautions are applied according to national and/or international regulations,
- appropriate Standard Operation Procedures (SOP) are established and followed,
- there is a quality assurance program including designated personnel,
- the study plan, where appropriate, is agreed upon in conjunction with the sponsor,
- amendments to the study plan are agreed upon and documented,
- copies of all study plans are maintained,
- a historical file of all SOP is maintained
- there is for each study a sufficient number of personnel available for its timely and proper conduct,
- there is designated for each study an individual as study director with the appropriate qualifications, training and experience before the study is initiated. If it is necessary to replace the study director during a study, this should be documented.
- an individual is identified as the responsible for the management of the archives.

Responsabilities of the study director.

The study director has the responsability for the overall conduct of the study and for the reporting of result(s).

The responsabilities include as a minimum the following functions:

- Accordance should be ensured between the performed study and the study plan.
- The procedures specified in the study plan are followed and any authorized modification which is accepted and incorporated is documented together with the reasons for the modifications.
- All data generated are fully documented.
- A signature and date on the final report as an indication for the validity of the data and confirmation of compliance with these to the principles of GLP.
- The plan(s) of the study, the final report, the raw data, any supporting material are kept properly in the archives after the termination of the study.

Responsabilities of the personnel.

The personnel should keep themselves abreast of safe working practice and adhere to these. Chemicals should be handled with suitable caution until their hazard(s) has been established.

The personnel should adhere to health precautions to avoid unnecessary risk to themselves and to ensure the integrity of the study.

The personnel who is aware of having a health or medical condititon which is likely to have an adverse effect on the study in any way should be excluded from operations that may affect the study.

QUALITY ASSURANCE PROGRAMME

General.

The test facility should have a quality assurance programme which is documented to ensure that studies performed are in compliance with the principles of GLP.

A quality assurance programme should be carried out by designated individual(s) who are familiar with the test procedures and are directly responsible to the management.

The same individual(s) should not be involved in the conduct of study which is being assured.

The individual(s) should report al findings in writing directly to the management and to the study director.

Responsabilities of the quality assurance personnel.

The responsabilities of the quality assurance personnel should at least include the following functions:

-    ascertain that the study plan and the SOP are available to the personnel conducting the study,
-    ensure that the study plan and SOP are followed by periodical inspections and/or auditing the study in progress,
-    retain reports of inspections and/or audits of the study in progress,
-    report immediately to the management and to the study director on observed deviations from the study plan and from the SOP,
-    review the final reports to confirm that the methods, the procedures and the observations are accurately described. It should be ensured that the reported results reflect correctly the raw data of the study,
-    prepare and sign a statement which should be included into the final report. This statement should specify the dates of any findings which were reported to the management and to the study director.

FACILITIES

General.

The test facility should be of a suitable size, location and design as to meet the requirements of the study and minimize disturbances which could interfere with the validity of the study.

The design of the test facility should provide a sufficient degree of separation for the different activities to ensure each individual study.

Facilities for test systems.

The test system facility should be provided with a sufficient number of rooms to ensure the isolation of test systems. This should especially be the case when substances known or suspected to be hazardous are involved.

Suitable facilities should be available for the proper diagnosis, treatment and control of diseases. This should ensure that there is no unacceptable degree of detorioration of the test system.

Sufficient storage area should be available for supplies and equipment. This area should be separated from the areas housing the test system(s) and should be adequately protected against contamination. Refrigeration should be provided for perishable commodities.

Facilities for handling of substances for test and reference.

Separate areas should be provided for receipt and storage of substances for the test and reference and for the mixing of the test substances with a vehicle to prevent individual and mutual contamination.

Areas for storage of the test substances should be separate from the areas housing the test systems. The facilities should be adequate to preserve identity of the test systems. Safe storage for hazardous substances should also be provided.

Facilities for archives.

Space should be provided for archives for the storage and retrievel of raw data, reports, samples and specimens.

Facilities for waste disposal.

Handling and disposal of waste should be carried out in a safe way so the integrity of the studies in progress is not jeopardized.

Handling and disposal of waste generated during the study performance should be carried out in accordance with pertinent regulatory requirements. These should include provisions for appropriate collection, storage and disposal facilities. They should also include provisions for decontamination and transportation of the waste.

Adequate records should be maintained on the handling and disposal of waste relating to the study.

APPARATUS, MATERIAL AND REAGENTS

Apparatus.

Apparatus used for the generation of data and for controlling environmental factors relevant to the study should be appropriately located and of adequate design and capacity.

Aparatus used in the study should be inspected periodically, cleaned, maintained and calibrated according to SOP. Records of procedures should be maintained.

Apparatus used in studies should not interfere with the test systems or only interfere in a well defined manner to ensure the quality of the data.

Material.

Material used in studies should not interfere with test systems or interfere in a well defined manner to ensure the quality of data.

Reagents.

Reagents should be labelled in an appropriate manner to indicate source, identity, concentration and information on stability. The labelling should include the date for preparation, the earlist expiration date and specific instructions for storage.

TEST SYSTEMS

Physical/chemical.

Apparatus used for the generation of physical/chemical data should be suitably located and of appropriate design and of adequate capacity.

Reference substances should be used to ensure the integrity and validity of the apparatus in the physical/chemical test systems.

Biological.

Proper conditions should be established and maintained for the housing, handling and care of animals, plants, microbial and other cellular/subcellular systems to ensure the quality and validity of the data.

The conditions should comply with the appropriate national regulatory requirements including the import, collection, care and use of animals, plants, microbial as well as other cellular/subcellular systems.

Animal/plant test systems which just have been received should be isolated until their health status has been ensured. The lot should not be used in the studies if any unusual mortality or morbidity occurs and - when appropriate - the lot should be destroyed in a humanely manner.

Records of source, date of arrival and condition at arrival of the animal/plant test systems should be maintained and preserved.

Animal, plant, microbial and cellular systems should be acclimatised to the test environment for an adequate period before the start of the study.

All necessary information for proper identification of the test system should appear on their housing or containers.

The diagnosis and treatment of any disease before or during a study should be adequately recorded.

TEST SUBSTANCES AND REFERENCE SUBSTANCES

Receipt, handling, sampling and storage.

Records should be maintained on substance characterization, date of receipt, quantities received and used in studies.

Handling, sampling and storage procedures should be identified to ensure the highest possible degree of homogeneity and stability and to prevent contamination and/or mistaken identity.

Storage container(s) should carry tags with information of identification, earliest expiration date and specific storage instructions.

Characterization.

Each test/reference substance should be appropriately identified (by code, name, batch number etc.).

The identity including batch number, purity, composition, concentration and other characterizations should be known for each study to define appropriately each batch of the test/reference substance.

The stability of test/reference substances should be known for all studies under conditions of storage.

The stability of test/reference substances should be known for all studies under the test conditions.

SOP should be established if the test substance is administred in a vehicle for testing the homogeneity and stability of the mixture.

A sample for analytical purposes should be retained from each batch of test substance, especially for studies in which the test substance is tested longer than four weeks.

STANDARD OPERATING PROCEDURES

General.

A test facility should have a written set of Standard Operating Procedures (SOP) which are approved by the management. They are intended to ensure the quality and integrity of the data generated during the study.

Each separate laboratory unit should have close-at-hand a copy of the SOP relevant to the activities being performed in that laboratory. Published text books, articles and manuals may be used as supplements to which references should be made.

Application.

SOP should at least be available for the following types of laboratory activities. The catch words mentioned below should be considered as illustrative examples and not as guidelines or as limitations.

Test/Reference substances:
Receipt, identification, labelling, handling, sampling and storage.

Apparatus and reagents:
Use, maintenance, cleaning, calibration of measuring apparatus and environmental control equipment. Preparation of reagents.

Record keeping, reporting, storage and retrieval:
Coding of studies, data collection, preparation of reports, indexing systems, handling of data including computerised data systems.

Test systems:
Preparation of conditions in room and environmental for the test system. Establishment of procedures for receipt, transfer, proper placement, characterization, identification and care of test system. Preparation of test system, observations/examinations before, during and at termination of the study. Procedures for handling of test system individuals found dying or dead during the study. Procedures for collection, identification and handling of specimens including necropsy and histopathology studies.

Quality assurance procesures:
Operation of quality assurance personnel in performing and reporting study audits, inspections and final study report reviews.

Health and safety precautions:
As required by national and/or international legislation or guidelines.

PERFORMANCE OF THE STUDY

Study plan.

A plan should exist in a written form prior to initiation of each study.

The study plan should be retained as raw data.

Any changes, modifications, revisions or amendments of the study plan agreed upon by the study director including justification(s) should be documented, signed and dated by the study director and retained with the study plan.

Content of the study plan.

The study plan should at least contain the following information:

- Identification of the study, the test and reference substance as:
  - a descriptive title,
  - a statement on the nature and purpose of the study,
  - an identification of the test substance by code or name (e.g. IUPAC: CAS number),
  - the reference substance to be used.

- Information concerning the sponsor and the test facility:
  - the name and address of the sponsor,
  - the name and address of the test facility,
  - the name and address of the study director.

- Dates:
  - the date of the agreement to the study plan together with signature of the study director and, when appropriate, of the sponsor and/or of the test facility management,
  - the proposed starting date and time schedule.

- Test methods:
  - the reference(s) to appropriate national or OECD test guidelines or other test guidelines to be applied.

- Issues:
  - the justification for the selection of the test system,
  - a characterization of the test system (e.g. the species of supply, number, body weight range, sex, age and other pertinent information,
  - the applied method of administration and the reason for its choice,
  - the dose levels and/or concentration(s), frequency, duration of administration,
  - a detailed information on the experimental design including description of the chronological procedure of the study, all applied methods, materials and conditions, type and frequency of analysis, measurements, observations to be performed.

- Records:
  - a list of records derived from the test should be retained.

Conduct of the study.

An unequivocal identification should be given to each study. All items concerning this study should carry this identification.

The study should be conducted in accordance with the study plan.

All data generated during the conduct of the study should be recorded directly, promptly, accurately and legibly by the individual who enters the data. These entries should be signed or initiated and dated.

Any change in the raw data should be made in a way that does not obscure the previous entry and should indicate the reason - if necessary - for change and should be identified by date and signed by the individual making the change.

Data generated as a direct computer input should be identified by the date and the time of data input by the individual(s) responsible for direct data entries. Corrections should be entered separately with the reason for the change, with the date and time and the identity of the individual making the change.

REPORTING OF STUDY RESULTS

General.

A final report should be prepared for every study.

The use of the international system of units should preferably be applied.

The final report should be signed and dated by the study director.

If reports by principal scientists from cooperating disciplines are included in the final report, they should sign and date them individually.

Corrections and additions to the final report should be in the form of an amendment. The amendment should clearly specify the reason for the correction or the addition and should be signed and dated by the study director and by principal scientists from each discipline involved.

Contents of final report.

The final report should include at least the following information:

- Identification of the study, the test and reference substance:
  - a descriptive title,
  - identification of the test substance(s) by code or name,
  - identification of the reference substance by chemical name (not just placebo),

- • characterization of the test substance including the purity, the stability and homogeneity.

- Information on the test facility:
  - • name and address
  - • name of the study director
  - • name of other principal personnel having contributed reports to the final report.

- Dates:
  - • dates on which the study was initiated, the time plan and date for completion.

- Statements:
  - • a quality assurance statement certifying the dates for inspection and the dates for any findings which were reported to the management and to the study director.

- Description of materials and test methods:
  - • description of test methods and materials used,
  - • reference to OECD test guidelines or other relevant test guidelines.

- Results:
  - • a summary of the results and conclusions,
  - • all information and data required in the study plan,
  - • a presentation of the results including calculations and description of the statistical methods,
  - • an evaluation and discussion of the results, perhaps with conclusion(s).

- Storage:
  - • description of the location where all samples, specimens, raw data and the final report are to be stored.

STORAGE AND RETENTION OF RECORDS AND MATERIAL

Storage and Retrieval.

Archives should be designed and equipped for adequate accomodation and secure storage of:

- the study plans,
- the raw data,
- the final reports,
- the reports of laboratory inspections and study audits performed in accordance to the Quality Assurance Programme,
- samples and specimens.

Material retained in the archives should be indexed to facilitate orderly storage and rapid, correct retrieval.

Only personnel authorized by the management should have access to the archives. Transfer of material in and out of the archives should be recorded properly.

Retention.

The following should be retained for the periods specified by the appropriate authorities:

- the study plan, raw data, samples, specimens and the final report of each study,
- records of all inspections and audits performed by the Quality Assurance Programme,
- summary of qualifications, training, experience and job descriptions of personnel,
- records and reports of the maintenance and calibration of equipment,
- the historical file of Standard Operating Procedures.

Samples and specimens should be retained as long as the quality of the preparation permits proper evaluation. If a test facility or an archive of a contracting facility goes out of business and has no legal successor, the archive material should be transferred to the archives of the sponsor(s) of the study(ies) and retained there.

Rules for Good Manufacturing Practice (GMP)
General Considerations

Author: S. Dawids

Institute of Engineering Design, Biomedical Section, Techn. Univ. of Denmark, DK - 2800 Denmark

INTRODUCTION

The purpose of the guidelines for GMP is to provide a framework which enables the producer to establish appropriate and necessary structures to ensure products of the intended quality and uniformity.

The following description of Good Manufacturing Practice which aims at medical devices has no statutory force and cannot be regarded as an interpretation of the requirements of any act, regulation or directive. It is intended as an aid to the manufacturers who -apart from the cited main guidelines - must acquire additional knowledge of national requirements.

The general guidelines for GMP are particularly concerned with the aspects of manufacturing conditions and processes which may affect the quality, safety and intended performance of the product. However, complementary specifications, monographs and guides or codes exist in many countries and detail specific requirements for particular products or product groups.

In general, GMP guidelines with minor deviations from those described in this chapter will be accepted as equally usable if they ensure the same end results.

Basic concept and principles of Good Manufacturing Practice.

The principles represent the sum of the activities necessary to ensure that the medical devices meet the requirements of the customers and of the authorities for quality, safety and performance.

The basic principle encompass:

S. Dawids (ed.), Test Procedures for the Blood Compatibility of Biomaterials, 615–643.
© 1993 *Kluwer Academic Publishers. Printed in the Netherlands.*

- an integrated system of manufacture and quality assurance,
- allocated management responsabilities for both production and quality assurance,
- suitable, clean, well kept premises, equipment and pure raw materials,
- properly trained personnel,
- documented procedures for the manufacture and quality assurance,
- appropriate batch and product records,
- adequate transport and storage,
- a recall system
- good manufacturing practice throughout the production unit.

GLOSSARY OF TERMS

Below are given definitions which are used in the GMP rules:

Advisory Note
A notice issued to advise what action should be taken in the use, disposal or return of a product where a recall would not be appropriate.

Ancillary Materials
Solvents and other aids in manufacture.

Approval/Registration Scheme
A system by which manufacturing and quality control operations are assessed and approved/registrered by a regulatory authority.

Batch (Lot)
A defined quantity of a raw material, intermediate products, work-in-progress or finished products. The quantity in a batch can be governed for example by a sterilizer load, a period of manufacture or other parameters which will allow effective control.

Batch Manufacturing Records
Written and where necessary authorized records relating to individual batches comprising quality control and production records including details of raw materials and/or components, intermediate products, labels and any production conditions. All documents should relate to the batch number.

Batch Number **(or Lot Number)**	The designation of a batch by means of a distinctive combination of numbers and/or letters which identifies it and permits its history to be traced. A batch may be a single product in which case this may be referred to as a serial number.
Clean Room	A room with environmental control of particulate contamination, temperature and where necessary humidity, constructed and used in such a way as to minimize the introduction, generation and retention of particles inside the room and in which special attention is paid to the control of microbial contamination.
Code Number	A unique identity number for a type and/or size of a finished product. A catalogue or reference number.
Controlled Area	Any work space or room which cannot be classified as a clean room, but in which the air is required to be cleaner than that of the outside environment. No environmental conditions are specified. It may, however, be necessary for selected environmental conditions to be used as appropriate for the nature of the product.
Finished Product	A sterile product in its unit, correctly labelled.
Intermediate Product	Components, work-in-progress, sub-assemblies or any part-finished product.
Manufacturing Environment	The premises, environmental conditions, equipment and ancillary materials used during manufacture.
Manufacturing Procedure	The manufacturing steps required to produce an intermediate or finished product.
Master Documents	Approved manufacturing and quality control specifications defining a product and its method of manufacture and control. It is from these documents that operational documents required for production and quality control are derived.

Medical Device	An instrument, apparatus, implement, appliance, implant or other similar or related article which is intended for use in the treatment of humans, contraception or in diagnosis. A device achieving its principal intended purpose through chemical action within or on the body is excluded from this guide.
Outer or Transit Container	A package, carton or other protective container which may contain one or more unit or shelf containers and which provides additional protection for storage and distribution.
Products	Used generally in the text to cover medical devices and surgical products.
Quarantine	The status of materials or products awaiting release by the quality controller.
Quarantine Area	An area isolated by physical barriers or other effective means which is used for the storage of any materials or products whilst subject to quarantine.
Raw Materials	Any material or fabricated component used singly or in conjunction with other raw materials and/or components in the assembly or fabrication of part or in total production of products.
Serial Number	See Batch number.
Shelf or Multi-Unit Container	A package, carton or other container for one or more unit container.
Single-Use	The use of a product for one patient for one procedure only.

Sterile Medical Devices and Surgical Products (or Sterile Products)	Medical devices, surgical products or any combination thereof which are supplied sterile.
Surgical Product	An adhesive or non-adhesive, non-medicated surgical dressing or other similar or related article which is used during treatment of humans.
Unit Container	A pack containing a single item or a combination of procedure related components or products. The pack is designed to maintain the sterility and integrity of the contents up to the time of use, and to permit, where appropriate, aseptic removal of its contents, and such that once opened it cannot be resealed easily, and clearly reveals that it has been opened.

MANUFACTURING ENVIRONMENT

The environment is one of the three main sources of contamination of a product - dust and microbial - prior to its sterilization (the others being direct contact with personnel and contaminated raw material/components). If the end product is intended to be sterile, Good Manufacturing Practice requires that adventitious contamination from all sources is minimized by identifying the sources and adjusting the conditions by all practicable means.

Basic consideration for premises.

At best premises should be purpose-designed and purpose-built. Both newly constructed or modified premises require the following basic considerations:

- The essential part of, or all of the production should take place in a controlled area or clean room.
- Passage of all materials, components and personnel should be controlled.
- Correctly designed and situated cloakroom and toilet facilities should be provided.
- Special storage areas and conditions are generally necessary.
- Special facilities for cleaning are generally required.

- Maintenance, repair, building activities, pest control and other necessary services need special provision.

Production area requirements.

Production areas - large or small - in which components, subassemblies and/or finished products are exposed for considerable periods to the environment and/or are handled, should be situated in controllled facilities. In this respect special attention must be paid to microbial contamination levels by regular testing of air and of the hands and clothing of the personnel.

Although no general specification can be given on the class of clean room or controlled area, it is important that the cleanliness of the air should be determined during workhours. It should be noted that the determinations depend on the nature of the work, the product and the degree of handling and exposure of the product. The level of acceptable contamination depends on the end application of the product.

Within the controlled areas it can be desirably to use locally contained work stations (e.g. laminar flow benches) which provide a protective air flow of a high standard.

The following features are common to all clean rooms, and a majority of them also apply to controlled areas:

- Clean rooms must be supplied with a flow of filtered air. Air pressure of the areas should be maintained above that of surrounding areas to prevent ingress of unfiltered air from surrounding areas. Windows should be kept closed and sealed. Doors should be self-closing and tight-fitting. Air locks are often necessary to maintain the necessary positive pressure in the area. Doors or other openings (apart from emergency exits) should be located as far as possible from clean work stations or laminar flow benches.
- The interior of work rooms should be designed to avoid dust traps and to permit easy cleaning. Junctions between floors, walls and ceilings should preferably be coved. Walls, floors, ceilings and other surfaces should be of a smooth finish able to withstand frequent cleaning.
- Furniture and work benches should be made of non-shedding materials with a smooth resistance finish without cracks or chinks.

- Fitments should have smooth surfaces preferably free from sharp corners.
- Waste material should be collected in suitable containers for frequent removal. Care should be taken that waste containers do not act as contamination sources.
- The disposal of raw materials and rejected products should be controlled, and reuse of raw materials and rejected products should only be considered under very strict precautions.
- Production areas (clean areas) should not be used for storage or as travelling route for transport of materials or throughway of personnel.
- Cloakrooms and toilets must be segregated from the production areas. Changing rooms and washing facilities for the clean room or the controlled area(s) should be adjacent. Personnel should only be able to enter the production area after provided facilities have been used. Changing rooms and washing facilities should be at the down stream end of any air pressure gradient.
- Changing chambers, washing facilities, toilets etc. should be maintained in a very clean and tidy condition. Personnel should not enter toilet facilities wearing their outer protecting clothing. The toilet should be provided with:
  - dispersable hand detergents,
  - hot air dryers and/or single use towels,
  - waste bins with foot-operated lids,
  - taps or wash basins fitted with foot or elbow-operated controls, mirrors should be available to assist correct adjustment of head covering,
  - nailbrushes - when considered necessary - should be either clean single-use brushes or brushes maintained micro-biologically clean.

Equipment and materials.

Equipment used in the production of sterile items should be designed for easy cleaning. Matter or lubricants from the machinery must not come into contact with components of the production and the machine should be adequately shielded, perhaps sealed.

The equipment should be designed (or modified) to perform the production processes with ease, and should demonstrate a capability of carrying out the processes for which it is intended in an adequate hygienic standard.

Equipment should be adequately spaced to avoid congestion, most contamination and accidental mixing of different products.

Containers used for temporary storage and handling should be constructed from non-shedding materials and designed for easy cleaning. Care should be taken that they do not act as sources of contamination.

All ancillary materials should be adequately identified and labelled. Fluids should be contained in properly labelled dispensers. All labels should be easy to read and clearly specify the contents and preferably display appropriate hazards warnings and safety precautions.

All fluids and any other materials which could be considered capable of contaminating the product should be carefully monitored micro-biologically as well as for any other hazard.

All filters for environmental and compressed air suppliers should be regularly serviced. Care should be taken that filters are well fitted to avoid inflow of unfiltered air.

Cleaning and cleaning schedules.

A (preferably typewritten) cleaning schedule should exist for all areas and equipment. The schedule should specify methods and materials to be used, the frequency of cleaning and the persons with the allocated responsability for carrying it out. The schedule should be approved by a competent person, e.g. the production manager.

Cleaning equipment should be stored in a clean, dry and tidy manner and be well maintained. Cleaning equipment used in clean rooms should only be used for this area and not used in any other area. It should be stored in an enclosed area adjacent to the clean room.

Most surfaces which are required to be cleaned can be washed satisfactorily with free-rinsing detergent alone. For working surfaces the subsequent use of hypochlorite solution or 70% v/v alcohol or suitable equivalent is acceptable.

To avoid resistant organisms, more than one disinfectant should be applied and interchanged periodically.

Solutions of detergents and disinfectents should be made afresh before cleaning operations as they may loose their effectiveness and thereby become a microbiological hazard.

Vessels and pipelines used for water should be cleaned regularly and maintained in good condition. Stored water quickly becomes contaminated and should be avoided. If necessary, the equipment for fluids should be sterilized or disinfected regularly.

Containers for solvents and other solutions which may support bacterial growth should not be stored after having been opened. They should be cleaned regularly and thoroughly and unused contents discarded.

Cleaned equipment which raises dust must not be used. Positive pressure airlines should be avoided for cleaning purpose of equipment.

Storage area(s).

Suitable storage areas should be provided for raw materials and for finished products.

A specific area should be provided for the quarantine products awaiting release by the quality controller and/or the microbiologist.

A specific area should be provided for the segregated storage of rejected materials and discarded products.

A specific area should be provided for the storage of customer returns.

These areas should be maintained in a clean and tidy condition. They should be designed and controlled to avoid contamination from outside or from other areas.

Potential contamination of the manufacturing environment e.g. from transferring items from the storage area should be avoided by adequate cleaning or other handling procedures (e.g. storing in containers or boxes).

Appropriate records must be maintained to control the identity and movement of each item (batch of items) transferred.

PERSONNEL, TRAINING AND HYGIENE

The quality of a product is to a large extent dependent on how well the personnel involved in the manufacture is aware of critical aspects of the production. Each member carries a responsability for performing correctly and satisfactorily at each stage or stages in the manufacturing process.

Adequate training is of paramount importance to ensure that each one understands the nature of the work for which he/she is responsible and the possible consequences of failure to observe good manufacturing practice.

Personnel.

At each step in the production there should be sufficient personnel with the ability, training experience and other qualifications (professional and technical) appropriate for the tasks assigned to them. Their duties and responsabilities should be made clear to them and recorded e.g. as an accepted job description.

In the context of the guide the key personnel are the production manager and the quality controller. They should be different persons neither of whom is responsible to the other, but who are jointly responsible for ensuring quality. The persons in these responsible positions should have sufficient authority to discharge their duties.

Persons engaged in a consultative capacity or on a part-time basis should not be appointed as the key personnel.

Designated deputies should be capable of assuming the responsabilities of key personnel in their absence.

Key personnel should be given adequate supporting staff to ensure that requirements are met for quality and production.

Training.

All personnel should be adequately trained in the principles of good manuafacturing principles at least relevant to the tasks assigned to them. The training should be in accordance with written programmes approved by the production manager and ,when appropriate, by the quality controller.

Personnel only intermittently employed in production areas e.g. maintenance staff and cleaners should also be trained.

Personnel working in a clean room should be given specific training appropriate to their tasks.

Personnel should be educated in the intended use of the end product(s) thereby achieving greater motivation towards quality requirements.

Training should be given as part of a general introduction to all new employees and for all personnel at regular intervals.

Hygiene.

Individuals liberate both microorganisms and non-viable particles which constitute contamination risks. Surfaces of the human body which shed particles are ideal environments for multiplication of microorganisms including virus. Important locations are the skin, the hair, the nose, the oral cavity and especially the large intestine. Contact with these areas to e.g. the hands can transmit millions of microorganisms, their spores, virus etc. to the products. An individual can liberate millions of bacteria-carrying particles per minute, and movement of people creates aircurrents and turbulence which stir up particles and delay sedimentation. Even sitting persons cause air convection currents, thus liberating particles to the surrounding air.

Personnel employed in production areas should be medically examined and provide a medical certificate before employment.

No person should be employed on production processes who are known to have skin lesions on exposed surfaces of the body, a contagious disease or to be the carrier of such a disease or in other way likely to constitute a contamination hazard.

Personnel in contact with the production or its environment shall be clean, healthy and suitably attired as well as performing an adequate conduct.

Any personnel who by observation or by medical examination appear to have a condition which could adversely affect the product should be excluded from the sensitive operations for the necessary period.

Personnel should be instructed and encouraged to report such conditions to their supervisor.

A list of rules should be issued to the personnel, and steps should be taken to ensure that they are read and understood.

Those rules which apply to clean rooms should also be prominently displayed at their entrance.

The rules and procedures for entry into and behaviour during the presence in clean rooms and controlled areas should rigorously be followed by **all** persons entering the areas (including visitors, maintenance staff and cleaners).

Non-essential personnel and visitors should be discouraged from entering these areas.

The appropriate supervisor should have the responsability to ensure that the correct procedures are adhered to at all time.

In the rules for clean rooms and if necessary for controlled areas the following further points are important:

- Access must be restricted to authorized persons only. Visitors should obtain prior authorization before they are allowed to enter..
- No persons are allowed to enter except through a changing room in which their regular outdoor clothing should be left in an enclosed area separate from that of protective clothing.
- All persons must be required to wash their hands immediately before entering.
- Protective clothing and special foot-wear must be worn by all persons and must not be worn outside the clean room or controlled area.
- Eating, drinking, smoking, chewing etc. is not permitted in clean room or controlled area.
- Use of cosmetics should be discouraged. Those which can shed particles e.g. powder based cosmetics should not be worn.
- No personal belongings e.g. purses, handbags and easily removable jewellery should be taken into these areas, but lockers should by available for storing of personal belongings or valuables.

Protective clothing: Clean rooms.

A protective clothing for personnel and for visitors should be of essential non-linting material and designed to cover the wearer and every-day clothes effectively. It should be noted that even the use of protective clothing will fail if the person rubs or itches exposed skin areas with the hands. External pockets should be avoided.

The head covering should be designed to cover the hair completely and supervision should ensure that everyone wears it correctly. Beards and moustaches should not be permitted for personnel in clean rooms unless adequately covered.

Used protective clothing should be cleaned regularly and be of essentially non-linting material. Single use foot-wear protection may be used.

Where gloves or other hand-coverings are to be worn while handling components or products they should be of essentially non-linting and non-shedding materials and should be discarded when leaving the clean room. Clean gloves should be available for personnel when re-entering the room and at any time during the working day when gloves need to be changed.

Protective clothing should be stored in a separate area from outer clothing and should be maintained in a good and clean condition. The garments should be regularly and frequently replaced by clean sets.

The type of protective clothing must be appropriate for the area and operation in which it is used.

Protective clothing: Controlled area.

The protective clothing provided for personnel and visitors to controlled areas should be of essentially non-linting material and designed to cover the wearer and everyday clothes effectively.

The coat or over-all should fasten throughout its length.

Hair covering should be designed to cover the hair completely, and supervision should ensure that they are worn correctly.

The protective clothing should be regularly and frequently replaced by clean sets.

QUALITY ASSURANCE

The assurance of quality of a finished product requires an integration of the manufacturing and quality control procedures to ensure that the products are correctly made and comply with the appropriate specifications.

Quality controller.

Several recommendations are given here concerning the personnel structure of the manufacturer's organization. These are included because it is considered essential that the quality controller is independent of that part of management which has a first duty towards purchasing, production or marketing. A quality controller should therefore be responsible to a senior executive who has no direct responsability for the production.

In small organizations where this structure is not possible, the management must ensure that quality and safety are not compromised by considerations of increased production or by market pressure.

The quality controller should be given the authority to establish all necessary procedures. He should have the authority independently of production to approve raw materials, packaging materials and to reject raw materials, products and packaging materials which do not comply with the relevant specifications or which have not been manufactured in accordance with the approved methods or under the prescribed conditions.

Quality control.

The responsability of the quality controller should encompass:

-    the specifications of the system of inspection, sampling and testing procedures with defined limits for acceptance and rejection, and the provision and control of all relevant documentation,

- the provision of specifications which describe the standards to be met by the materials and components used in the manufacturing process and by the finished products,
- the provision of control and information data. There should be facilities adequately staffed and equipped for performing all control tests without undue delay,
- the provision for the calibration and maintenance of the test equipment,
- the approval of all raw materials, intermediate products and finished products. Materials should not be passed on to the next stage of processing before approval. Neither should finished products go to distribution without the signed agreement of the quality controller. There should be a system to identify and segregate accepted and rejected material,
- the taking of samples from any part of the premises at any time. Samples for testing should be taken in accordance with the sampling procedure which will ensure that the result of the test is representative of the batch. Other sampling procedures may be defined for specific purposes. Samples for testing should be taken by the quality control personnel. Samples for in process control should be taken by authorized production personnel according to guidelines approved by the quality controller. In case of inspection of every item by the production staff, the quality controller is still responsible for the approval of the test methods and the acceptance criteria,
- the preservation of appropriate records and samples relating to each batch as a normal procedure. The length of time of storage of records and samples should be determined by the quality controller. This should normally be minimum 5 years even if expiry dates qualify for a shorter period. The records should include:
  - key manufacturing, quality control, sterilization, microbiological control tests and other appropriate historical records,
  - the signature of persons who have carried out the sterilization process and tests,
  - a dated and signed statement that the batch in question has been approved for release or has been rejected. A rejection note should include instructions for disposal,
- the examination and documented control of all returned products to determine whether they may be released, reprocessed or destroyed.

Microbiological control.

The responsability for all microbiological aspects of production and control necessary to ensure the sterility of the product shall be allocated to a person qualified and experienced in microbiology. Competent knowledge on production hygiene is also important. The use of an outside laboratory service covering the particular needs of the company is an alternative.

The laboratory for control should preferably be on the manufacturer's premises, in fact this is strongly recommended if ethylene oxide sterilization or other types of chemical sterilization are used. The laboratory staff should be capable of carrying out routine tests using aseptic procedures.

A competent person should be responsible either to the quality controller or senior executive. The person should be independent of production at all levels of management. The competent person may be identical with the quality controller.

No product labelled as "sterile" should be released for distribution until the prescribed microbiological control procedures and adequate monitoring of sterilisant residues have been satisfactorily completed and the result approved by the competent person.

The responsabilities of the competent person should include:

-    the commissioning of sterilizing plant and the defining of effective sterilization and quarantine procedures, based on measurement of sterilisant residues
-    the operation or supervision of the sterilization plant and/or the monitoring of the sterilization procedures,
-    definition and routine monitoring of the performance of air conditioning and filtration equipment in the production areas,
-    approval of written hygiene regulations and the monitoring of their correct implementation,
-    approval of written cleaning schedules for all areas and equipment,
-    investigation of the level of presterilization microbial contamination of the product. The elimination or reduction of suspected sources of contamination where possible,

- monitoring and recording of environmental contamination (e.g. with split-sampler) at an appropriate frequency; isolating and identifying the pattern of contaminants and attempting to determine their sources,
- the microbiological monitoring of materials etc.

Customer complaints.

A system for the precise recording, evaluation and processing of product quality complaints should be in operation. All such complaints should be presented to the quality controller.

Auditing.

All aspects of GMP should be subject to planned and documented internal audits.

MANUFACTURE

The process of manufacture begins with the raw materials and provision of equipment. All subsequent stages of production are included until the finished product is at a stage when subject to approval. After passing this stage it can be released for sale.

Process capability.

In all stages of a new design of product the manufacturing process should be closely evaluated to establish its efficacy.

Similar procedures should be followed where any important change in the processing occurs which should involve new or modified equipment, major overhauls of equipment, change of location, change of positioning of production line, changes of raw material or change of suppliers to raw materials.

Processes should continuously be subject to critical appraisals to ensure that they maintain capability of achieving the intended results.

Manufacturing procedures.

A documented procedure should be agreed upon by the production manager and the quality controller for each stage of a manufacturing operation.

No deviation from the defined procedure should be made without the prior agreement of the production manager and the quality controller. An authorized amendment to the documented procedure should be made prior to the deviation.

A defined system must exist to ensure that the work area and equipment are uncontaminated by all raw materials, components, sub-assemblies and other products not required for the product to be made before manufacturing begins.

Line cleaning procedures should be carried out when considered essential between production runs to prevent the possibility of cross-contamination.

Bulk containers and equipment used in processing should be identified where appropriate as relating to the product and batch being processed. This identification is not necessarily the code used on the finished product label, but should on the other hand be easily related to the product code.

Prior to labelling of the containers or equipment, all labels previously used thereon should be removed and maculated.

To avoid accidental mixing of products work-in-progress should be segregated and identified.

The finished products should be stored until released by the quality controller.

Production control.

The production manager is responsible for manufacturing intermediate and finished products in the required quality.
Production resources should be available and adequately staffed and equipped to ensure an effective control of the production.

The production manager shall be responsible for ensuring that:

-    production schedules are provided,

- production batch manufacturing records are prepared and kept,
- raw materials are issued through the manufacturing processes,
- premises and equipment are kept clean and well functioning,
- in-process controls are operated according to the methods agreed upon with the quality controller,
- appropriate in-plant transport and storage is provided.

Raw materials, intermediate and finished products.

Appropriate specifications with relevant sampling and test procedures should exist. The records of such tests should be included in the batch documents where appropriate.

Raw materials and products should be quarantined until released for use by the quality controller. Those items which have no direct influence or cannot jeopardise the finished product do not require such control.

Raw materials and products should be correctly identifiable at all times. Unauthorized descriptions should not be permitted.

Each delivery of a batch of raw materials should be allocated an identifying number which can be related to the material throughout storage and processing. Subject to approval by the quality controller, certain raw materials which cannot jeopardise the product does not require such restrictions.

Raw materials and products should be issued in rotation according to an approved and documented procedure.

Stock records should be kept to indicate each receipt and issue, enabling stock reconciliation.

Storage.

Raw materials and products should be stored in a manner approved by the quality controller.

The storage conditions should be such that deterioration or contamination as well as damage is minimised. Containers may afford adequate protection.

If special environmental storage conditions are required at any stage, such conditions should be established, monitored and controlled according to an approved procedure.

Storage conditions should be orderly to facilitate protection of stock, batch differentiation and ease of cleaning.

Access to material in quarantine area should be restricted to authorized persons.

Items which have been rejected, recalled or returned should be placed in a separate quarantine area to prevent mixing or mistaken for other materials or products. Release and disposition should be authorized only by the quality controller.

A specific area should be provided for material recovery operations. This may be designated in a production area. A potential contamination should be minimised by adequate cleaning procedures or isolation.

Labelling process.

Packaging of products of similar appearance in close proximity to one another should be avoided. Line identification and clearance procedures should be practiced between production runs to avoid confusion.

The risk of errors in labellling and packaging may be minimised e.g. by:

- the use of roll feed labels,
- the issue of a known number of labels and reconciliation of usage,
- on-line batch coding (as distinct from preprinted batch coding),
- the use of electronic code encoders/readers and label counters,
- labels and other printed materials designed to give a marked differentiation between products whereever possible.

After completion of a packaging procedure any unused batch coded labels or coded packaging materials should be destroyed. Uncoded materials not required should be returned to stock by a documented procedure to avoid contamination and mixing.

Calibration.

Equipment used to sterilize products should be provided with recording devices and/or indicators which should be calibrated initially and checked at specific intervals by adequate methods. The density of products in sterilizing equipment should not change from the calibration situation to the production situation.

Process control equipment should be calibrated and checked according to a planned maintenance schedule.

PACKAGING AND LABELLING

The packaging should be designed to contain and to protect the product physically and to maintain sterility. Normally this is obtained by establishing three protective layers between the outside and the sterile product. Maintenance of sterility is event related rather than time related. Products should maintain sterility if they are both transported and stored correctly.

Before any packaging method is adopted it should have been evaluated to establish its suitability for the intended purpose.

Packaging.

Requirements for unit containers:

- Sterile products must be contained in sealed units. In the case of devices where only the inside surfaces need to be sterile, the requirements for unit packaging may be less strict.
- The design should permit in situ sterilization of the contents without the need for further sealing of the pack. Where the chemical or physical characteristics of the contents preclude this, such contents should be sterilized and subsequently sealed under aseptic conditions. In these cases the quality controller should adequately ensure the sterile conditions through approved procedures. Normally such items should not be defined as "sterile", but the term "aseptically packed" should be preferred.

- It should protect the contents from contamination until just prior to use and afford adequately physical protection in conjunction with shelf storage when subjected to normal methods of handling, transit and storage.
- The integrity of the complete container should be verified by test procedures approved by the quality controller, e.g sample test.
- It should be designed to ensure that the product can be presented for use in an aseptic manner where appropriate. This can be achieved by e.g. double wrapping. Once the unit container has been opened, it should not be possible to reseal it without revealing that it has been opened.

Shelf containers:

- A suitable number of unit containers may be packed into a shelf container for ease of storage and to afford adequate physical protection when subjected to normal methods of handling, transit and storage.
- Further protection during transit should be provided by packing shelf containers into a transit container. If sterilized with these packaging facilities, the required three layers of protective packaging is achieved.

Labelling.

The unit and shelf containers should be labelled as indicated below. If it is impossible or impractical to label the unit container as specified below, this information must appear clearly on the shelf container.

Unit containers:

- A description of the contents.
- The word "sterile" (or "aseptically packed") where appropriate in prominent form. This may form part of the description. The quality word of sterile should be respected.
- The words "single use" or "single use only" where appropriate.
- The identity of manufacturer and/or supplier.
- A batch number using a distinctive combination of numbers and/or letters which identifies it and permits the tracing of the production history of the product.
- Instructions for use (if required).

- Special precautions (if any).
- Date (month/year) of expiry for products having a determined shelf life. Normally a shelf life is acknowledged for 2 years.

Shelf containers:

- A description of the contents including the size and the number of contents where appropriate.
- The word "sterile" where appropriate in a prominent form. This may form part of the description.
- Name and address of the manufacturer and/or supplier.
- A batch number using distinctive combination of numbers and/or letters which identifies it and permits tracing of the production history of the product.
- Date of sterilization (year, week and/or month).
- Date (year, week and/or month) of expiry for products having a determined shelf life.
- Instruction for storage, especially limitations (if necessary).

STERILIZATION

Sterilization is a critical part of the manufacturing process, and it is essential in the principles of GMP. They must be applied to all aspects of the sterilization processes depending on the chosen process.

Sterilization is used to inactivate viable microorganisms on a product according to the desired margin of safety. The margin has been discussed, and it is generally agreed that a reduction to a level of 10^{-6} is fully satisfactory. Whatever the nature and resistance of the microorganisms to the sterilization process is, it is clear that the fewer microorganisms exist on the product prior to sterilization, the greater is the probability of achieving sterility after the processing.

Adherence to the principles of GMP is therefore to keep the number of microorgannisms - the bioburden - on the product to a minimum at any time. It should be noted that even dead microorganisms may release proteinic substances which can lead to unwanted reactions in the exposed patient.

The selected sterilization processes must be properly validated and accurately controlled. Validation is an overall programme which is used to demonstrate that a specific product can be reliably sterilized by the designed process. Validation must ensure that the product is uniformly exposed to the process parameters.

Specific European guidelines and recommendations can be found in *"Sterilization of medical devices"* as considered by CEN (European Committee for Classification) in documents for different forms of sterilization (*Brit.Stand.Inst., 1991*).

DOCUMENTATION

The previous reference to the written procedures which are required to define the system of the manufacture and control is summarized in this section. The procedures of documentation reduce the risk of error which may easily occur when only verbal instructions or communications are used.

Care and attention to every detail in the procedure are essential in preparing all instruction documents to ensure that the instructions given are clear and unequivocal and can be readily understood by all concerned in carrying out the described tasks.

General.

An operator must have ready access to the appropriate documents and instructions. They should be available, whenever possible, near to the operation which is to be carried out.

Each document should clearly state its title and scope. It should be dated and signed by the person who issues it and the persons authorizing it. In case of copies or computerised documents, the original must be authorized.

The system should include provision for periodical review and revision.

Superseded documents other than master copies should be destroyed. Superseded master copies should be clearly marked and retained and if necessary filed in a manner which cannot lead to confusion with current documents.

All specifications should be approved by the quality controller and the production manager.

Raw material specification.

The specifications should include:

- a description of the material (composition) or component and drawings, where appropriate,
- designated name of the material and a code reference unique to the material,
- a reference to approved supplier(s) of the material,
- a description of the method of packing agreed with the supplier,
- any safety instruction for handling and use,
- sampling instructions,
- details of quality control tests to be performed (including analytical methods where appropriate), specification and classification of defects and acceptance limits,
- requirements for standard reference samples (standard reference samples and sampling should be agreed with the supplier),
- instructions for storage conditions,
- Frequency of retesting of the stored raw material where appropriate.

Manufacturing specifications.

The specifications should include:

- description of the equipment and materials to be used,
- description in detail of all precautions to be taken,
- step by step manufacturing instructions,
- description of the process of quality checks to be conducted by production personnel,
- instructions on the procedure to be followed in event of problems relating to the necessary product quality,
- instructions for the disposition of accepted or rejected intermediate or finished products (e.g. transfer to sterilization area, work-in-progress stores, rejection storage area or destruction).

Intermediate and finished product specifications.

The specifications should include:

- the accepted name of the product and its code number,
- a description, preferably including a photograph and/or drawing of the physical form of the product together with any standard reference samples,
- detailed instructions on sampling, frequency and method of testing,
- a classification of defects and acceptance limits of the product.

Batch manufacturing records.

Batch manufacturing records should be prepared in accordance to the currently approved version of the appropriate master documents. The system should be designed to avoid errors in the lay-out of the records. Whenever possible photocopying or other methods of reproduction are to be preferred in the preparation of the records. An unambiguous batch number should be issued.

If the batch manufacturing record does not include complete details on the process, the operator should have access to the relevant sections of the approved manufacturing specifications.

During manufacture the following should be entered on to the batch manufacturing record:

- the quality of the raw materials, components and intermediate products and their batch number (if considered appropriate by the quality controller),
- the quantity of the manufactured product should be given the batch number and where appropriate also the number of bulk containers or finished product containers,
- the results of all in-process controls and the initials of personnel taking samples,
- designation of the production line on which the product was manufactured together with the date(s),
- details of any deviation from the manufacturing specifications which should be approved by the quality controller and the production manager,

- detailed sterilization records.

Other documents.

In general, procedures covering other aspects of GMP should be recorded even if they are not dealt with in the above paragraphs.

Written records should normally be kept for an appropriate time and include:

- cleaning and maintenance of buildings and equipment,
- commissioning, maintenance and checking of all equipment,
- monitoring and control of the environment,
- training of personnel, particularly with regard to understanding of relevant procedures and hygiene,
- the return of unused material to storage,
- relevant action instructions for materials and products which do not comply with their specifications,
- the handling of return products.

References to such documents are made in other sections of the guide.

PRODUCT RECALL AND ADVISORY NOTICES

There should be a defined policy on the action to be taken in the event of a product defect. This action must take into account the nature or the product and the potential risk for the user (patient).

If a product is defective or is subject to adverse reports, it may be necessary to issue an advisory notice or to instigate a recall.

The nature and seriousness of the fault will determine whether it will be necessary to issue an advisory notice or toinstigate a recall, the extent of the action and the speed. In many instances contact to authorities is advisable.

There should exist a written recall or advisory notice procedure, approved by the quality controller and known to all persons concerned. It should be capable of being pur into operation at any time (also outside normal working hours, during holidays etc.).

A person should be allocated the task to initiate and coordinate the procedure. A designated deputy should be appointed to act for the nominated person during periods of absence.

Records should be kept which will facilitate the issue of advisory notices and effective recall.

The procedure should specify:

-   the methods to be initiated in order to prevent the distribution of a batch or batches in question,
-   how and which regulatory authorities should be informed
-   the notification methods.

The notification for preventing distribution, for informing regulatory authorities, for the issue of advisory notices and for recall notices should include:

-   the description of the product and the code number (if necessary),
-   the product batch number or serial number,
-   the detailed action to be taken,
-   the eason(s) for the necessary action and the advise on any associated hazards.

The progress of a recall and reconciliation of the amounts received should be monitored accurately.

All returned products should be placed in quarantine for subsequent inspection by the quality controller and product manager.

CONTRACT MANUFACTURE

The manufacture or processing of raw materials, components, intermediate and finished products may be delegated wholly or in part by contract.

Arrangement should be made to ensure that the standards prescribed in this guide are followed by the contract acceptor and that agreed quality standards are specified and maintained.

A contract acceptor should not pass a contract or any part therein to a third party without the prior concent of the contract giver.

The contract should be in clear writing and include arrangements for technical information as appropriate. The limits of the responsabilities accepted by each of the parties of the contract should be clearly defined.

A contract giver bears the ultimate responsability for ensuring that:

-   the product of the contract acceptor or manufacturing process complies with the specifications,
-   the required conditions for storage, transport and distribution are fulfilled.

The contract giver should ensure that the contract acceptor has adequate premises, equipment and staff with sufficient knowledge and experience to carry out the work in a satisfactory way.

Any changes in the arrangement for manufacture should be agreed by both parties to the contract and should be in writing.

The contract should permit the contract giver to have the premises of the contract acceptor visited by a third party acceptable to both parties whenever the product is bein manufactures to ensure that the good manufacturing practice and the agreed standards of quality are being maintained.

The quality controller of the contract giver and a competent person with the required authority nominated by the contract acceptor should:

-   agree on specifications and conditons of manufacture,
-   agree on the productions for in-process control tests and for testing of finished products,
-   agree on the procedure for batch release (after acceptance by the quality controller),
-   agree on the system of documentation.

The contract giver normally implies the company.

SUBJECT INDEX

A

B

C

F

G

H

I

K

L

M

N

O

P

Q

R

S

W

X

Y

Z

AUTHORS INDEX